American Tropics

FLOWS, MIGRATIONS, AND EXCHANGES

Mart A. Stewart and Harriet Ritvo, *editors*

The Flows, Migrations, and Exchanges series publishes new works of environmental history that explore the cross-border movements of organisms and materials that have shaped the modern world, as well as the varied human attempts to understand, regulate, and manage these movements.

MEGAN RABY

American Tropics

The Caribbean Roots of Biodiversity Science

The University of North Carolina Press *Chapel Hill*

This book was published with the assistance of the Wells Fargo Fund for Excellence of the University of North Carolina Press.

Set in Arno Pro by Westchester Publishing Services
Manufactured in the United States of America

The University of North Carolina Press has been a member of the Green Press Initiative since 2003.

Library of Congress Cataloging-in-Publication Data
Names: Raby, Megan, author.
Title: American tropics : the Caribbean roots of biodiversity science / Megan Raby.
Other titles: Flows, migrations, and exchanges.
Description: Chapel Hill : University of North Carolina Press, [2017] | Series: Flows, migrations, and exchanges | Includes bibliographical references and index.
Identifiers: LCCN 2017009456| ISBN 9781469635590 (cloth : alk. paper) | ISBN 9781469635606 (pbk : alk. paper) | ISBN 9781469635613 (ebook)
Subjects: LCSH: Biodiversity—United States—History—20th century. | Biology—Fieldwork—Caribbean Area—History—20th century. | Biology—Fieldwork—Political aspects—Caribbean Area. | Biology—Fieldwork—United States—History—20th century. | United States—Foreign relations—1897–1901. | United States—Foreign Relations—20th century.
Classification: LCC QH541.15.B56 R33 2017 | DDC 333.95—dc23
LC record available at https://lccn.loc.gov/2017009456

Cover illustration: Scientists at the Canal Zone Biological Area, Barro Colorado Island, Panama, pose at the foot of *Bombacopsis fendleni*, a.k.a. the "spiny cedar" (Smithsonian Institution Archives, Image SIA2011-0102).

Portions of this book were previously published in a different form. Chapter 1 includes material from "A Laboratory for Tropical Ecology: Colonial Models and American Science at Cinchona, Jamaica," in *Spatializing the History of Ecology: Sites, Journeys, Mappings*, ed. Raf de Bont and Jens Lachmund (New York: Routledge, 2017), 56–78. Chapters 2 and 3 include material from "Ark and Archive: Making a Place for Long-term Research on Barro Colorado Island, Panama," *Isis* (2015): 798–824. Used here with permission.

Contents

Illustrations, Maps, and Table

ILLUSTRATIONS

MAPS

TABLE

Acknowledgments

This book is a product of the time for research and writing generously provided by fellowships at the Smithsonian Institution Archives and National Museum of American History and the Institute for Historical Studies, University of Texas at Austin. Travel for research and sharing ideas was also supported by JSTOR Global Plants and the Andrew W. Mellon Foundation; the National Science Foundation CHANGE-IGERT, Nelson Institute for Environmental Studies, University of Wisconsin Madison; and the College of Liberal Arts, University of Texas at Austin.

Although it sometimes feels like it, no book is a solitary endeavor. Historical research is possible only because of the dedicated work of archivists and librarians. I have many to thank, but my time at the Smithsonian Institution Archives was certainly the foundation of this project. This is not only because of the tremendous resources available there but also because of Pamela Henson's invaluable mentorship from the very beginning of my career as a historian. Thank you also to Courtney Bellizzi, Ellen Alers, Tad Bennicoff, and the whole archives staff. I have been fortunate in finding enthusiastic and skilled guidance at every collection I consulted—indeed, several individuals went far beyond the call of duty in helping me. My sincerest thanks go to Mai Reitmeyer and Colin Woodward at the American Museum of Natural History Research Library; Mary LeCroy of the American Museum of Natural History, Department of Ornithology; Lizeth Zepeda at the Arizona Historical Society; Sheila Connor, Lisa Pearson, and Larissa Glasser at the Arnold Arboretum Archives of Harvard University; Amy McDonald at the Duke University David M. Rubenstein Rare Book and Manuscript Library; Nancy Korber at the Fairchild Tropical Botanic Garden Special Collections; the staff of the Harvard University Archives; Jim Stimpert at the Johns Hopkins University Ferdinand Hamburger Archives; Andrew Colligan at the Missouri Botanical Garden Archives; Janice Goldblum at the National Academies Archives; Chris Killillay and Tim Nenninger at the National Archives and Records Administration; the staff of the New York Botanical Garden Mertz Library Archive; the Organization for Tropical Studies fiftieth anniversary oral history project; the staff of Princeton University Library's Manuscripts Division; Tom Rosenbaum at the Rockefeller Archive Center; Vielka Chang-Yau,

Angel S. Aguirre, Marialuz Calderon, Ron Herzig, and Alvin Hutchinson at the Smithsonian Tropical Research Institute Earl S. Tupper Tropical Sciences Library; and Kerry Prendergast and Sana Masood at the Wildlife Conservation Society Library and Archives. I also thank the staff of the University of Wisconsin Libraries, University of Texas Libraries, and interlibrary loan librarians everywhere.

Many people have contributed conversation, camaraderie, or other forms of key assistance or insight at just the right moment. It is a great pleasure to thank Anabelle Arroyo, Mitch Aso, Robert Beattie, Erika Bsumek, Guillermo Castro Herrera, Catherine Christen, John Christy, Thomas Croat, Fritz Davis, Raf de Bont, Peter Del Tredici, Carmela Diosana, Laura Flack, Arturo Flores, Seth Garfield, Fred Gibbs, Lawrence Gilbert, Allen Herre, Dr. Jones, Rachel Koroloff, Jens Lachmund, Matthew Larsen, Marixa Lasso, Egbert Leigh, Dan Liu, Jonathan Losos, Scott Mangan, Alberto Martinez, Katherine McLeod, Courtney Meador, Samantha Muka, Rachel Page, Germán Palacio, Steven Paton, Pedro Pruna, Lynnette Regouby, Ann Saudek, Tony Saudek, Robin Scheffler, David Singerman, Jeffrey Stine, Jenna Tonn, Greg Urwin, Sonja Walch, William Wcislo, Edward O. Wilson, Mike Wise, Anna Zeide, and Katie Zimmerman. My deepest apologies to those I have accidentally neglected to include.

As this project has developed, I have also benefited greatly from the questions asked and suggestions made by participants at a variety of conferences and workshops, including the Arizona State University–Marine Biological Laboratory History of Biology Seminar; History and Philosophy of Science Colloquium, University of Texas at Austin; History of Science Colloquium, University of Wisconsin–Madison; and Spaces of Ecological Knowledge Workshop, Faculty for Arts and Culture, Maastricht University; as well as events sponsored by the American Historical Association; American Society for Environmental History; Center for Culture, History, and Environment, University of Wisconsin–Madison; Ciudad del Saber, Panamá; Columbia History of Science Group; Ecological Society of America; History of Science Society; Institute for Historical Studies, University of Texas at Austin; Smithsonian Tropical Research Institute; Sociedad Latinoamericana y Caribeña de Historia Ambiental; and Southern Historical Association. I have also been immensely privileged to be a member of the History Department of the University of Texas at Austin as I completed this book and to enjoy the support of this amazing community of scholars.

I am most deeply indebted to those individuals who read various parts, iterations, and offshoots of this project. This first of all includes the mentors who have shaped my thinking about the history of science and the environment. Thank you Gregg Mitman, Lynn Nyhart, and Bill Cronon. I would not be in this field without the encouragement of Michael Reidy. This would also be a much poorer book without critical feedback from Ashley Carse, Andrew Case, Brent Crossen, Lina Del Castillo, Jennifer Holland, Charles Hughes, Christine Keiner, Doug Kiel, Christine Manganaro (who also deserves special thanks for reminding me to breathe), Stuart McCook, Meridith Beck Mink, Crystal Marie Moten, Abena Osseo-Asare, Franco Scarano, Blake Scott, Andrew Stuhl, Paul Sutter, Libby Tronnes, Jeremy Vetter, Sam Vong, Emily Wakild, Don Waller, Robert-Jan Wille, Amrys Williams, and Kristin Wintersteen. For their generosity in providing thoughtful, detailed comments on the entire book manuscript, I am forever grateful to Elizabeth Hennessy, Bruce Hunt, Jerry Jessee, and the anonymous reviewers. My thanks also go to Brandon Proia, the series editors Mart Stewart and Harriet Ritvo, and everyone at UNC Press for making the publication of this book such a pleasant experience.

My family has had to put up with many years apart and many years of my working through holidays. Thank you for your love and patience, Stephen and Nancy Raby. Kristy and Terry Williams, thank you also for your son.

Without Eric Williams I would never have been able to complete this book. He has kept me fed, clothed, and proofread. You mean the world to me, Eric. Thank you for being my partner in this and all things.

Abbreviations in the Text

AEC	Atomic Energy Commission
ATB	Association for Tropical Biology
BCI	Barro Colorado Island
CZBA	Canal Zone Biological Area
ESA	Ecological Society of America
IBP	International Biological Program
IRTA	Institute for Research in Tropical America
MCZ	Museum of Comparative Zoology (Harvard)
NRC	National Research Council
NSF	National Science Foundation
NYBG	New York Botanical Garden
NYZS	New York Zoological Society
OTS	Organization for Tropical Studies
STRI	Smithsonian Tropical Research Institute
UNAM	Universidad Nacional Autónoma de México
USDA	U.S. Department of Agriculture

American Tropics

INTRODUCTION

From Tropicality to Biodiversity

The problems of human beings in the tropics are primarily biological in origin: overpopulation, habitat destruction, soil deterioration, malnutrition, disease, and even, for hundreds of millions, the uncertainty of food and shelter from one day to the next. These problems can be solved in part by making biological diversity a source of economic wealth.

—Edward O. Wilson, 1988

In September 1986, sixty scientists and policy makers convened for the "National Forum on BioDiversity" in Washington, DC. The conference, organized under the auspices of the Smithsonian Institution and the National Academies of Science, included some of the biggest names in the U.S. science and conservation communities, Edward O. Wilson, Thomas Lovejoy, Paul Ehrlich, Peter Raven, Stephen Jay Gould, and Michael Soulé among them. Although it was a U.S. "national" forum, its ambitions were decidedly global. As each speaker came to the podium, a picture of a worldwide extinction crisis emerged. Together, they made a forceful, and very public, case for the need for more scientific research in support of conservation around the globe. Species were being lost, they warned, before they could even be discovered.[1]

To articulate their cause, the conference organizers coined the term *biodiversity*, which quickly became the rallying cry of the emerging field of conservation biology. As the forum was telecast and participants interviewed by news agencies nationwide, it even became a household word. In a narrow sense, *biodiversity* refers to the number and variety of species in a given area. Although most definitions also include variation within species (genetic diversity) and at the level of ecosystems, the term is often used as a synonym for *species diversity*—the number and relative abundance of species in an area. As a scientific measure, biodiversity offered an important tool for making conservation priorities. The discourse surrounding the term *biodiversity*, however, also helped reinforce the global nature of the conservation problem at hand. At stake, conservationists argued, was not just particular wild places or even individual endangered species; the threat was to the diversity of life on Earth itself.[2]

Nevertheless, one global region dominated the dire stories and statistics that the participants cited: the tropics. A whole session of the forum (the only one delimited geographically) focused on tropical problems. Many of the organizers and participants were specialists in tropical biology, and the institutions hosting the forum were among the country's most important supporters of tropical research. Tropical imagery dominated even the poster that loomed onstage beside each speaker, and that afterward became the cover of the forum's proceedings. On it, water droplets cling to a lush background of greenery, each reflecting images that represent Earth's diverse biomes; the largest drop contains a rainforest whose colorful animal inhabitants peer out at the viewer. This tropical emphasis was not accidental. Wilson reminded his audience that, although tropical rainforests "cover only 7% of the Earth's land surface, they contain more than half the species in the entire world biota."[3] It was also the region most in peril, as deforestation and population growth ran rampant "especially in tropical countries."[4] For this reason, Wilson argued, "rain forests serve as the ideal paradigm of the larger global crisis."[5] The biodiversity crisis might be global, but it was both centered on and symbolized by the tropics, particularly tropical forests.

Significantly, Wilson and the other participants cast tropical diversity not only as threatened nature but also as a natural resource. The uncatalogued species of the world, the vast majority of which lay in the tropics, figured as humanity's most irreplaceable and untapped asset. Tropical nature, they suggested, could be transformed into a source of salvation rather than suffering, thus addressing some of the world's most pressing social and economic problems at their root. With biologists' expertise, the diversity of life could be recognized and valued as "a source of economic wealth," thus effecting development in harmony with conservation at a global scale.[6] This move linked a need for basic research in tropical biology—long an obscure and underfunded field—to some of the most politically significant issues of the day. But how had scientists come to connect the abstract and technical concept of species diversity to the problems of international development? And why did a group of biologists from the north temperate zone insist on moving the tropics to the center of global debates at this historical moment?

The suddenness with which *biodiversity* landed on the lips of policy makers and "Save the Rainforest" appeared on bumper stickers belies a deeper intellectual and political history. Consciousness of tropical biodiversity exploded onto the scene in the mid-1980s, but it was not a new concept to biologists. U.S. scientists' engagement with life in the tropics already stretched back a century. During this time, scientists had struggled with questions of

Cover of the proceedings of the 1986 National Forum on BioDiversity. Reprinted with permission from *BioDiversity*, copyright 1988, National Academy of Sciences, courtesy of the National Academies Press, Washington, DC.

the biological differences of the tropics—especially its richness in species—and at the same time entangled themselves in U.S. corporate and government efforts to exploit tropical resources. This book argues that both the key scientific concepts and the values embedded in the modern biodiversity discourse had significant precedents in biologists' involvement in U.S. encounters with the tropical world over the course of the twentieth century, centered on the circum-Caribbean region. From the era of the Spanish-American War and the construction of the Panama Canal through the revolutionary 1960s and

1970s, U.S. biologists sought ongoing access to research sites in the tropics. They found it through a complex set of partnerships and the intertwining of intellectual and economic agendas. *American Tropics* argues that the ideas, attitudes, and institutions forged at field sites in the colonies and neocolonies of the circum-Caribbean are crucial for understanding the emergence of this new paradigm in biology and conservation at the end of the century. Long before the 1986 BioDiversity Forum extended such ideas to the globe, U.S. biologists had begun both to articulate fundamental biological questions raised by the diversity of tropical life and to argue for its potential as a resource.

Biology, Diversity, and the Tropics

Although the word *biodiversity* was new when it appeared in the title and throughout the speeches of the 1986 forum, its organizers did not feel they were doing anything radical by contracting a conventional phrase. *Biological diversity* was already in use, not to mention long-standing formulations like *organic diversity*, or the host of specialized quantifications of species richness and species diversity.[7] Historians and philosophers of biology have referenced such precursors, but their primary focus has been on the use of the term *biodiversity* in conservation circles since the 1980s.[8] In fact, the contemporary conception of biodiversity emerged from much longer-standing scientific efforts to understand the numbers and distribution of species on Earth, and particularly in the tropics. In part, this book argues that tropical biologists played a much more central role in shaping this intellectual history through the twentieth century than scholars have previously acknowledged.

To understand why tropical biology was central to formulating modern concepts of biodiversity, however, we must first understand what constitutes this field of study. What is tropical biology? In one sense, it is simply the study of living things in the tropics—the global region surrounding the equator between 23° 26′ 16″ north and south latitudes. Geographically, the tropics are marked by a near-constant day length throughout the year and climates free from frost (except at high elevations). The nature of biological adaptations to such environmental factors is among the problems that interest tropical biologists. But if tropical biology is a regional specialization, it is also an interdisciplinary field. The biologists who publish in journals like *Biotropica, Tropical Ecology*, or *Revista de Biologia Tropical*, who attend the meetings of the Association of Tropical Biology and Conservation or the International Society for Tropical Ecology, or who work at organizations like the Smithsonian Tropical Research Institute or the Organization for Tropical Studies come from the full range of subdisciplines within biology. They may study particular groups of organisms or focus on biological processes. Their approach may emphasize taxonomy, behavior, physiology, evolution, or ecology—although ecology becomes the touchstone when confronting the complexity of many tropical ecosystems.[9] Tropical biologists are driven by a range of motivations, including curiosity about the puzzles of theoretical biology, a love of fieldwork, or concerns about conservation and resource use. The work that tropical biologists do is nearly as diverse as the ecosystems they study.

Tropical biology is more than a geographic pigeonhole, however. Why are there scientific journals, professional associations, and research institutions devoted to tropical biology while "temperate biology" remains an unmarked category? Historically, the field of tropical biology was demarcated by outsiders. From the European voyages of exploration of the sixteenth century to the overland expeditions of nineteenth-century naturalists, northern visitors to the equatorial regions tended to set the plants, animals, and people they encountered there in comparison with those of their temperate homelands. During the same period, tropical countries developed their own local scientific communities, but whether these chose to designate their research as specifically "tropical" varied. An outsider's perspective thus left its mark on the kinds of questions the emerging community of tropical biologists asked: How does tropical life compare with life elsewhere? Does life in the tropics share qualities that make it unique? Conversely, might the study of living things in the tropics reveal something fundamental about the phenomena of life everywhere?[10]

These are questions with roots that extend far beyond the scientific community. The tropics are loaded with powerful meanings and imagery for outsiders. In the imaginations of Europeans and North Americans, the equatorial regions have figured as an earthly paradise or a green hell. This discourse of tropicality, as identified by cultural geographers and historians, has pictured the tropics as an exotic Other—a place both attractive and dangerous, a realm of unbounded abundance and riotous growth as well as disease and decay.[11] Like Orientalist discourses, this exoticized imaginary geography of the equatorial regions has functioned to justify the colonization and exploitation of its people and environments.[12] In it, tropical people appear backward and lazy, spoiled by the abundance offered by tropical nature, and unable to efficiently develop the resources of their own lands. In popular imaginings, the tropics were—and largely remain—a region of untapped potential. Thus, while *tropical biology* may be defined as "biology in the tropics," the complex, contested cultural meanings of *the tropics* make this far from a simple declaration.

Scientists' views of the tropics have run in tandem with these broader cultural understandings. The narratives of naturalists and explorers like Alexander von Humboldt, Henry Walter Bates, Richard Spruce, Thomas Belt, and Alfred Russel Wallace worked alongside those of other foreign travelers to construct and reproduce the familiar tropes of the tropical. Traveling naturalists saw with "imperial eyes," carrying with them on their journeys prejudices and assumptions shaped by generations of previous visitors.[13] Working within the naturalist tradition, however, they sought to systematize and explain the phenomena they confronted—in ways that often reinforced but sometimes challenged their own preconceptions. Encounters with nature in the tropics sparked some of the most important biological syntheses of the nineteenth century, from Humboldt's plant geography to Darwin and Wallace's theory of evolution by natural selection. To varying degrees, some naturalists were aware of the ways their expectations about life in the tropics were shaped by forerunners' accounts. Darwin acknowledged how Humboldt's descriptions loomed in his mind when he "first saw a Tropical forest in all its sublime grandeur."[14] For others, heightened anticipation of superabundant life led to disappointment when lush vegetation appeared "monotonous" and parrots failed to adorn every tree.[15]

This heavy burden of prior writing led Wallace to open his influential 1878 essay *Tropical Nature* with the admonition, "The luxuriance and beauty of Tropical Nature is a well-worn theme, and there is little new to say about it." What was sorely lacking, he argued, was a serious effort to "give a general view of the phenomena which are *essentially* tropical, [and] to determine the causes and conditions of those phenomena."[16] To replace the "many erroneous ideas" in circulation, Wallace synthesized his own extensive field observations in Malaysia and the Amazon with other reports. As evolutionary theories increasingly provided a framework for such analyses from the late nineteenth century on, this essay became a new starting point for investigations into tropical difference. The basic patterns Wallace described in *Tropical Nature* are still recognized and studied by tropical biologists today; thus, they provide a convenient introduction to the phenomena that would continue to populate the conceptual space of the tropics for biologists during the following century.

First among the general features of tropical vegetation that Wallace recognized was its lushness of growth—in modern terms, the sheer biomass of plant life. In contrast, animal life was inconspicuous and widely dispersed. For both animals and plants in the tropics, however, the "best distinguishing features are the variety of forms and species."[17] He noted the prevalence of certain types of plant adaptations, including climbing, epiphytic, and parasitic habits; the division of mature forests into layers of canopy and undergrowth forms; and a variety of trunk structures, such as buttressing, that were highly unusual outside of the tropics. Among animals, Wallace found "such a diversity of forms, structures, and habits, as to render any typical characterisation of them impossible," but he did note, for example, some particularly close interrelationships of plants and insects.[18]

Wallace not only enumerated these "peculiar" features of plant and animal adaptation but also described certain broad patterns in the distribution of species at local and global scales that have remained among the central puzzles of tropical biology.[19] Within tropical forests, Wallace contrasted the high number of species with the low abundance of individuals of any one species. This was in no case more evident than among trees: "If the traveller notices a particular species and wishes to find more like it, he may often turn his eyes in vain in every direction. Trees of varied forms, dimensions, and colours are around him, but he rarely sees any one of them repeated. Time after time he goes towards a tree which looks like the one he seeks, but a closer examination proves it to be distinct. He may at length, perhaps, meet with a second specimen half a mile off, or may fail altogether, till on another occasion he stumbles on one by accident."[20] While temperate forests are dominated by one or a few key species, tropical forests are in general characterized by diversity and rarity—no one species dominates. At the same time, Wallace noticed a curious global pattern in species distribution. For most groups of plants and animals, more member species inhabited the tropics than the temperate zone. A genus or family might contain many times more tropical than temperate species. Moreover, the tropics harbored a wide array of groups rarely found or entirely absent in the North—parrots and palms, for example, as well as whole families of insects, reptiles, and amphibians. Oddly, the reverse pattern was extremely uncommon. Whereas pre-Darwinian biogeographers used such regional concentrations of species numbers to posit "centers of creation," Wallace offered evolutionary explanations, notably a "less severe" struggle for existence against "physical conditions" and the historical absence of glaciation in the tropics.[21] This, he argued, had allowed the "comparatively continuous and unchecked development of organic forms" at the equator, leaving the region a "more ancient world."[22] Biologists would continue to debate these proposed evolutionary and ecological causes of tropical diversity over the next century.

The way that Wallace and subsequent researchers imagined the tropics resonated powerfully with broader ideas about tropical exuberance, excess, and primitiveness. Nevertheless, biologists' views cannot be reduced to such tropes. Tropical biology emerged around a scientific subdiscourse about the characteristics and qualities of tropical life that was also part of broader currents in ecological and evolutionary thought. The specificity of these concerns distinguished (although never freed) scientists' conceptions from the broader discourse of tropicality. Among their key concerns about tropical life was its diversity—its great numbers and variety of species.

This book argues that the articulation of biological ideas about diversity and tropicality went hand in hand through the twentieth century. It examines this relationship particularly as biologists from the United States began to work in the circum-Caribbean and form a professional community of tropical biologists. Biological concepts of diversity have many roots; those that lie in tropical research are especially significant and have been surprisingly neglected by historians. This neglect has obscured the ideas and institutions that laid the foundations for the explosive rise of *biodiversity* at the end of the century.

A Place-Based Science

Tropical biology is a place-based science that has historically been practiced by people from outside that place. This central irony has played a profound role in determining the development of the field and its major institutions. For North Americans and Europeans, studying tropical life meant traveling to the tropics. Expeditions allowed northern naturalists to pass through the tropics, bringing back observations and collections of specimens to the metropolitan scientific centers of their home countries. Expeditionary science could reveal the global biogeographic patterns that Humboldt, Wallace, and others commented upon, but it did not permit extended, in situ observations of living tropical organisms and their complex interactions. Understanding the ecological and evolutionary processes that caused the great differences and diversities of tropical life meant not just traveling through but staying in the tropics. It required a more permanent institutional basis, in the form of tropical research stations.[23]

By not only tracing the flow of U.S. scientists into the tropics but also examining how they were sometimes able to take root in place, *American Tropics* contributes to understandings of mobility and knowledge production.[24] Field stations became nodes of scientific migration, giving form to a community of tropical biologists in the United States and shaping its research practices and priorities. As the United States made its first foray into tropical empire at the turn of the twentieth century, so did the nation's biologists, founding such stations as Harvard's Atkins Institution at Soledad, Cuba (1899); Cinchona Botanical Station in Jamaica (1903); William Beebe's stations in British Guiana (Guyana, 1916 and 1919); and, most notably, the station at Barro Colorado Island (BCI), Panama (1923). While the first professional associations, university programs, and specialty journals in tropical biology did not appear until the 1960s, they emerged directly from a ready-made community centered on stations founded at the beginning of the century. Indeed, we will see in chapter 5 that the final stage in the professionalization of tropical biology was catalyzed by political and environmental threats to long-established research sites.

The U.S. movement to establish tropical stations was part of a larger, international groundswell for stations in the field. The influential marine stations Stazione Zoologica in Naples, Italy, founded in 1872, and Marine Biological Laboratory in Woods Hole, Massachusetts, founded in 1888, have played a significant role in the historiography of biology and American science, signaling the rising prestige of the laboratory and the professionalization of the discipline.[25] Investigations into the relationship between place and practice at field stations were pioneered by Robert Kohler, who approached them as border zones between the scientific cultures of the laboratory and field.[26] Historians of biology have recently begun to examine stations in relation to not only laboratory science but also other institutions and spaces, such as museums, aquaria, parks, and gardens, as well as in the context of a broader set of goals, priorities, and practices.[27]

If, as Raf de Bont notes, field stations have too often been seen primarily as extensions of urban laboratories, then tropical stations have doubly been cast as mere appendages to northern scientific institutions.[28] The Cinchona station, for example, has been discussed as part of the expanding scientific "empire" of the New York Botanical Garden (although its institutional affiliations changed over time), but how the tropical and Jamaican context shaped day-to-day scientific practice there has remained unexamined.[29] Likewise, BCI has been cast as a "failed" attempt to replicate the Naples and Marine Biological Laboratory model, finally saved only by coming under the aegis of the Smithsonian Institution after World War II.[30] Tropical stations have been approached as decidedly peripheral and subordinate to metropolitan, temperate-zone institutions. A significant exception is the pioneering Dutch tropical botanical station known as the Treub Laboratory of the Buitenzorg Gardens, Java, established in 1884, which Eugene Cittadino has argued played a central role in the emergence of Darwinian plant physiology.[31] Buitenzorg, in fact, stood out as the prime exemplar of the possibilities of tropical research

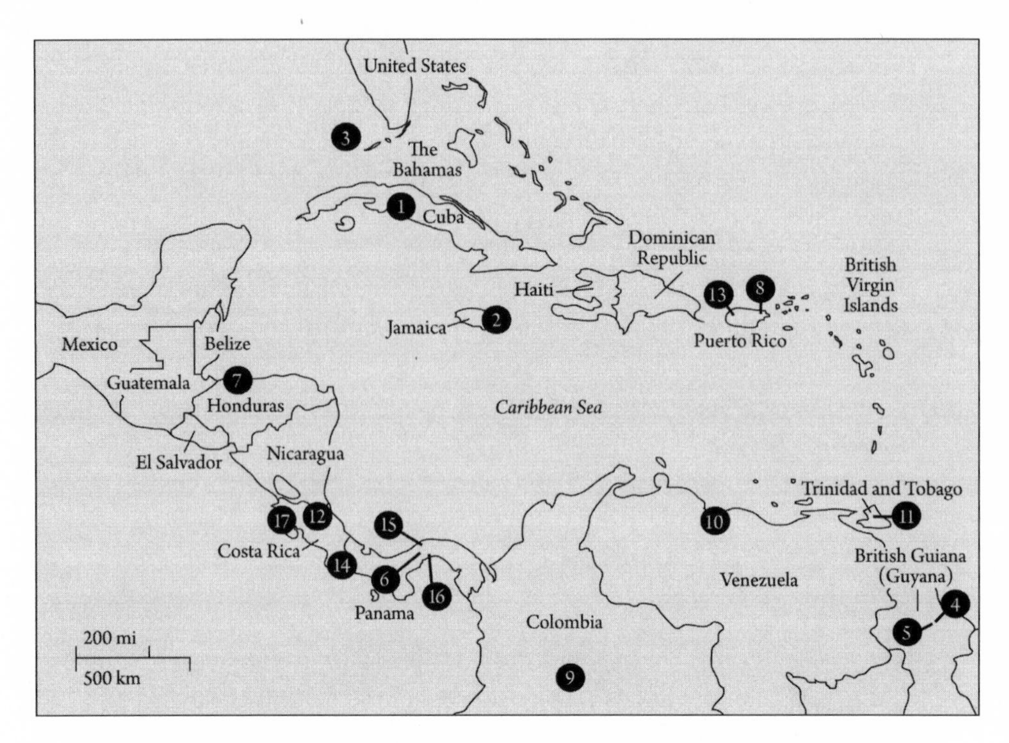

U.S. tropical biological stations, 1899–1969. This map depicts biological stations operated by U.S.-based institutions. Here, *station* denotes a scientific site with facilities and accommodations for permanent or prolonged and repeated research outdoors—not including temporary camps, parks, or other types of research lands—and *biological* research means basic research on living organisms. This does not include strictly agricultural or forestry experiment stations, botanical gardens, or medical laboratories but does include biological stations that contained or emerged from such institutions. Dates indicate when they were operated by U.S.-based institutions. If a station or other institution exists at this site today, its current name is noted.

for U.S. scientists in the early twentieth century, as explored in chapter 1. This history further complicates narratives centered on the dichotomy of lab versus field, because tropical research stations often emerged in relation to (or even on the grounds of) another distinct and venerable institution: colonial botanical gardens.[32] Thus, *American Tropics* examines U.S.-run stations not only in connection with a variety of U.S. metropolitan institutions but also in context with existing scientific institutions in tropical countries.

MAP KEY U.S. Tropical Biological Stations, 1899–1969

1. Soledad station (Also called Harvard Botanical Station for Tropical Research and Sugarcane Investigation/Atkins Institution of the Arnold Arboretum/Atkins Garden and Research Laboratory), Harvard, 1899–1961. (Today the Jardín Botánico de Cienfuegos.)
2. Cinchona Botanical Station (New York Botanical Garden, 1903–1914; Smithsonian Institution, 1916–1921), 1903–1921. (Today the Cinchona Botanical Gardens.)
3. Dry Tortugas Marine Biology Laboratory at Loggerhead Key, Carnegie Institution, 1903–1939.
4. "Kalacoon" Tropical Research Station, New York Zoological Society, 1916–1917.
5. "Kartabo" Tropical Research Station, New York Zoological Society, 1918–1926.
6. Barro Colorado Island (Institute for Research in Tropical America, 1923–1946; Smithsonian Institution 1946–present. Part of the Canal Zone Biological Area, 1940–1966; STRI 1966–present), 1923–present.
7. Lancetilla Experiment Station, United Fruit Company, 1925–1974. (Today El Jardín Botanico y Centro de Investigacíon Lancetilla.)
8. El Verde Field Station, U.S. Forest Service, 1937–present.
9. Villavicencio Yellow Fever Laboratory, Rockefeller Foundation, 1938–1948. (Today the Estación de Biología Tropical Roberto Franco.)
10. Rancho Grande Station, New York Zoological Society, 1945–1948. (Today the Estación Biológica Fernández Yépez.)
11. "Simla" Tropical Research Station, New York Zoological Society, 1949–1974. (Today William Beebe Tropical Research Station.)
12. La Selva Biological Station (OTS, 1968–present), 1954–present.
13. Institute of Marine Biology Laboratory/Isla Magueyes Laboratories, University of Puerto Rico, 1954–present.
14. Las Cruces Biological Station (OTS, 1973-present), 1962–present.
15. Galeta Marine Laboratory, STRI, 1965–present.
16. Naos Island Laboratories, STRI, 1965–present.
17. Palo Verde Biological Station, OTS, 1969–present.

Although they borrowed from established models, tropical stations were varied and heterogeneous. Some emphasized the study of plants over animals or vice versa. They focused on applied or basic science to varying degrees. They were sponsored by universities, zoos, botanical gardens, government agencies, and corporations, as well as through collaborations among such organizations. They incorporated a variety of scientific spaces on site: Soledad housed experimental plots, botanical gardens, and greenhouses, as well as a biological laboratory, and had access to uncultivated land. BCI comprised a forested island nature reserve, laboratory space, herbarium, dark room, and library, as well as buildings and plots of land that might be turned to a variety of experimental or observational purposes. Depending on the moment or context, these stations might be elided as *laboratories*, *gardens*, or *institutions* or just denoted by locality—naming practices that suggest their flexibility as scientific sites.[33] Rarely, however, did they appear without the modifier *tropical* close at hand. More than the promotion of any particular institutional model or set of research methods, then, tropical research stations shared a common purpose: to bring northern scientists into contact with nature in the tropics.

Of course, field stations elsewhere were also built to bring scientists "close to nature," but this goal carried very particular meanings for the twentieth-century U.S. scientific community regarding the tropics. During the nineteenth century, U.S. botanists and zoologists were occupied by the project of cataloguing the species of their nation's expanding continental empire.[34] Very few had experience working in the tropics. The new stations gave these temperate-zone scientists a tropical foothold, serving as outposts in the scientific colonization of lands to the south. Research stations removed much of the hardship of organizing a scientific expedition and ameliorated the real and imagined dangers to health and safety of travel in the tropics. By encouraging a much larger number and broader array of biologists to work in the region, stations built up a population of U.S. researchers with experience in tropical fieldwork during the twentieth century. Over time some stayed longer, using stations not just for brief tours but instead as a "home" in the field.[35] By midcentury, these researchers came to identify themselves as "tropical biologists."

Research stations not only offered the ability to access tropical organisms and environments while living in relative comfort, health, and safety, however. They also allowed a wider and qualitatively different set of research practices—practices that this book shows were crucial to the development of the study of species diversity. As discussed in chapter 1, as early as the beginning of the twentieth century scientists realized that the study of the ecologies, life histories, and behavior of tropical organisms lagged far behind those of temperate-zone organisms—comparatively few biologists worked in the tropics where, paradoxically, the greatest numbers of species existed. Whereas expeditions and surveys are best suited for the collection of specimens and measurements from a broad geographic area, permanent stations orient work toward the study of living organisms in place over longer periods of time. They permit repeated or ongoing experimentation and monitoring. Thus, tropical field stations opened up studies in ecology, physiology, and behavior where taxonomic and biogeographic work had long dominated. Taxonomic and distribution studies continued, however, and stations encouraged them to take place at a much more intensive, fine scale. Indeed, it seemed that the more closely biologists looked, the more species they found.

This concentration of research at stations in the tropics had the side effect of encouraging the accumulation of related studies over time. Biologists themselves have identified this as a significant quality of station work; BCI's Egbert Leigh argues for "the importance of 'unity of place,'" which allows "many different kinds of study, done in one small area [to] cohere into a unified picture."[36] Thus, even without the articulation of a formal research project, studies in a particular locality display increasing integrity, through the complex interconnections of the local ecology itself. For example, research on agouti behavior may provide background data for studies of the trees whose seeds they disperse, which in turn may raise questions about the effect of disease, or soils, or climate change on the diversity of the forest. Stations encourage a deep focus on particular places and organisms, fostering an approach that Ian Billick and Mary V. Price call the "ecology of place": "research that assigns the idiosyncrasies of place, time, and taxon a central and creative role in its design and interpretation, rather than as a problem to be circumvented through replication or statistical control."[37] This form of research shares qualities with the case-study approach of social scientists and is often long term and produced by loose groups of researchers from a variety of disciplines.[38] Such place-based research is not confined to the tropics, but, as Price and Billick argue, it has particular advantages for dealing with highly diverse and complex ecosystems. As we will see, it played a key role in the emergence of species diversity as a core concern of tropical biologists.

The history of tropical research stations is fundamental to understanding the intellectual development of biodiversity and tropical thinking, but it also helps contextualize persistent geographical biases in ecology. Despite the emergence and growth of tropical biology in the twentieth century, documented here, the tropics remain vastly underrepresented in current ecological research.[39] Most tropical research remains concentrated at a few well-established field sites, especially in Panama and Costa Rica, whose origins *American Tropics* examines.[40] Ecologists are only beginning to recognize such biases and assess how they affect understandings of tropical ecosystems and patterns of diversity; through histories of the earliest and most important tropical stations, this book contributes to understanding how and why these particular geographical patterns of tropical research came into being. *American Tropics* also explores how biologists historically attempted to deal with the epistemological problems of extending place-based research into general knowledge. For example, chapter 3 demonstrates how administrators of BCI at first argued for the representativeness of its tropical forest but later shifted toward encouraging comparative studies with other locations, as shown in chapters 4 and 5. As ecologists today struggle to establish new sites and networks of long-term research, to make data about tropical ecosystems more robust, this book also sheds light on the factors that permitted some stations to survive and thrive to the present day while others closed their doors after a few years or decades.

Just as these stations were founded by U.S. institutions, U.S. researchers themselves remain disproportionately prominent within the field of tropical biology—especially within research in Central America and the Caribbean. The number of tropical biology publications authored by U.S. researchers far exceeds those contributed by scientists from any other nation.[41] The geography of ecological research in the tropics continues to reflect colonial legacies, with U.S. scientists dominating studies in the Americas and Europeans tending to emphasize Africa and Asia.[42] Tropical biologists today recognize a need to increase the national, geographic, and racial diversity of their ranks if their work is to have a broader impact in the places they study; indeed, a common refrain among tropical conservation biologists is the need for more local researchers to catalogue, study, and protect biodiversity.[43] *American Tropics* thus provides important historical context for understanding the ongoing problem of unequal participation in the profession of tropical biology, as well as how it came to be recognized as a problem rather than accepted as a matter of course.

Caribbean Encounters

In the twentieth century, U.S. biologists argued for the need for "tropical" research, not specifically research in the Caribbean. Nevertheless, they focused their attention on a region they called the American tropics. Geographically, this category encompassed the tropical latitudes of the Americas, an area equivalent to the now more often used scientific term *neotropics*. In its historical context, however, the term *American tropics* carried additional connotations of colonial possession.[44] As is clear from the chapters that follow, U.S. biologists used the term not only to denote a particular region but also to strengthen rhetorical connections to U.S. colonial projects and advance the claims of their national professional community over the area. I have chosen this provocative actors' category as this book's title to emphasize these colonial dimensions. Within the text, however, I endeavor to use *U.S.* rather than *American* as an analytic category denoting people and organizations from the

United States, to avoid reinforcing this elision of the United States and the Western hemisphere.

In fact, the geographical frame of this book is the circum-Caribbean—the region including the islands of the West Indies and a broad swath of Central and South American land bordering the Caribbean sea.[45] The fundamental reason for this emphasis is that, historically, the few U.S. research stations that developed an emphasis on basic ecological research in the tropics—including investigations into questions of tropical species diversity—were indeed established in the circum-Caribbean and not elsewhere. U.S. biologists announced ambitions to universal knowledge and territorial claims on the "American tropics," but in practice the region they operated in did not actually encompass the entire neotropics. Research in tropical biology depends on access to tropical land, and the flows and migrations of U.S. scientists were constrained by imperial networks of infrastructure and patronage during the twentieth century.

Especially at first, they tended to work within British colonies, such as Jamaica or Guiana, where, as described in chapters 1 and 2, U.S. biologists found Anglophone populations, established scientific institutions, and existing infrastructure for permanent research stations. Increasingly, however, U.S. biologists accessed land and funding through the growing avenues of their own country's economic, political, and military hegemony. In 1898 Spain ceded Cuba, Puerto Rico, the Philippines, and Guam to the United States, and the United States annexed Hawaii. Following the U.S.-backed secession of Panama from Colombia in 1903, the Panama Canal Zone was created and came under U.S. administration. At the beginning of the twentieth century, the United States had thus acquired an "archipelago" of tropical colonies.[46] In the decades that followed, American agribusinesses—most notoriously the United Fruit Company—became major tropical landholders, exerting influence on economies, politics, and landscapes well beyond the borders of these outright colonies.[47] U.S. biologists could make inroads into tropical environments through these neocolonies—in the Caribbean, Latin America, and the Pacific—far beyond anything possible during the previous century.

For U.S. naturalists and ecologists from East Coast and Midwest universities, museums, zoos, and botanical gardens, however, transportation infrastructure and patronage networks favored research in the circum-Caribbean through most of the century. The tropical Pacific was also a significant area of U.S. tropical research, but for a variety of reasons institutions in the Philippines and Hawaii focused more heavily on agriculture, forestry, and medicine than on basic ecology.[48] At the same time, although the Amazon rainforest looms large in biodiversity and tropical discourse today, U.S. scientists worked there only sporadically until the latter part of the century—not coincidentally, after U.S. interests in Amazonian rubber burgeoned during World War II.[49] Until the 1970s, U.S. tropical biologists' involvement in South America remained largely expeditionary. Certainly, throughout the century some U.S. researchers accessed the tropics through direct collaboration with local scientists.[50] The oldest and most influential stations for basic ecological research, however, were established in the circum-Caribbean during the early twentieth century.

This book's focus on basic research in tropical biology should not obscure the importance of applied tropical science. In fact, this emphasis is intended to better illuminate the historical evolution of connections between basic biology and efforts to colonize and develop natural resources in the circum-Caribbean. The links between U.S. science and colonial and neocolonial ventures are certainly more obvious in the realm of tropical medicine and agriculture. The large numbers of entomologists, engineers, and medical doctors who worked to control disease-bearing mosquitoes in Cuba and the Panama Canal Zone provided the foundations of U.S. territorial control, as well as humanitarian justifications for invasion.[51] Scientific agriculture, chemistry, economic botany, and entomology also played key roles in enabling vast banana, sugar, and rubber monocultures, mitigating the effects of pests and disease that threatened the profits of corporations throughout the region.[52] Such examples of science, medicine, and agriculture working in the service of empire are relatively well known, although their role in the growth of U.S. hegemony has received far less attention than has been trained on European science and empire.

Nevertheless, the rise of a self-identified community of "pure" or "basic" tropical biologists was also deeply embedded in the expanding networks of U.S. corporate and government influence in the circum-Caribbean. Studies of butterfly mimicry, monkey behavior, or orchid taxonomy may seem rather esoteric, but they, too, depended on access to land, transportation, and patronage. Through much of the twentieth century, these came largely through government agencies and corporations—organizations with strong interests in controlling lands and resources in the region. Proponents of basic research on the natural history and ecology of tropical species often struggled to gain support, but gradually they found a variety of ways to make the case for the relevance of basic tropical biology to regional colonial interests. This book argues that significant among these strategies was the development of the diversity-as-resource argument, which would become a key component of biodiversity discourse at the end of the century. Indeed, at that time, many long-standing research areas in tropical biology—including questions of species numbers and distribution—that had previously been deemed basic were reframed as applied problems of central importance to conservation biology.

As well as capturing the sites of the most significant institutions for basic research in tropical biology, the circum-Caribbean also offers analytical advantages. As a linguistically, politically, and environmentally heterogeneous category, the Caribbean demands attention to movements across national and imperial boundaries. Caribbean historians have increasingly defined the region not by drawing sharp boundaries around it but through reference to material and cultural circulations—movements of people, commodities, and practices, mediated through shared histories of colonialism and plantation slavery.[53] The history of science in the region mirrors these flows. This book traces U.S. scientists' travels through and sojourns within the circum-Caribbean, which were highly determined by existing and developing networks of commerce. At the same time, concentrating on the Caribbean rather than on formal U.S. territories across the tropics helps more closely weave together a wider array of sites—including neocolonies, European colonial territories, and independent nations.

This regional and local focus brings to light the ways that experiences in place reshaped ideas about tropical environments and organisms and, ultimately, the diversity of life. U.S. scientists never simply went to "the tropics"; they worked at specific Caribbean localities that they framed as tropical. The view of tropical nature that each research site presented was quite distinct; scientists' consciousness of human-environment interactions, for example, could be influenced by the prevalence of agricultural land (as at Soledad) or forest (as at BCI) surrounding a station. Local conditions—of infrastructure, labor relations, environment, politics—shaped what science was done and how it could be done. At the same time, local conditions were often significantly structured by regional relationships and changes. Even as broader trajectories of the history of U.S. science played out, the specificity of locality and region—including and especially colonial contexts—made U.S. biology in the circum-Caribbean significantly different from research at coastal marine, agricultural, or montane biological stations in the continental United States. U.S. tropical field research has long been treated as peripheral in the historiography of twentieth-century science. This is in part because of the dominance of studies of European colonial science and in part because of a long-standing tradition of exceptionalism in the historiography of American science, and also due to a broader neglect of twentieth-century sciences of natural history and exploration.[54] At the same time, the discourse of biodiversity itself is highly globalizing—its nesting scales situate local species as part of a global biological heritage.[55] Biodiversity has been so successfully framed as a world resource that it is tempting to see its intellectual history as abstracted from place—or as the product of discussions in a Washington, DC, conference room. Countering this tendency, *American Tropics* argues that Caribbean contexts strongly shaped the emergence of biodiversity as a central paradigm in theoretical and conservation biology.

Overview

Centered on U.S.-Caribbean encounters, the following chapters examine the interconnected development of understandings of the tropics and the diversity of life. Each chapter pays special attention to the cultural and institutional contexts of this intellectual history. U.S. researchers' dependence on access to tropical land, however, means that this narrative also traces the outlines of the growth and transformation of U.S. political and economic hegemony in the region—an era bookended by the lead-up to the Spanish-American War at the close of the nineteenth century and the rise of nationalist and anticolonial movements during the Cold War era.

During this period, a growing trade and transportation infrastructure made tropical environments increasingly accessible to U.S. scientists. At the same time, both the interest in tropical agriculture and the rise of the "new botany"—which emphasized studying plants in their natural habitats—gave impetus to botanists to work in the tropics. U.S. botanists pointed with envy to the progress of European scientists, who had access to tropical colonies. Even before the Spanish-American War, as we will see in chapter 1, they pushed for the creation of an "American tropical laboratory" to study plants in their tropical home.

Early efforts to create institutions for ecological research were far more difficult to sustain financially than stations with agricultural goals, however. In the 1910s and 1920s, rival zoologists Thomas Barbour and William Beebe each drew on their wealth, corporate and political connections, and larger-than-life personalities to transform the landscape of basic tropical research. While differing in their spatial practices and relative emphases on taxonomy or ecology, both men argued that the study of life in the tropics was fundamental to a broad understanding of biology. Barbour argued furthermore that "tropical biology" was essential to solving the nation's growing practical problems in tropical agriculture and medicine. Chapter 2 examines the stations they developed—Beebe in British Guiana, Barbour at Soledad, Cuba, and Barro Colorado Island (BCI) in the Panama Canal Zone—and how they leveraged U.S. economic interests in the tropics to further basic science.

These new stations drew hundreds of U.S. biologists, few of whom would have attempted a rigorous tropical expedition on their own. BCI in particular, however, became a model tropical forest. Chapter 3 demonstrates how the station's location on an island nature reserve within the Panama Canal Zone enabled unprecedented control over space and scientific labor. There, biologists were able to develop practices to study living tropical organisms as part of a complex, dynamic ecological community. Moreover, the monitoring and censusing of BCI's ecology in the 1920s through 1940s allowed researchers like Frank Chapman, Robert Enders, Clarence Ray Carpenter, Warder C. Allee, and Orlando Park to investigate ecological change over time, planting the seeds of a challenge to the old idea of the primeval, changeless tropics.

Tracing the fieldwork and ideas of Robert H. MacArthur, Howard T. Odum, and Theodosius Dobzhansky, chapter 4 examines the post–World War II rise of efforts to capture the complexity of tropical nature using a simplified quantitative measure: species diversity. The new approaches were abstract but were shaped by U.S. biologists' experiences in an increasingly wide array of sites within and beyond the circum-Caribbean—facilitated by the U.S. government's interest in tropical warfare, demand for tropical products, and the growth in air travel. The rise of mathematical and systems approaches in ecology, along with the population perspective of the modern evolutionary synthesis, recast the old question of the biological difference of the tropics. The need for tropical data to solve biology's core theoretical problems was now unquestionable.

The 1960s and 1970s saw a wave of highly influential publications on problems of the distribution and ecological controls on species diversity, which drew heavily on data from key tropical field sites. Yet, at this same moment U.S. scientists' future in the tropics was thrown into question. Revolution swept Cuba and protests erupted in Panama against the U.S. occupation of the Canal Zone. U.S. tropical biologists confronted the loss of access to their most important tropical stations. They responded by realigning themselves, creating professional organizations, and taking new steps toward international collaboration. As chapter 5 explains, they also recast their justifications for the support of basic research. Tropical research was not merely in the U.S. national interest, they began to argue; understanding the biological diversity of the tropics was essential for sustainable global development.

American Tropics closes with an examination of the postcolonial situation of tropical research in the circum-Caribbean. Today, the institutions that are the most important and heavily used by U.S. biologists for tropical research and teaching are located in independent republics: the Organization for Tropical Studies (OTS) in Costa Rica and the Smithsonian Tropical Research Institute (STRI)—since the 1979 dissolution of the Canal Zone—in Panama. Key players in the move to bring "biodiversity" to the public stage in the 1980s were tropical biologists who had deep connections to OTS and STRI during the previous two decades of transition. The emergence of the modern biodiversity discourse, this book argues, is a direct product of the intellectual and political ferment of tropical biology during that revolutionary period. The significance of that moment, in turn, can be understood only in the context of the full twentieth century and its mixed legacies for tropical biology—the development of place-based research practices and a long-standing dependence on institutions supported by U.S. corporations and government agencies.

Through the twentieth century, tropical biologists developed ways to cast problems of deforestation, food and population growth, and species loss as justifications for the support of basic research on tropical species and the investigation of the ecological relationships that sustain them. With the emergence of biodiversity, the problems of the tropics became the problems of the world.

CHAPTER ONE

An American Tropical Laboratory

At our very door there lies a vast terra incognita, with excessively luxuriant vegetation, and inviting endless research. The duty seems to be laid upon American botanists to establish and maintain an international laboratory in the American tropics.

—"An American Tropical Laboratory," 1896

In November 1896, a radical statement appeared in an editorial of the *Botanical Gazette*. Botany was biased, it asserted. It "rest[ed] largely upon the results of researches carried on in the north temperate zone, in gardens and laboratories situated between the parallels of 40° and 55°." Botanists had neglected the tropics, deriving their knowledge of living plants almost entirely "from a study of the plants indigenous to a strip of territory fifteen degrees in width and extending across one continent and partly across another." Taxonomists might be in a position to make universal claims, it suggested; they could rely on metropolitan herbaria for access to specimens collected around the globe. But plant morphologists, physiologists, and ecologists—botanists interested in the forms, functions, and environmental relations of living plants—could not study tropical nature from a laboratory in North America or Europe. The editorial warned "that many of the conclusions reached under such circumstances are not capable of general application is becoming more and more apparent." Existing laboratories might be capable of producing regional knowledge but could not form the foundation of a universal understanding of plant life. "The various problems in which American botanists are interested . . . need tropical conditions for their proper investigation," it contended: "The time has come for the establishment of a laboratory in the American tropics."[1]

The editorial was calculated to touch nerves among the community of U.S. botanists of the turn of the twentieth century. It appeared in a prominent place. The *Botanical Gazette* was the most influential journal of plant science in the United States, founded and edited by University of Chicago botanist John Merle Coulter. Readers would likely have assumed his authorship of the unattributed 1896 editorial. In fact, he had condensed a text supplied by Daniel Trembly MacDougal, an ambitious thirty-one-year-old plant physiologist at the University of Minnesota, best known today for his administration of the Desert Botanical Laboratory at Tucson, Arizona, and support for mutation theory and experimental evolution.[2] MacDougal and Coulter were both avid promoters of the "new botany," an approach emphasizing the use of the latest experimental techniques to study living plants' physiological adaptations. This movement advocated for laboratory work in botanical education and research. Yet by proclaiming that much existing research was "not capable of general application," MacDougal and Coulter's editorial aimed to expose an epistemological problem at the root of laboratory botany.[3] Far from accepting laboratories as "placeless places" capable of producing universal knowledge, they provocatively drew attention to the global geography of laboratory science.[4] By concentrating their work in Europe and North America, they argued, botanists had confined themselves to the study of a narrow range of temperate-zone plants and a few tropical species "growing under abnormal conditions."[5] The climate and soils of the tropics could not be re-created in northern laboratories and greenhouses. Rather than call for refined experimental techniques or more representative model organisms, MacDougal and Coulter lobbied for a change of place. For a truly general science of botany, they argued, botanists must go to the tropics.

Why in 1896 did these U.S. botanists feel that the "time had come" to establish a tropical laboratory? Through the nineteenth century U.S. naturalists had shown little interest in the region. Apart from scattered tropical expeditions, U.S. botanists and zoologists had focused on creating an inventory of their country's expanding western territory. Historians of science have even characterized U.S. botanists as "indifferen[t] to tropical research" before 1898, suggesting that only after the acquisition of Spain's former colonies did U.S. government agencies and companies' demand for experts on tropical plants push U.S. botanists toward the South.[6] The extension of U.S. territorial control into the tropics would indeed provide naturalists new lands and sources of patronage. Nevertheless, tropical botany did not merely follow the flag. A deeper examination of the institutions and intellectual content of U.S. botany in the tropics at the turn of the twentieth century shows a more complex relationship than simply science in the service of the state.

This chapter traces the origins of the U.S. tropical laboratory movement; the resulting rental of the station at Cinchona, Jamaica; and the first decade of research there by members of the founding generation of U.S. ecologists. Most notably, this period produced Forrest Shreve's 1914 *A Montane Rain-Forest*, among the first monograph-length ecological studies of a tropical forest. This history reveals U.S. botanists' range of motivations for engaging in tropical research, from the period leading up to the Spanish-American War in 1898 through the outbreak of World War I and the opening of the Panama Canal in 1914. The study of tropical organisms—with their diversity of forms and adaptations so foreign to those familiar with temperate flora and fauna—seemed to offer a path to a truly general understanding of living things. At the same time, U.S. botanists saw tropical research as the key to a place on the international scientific stage. As the 1896 *Botanical Gazette* editorial indicates, U.S. botanists did not wait for state-sponsored colonial science. Driven by a distinct set of intellectual, cultural, and professional concerns, they were ready to filibuster for science to acquire an outpost for research in the Caribbean.

In the "Primitive Habitat of the Plant Kingdom"

In 1895, MacDougal sought a leave of absence from his position as instructor in plant physiology at the University of Minnesota. He did not set out for the tropics but began an intellectual journey that would lead there. MacDougal's path to the tropics demonstrates the significant role tropical botany played in the emergence of the discipline of ecology, as well as how U.S. botanists first became convinced of the need for place-based research in the tropics. Although MacDougal was in some ways exceptional in his enthusiasm, his story serves as an introduction to the botanical community at the turn of the twentieth century and helps explain why it was such fertile ground for his movement for an American tropical laboratory.

MacDougal had earned his master's degree four years earlier at Purdue while studying with J. C. Arthur, a major advocate for the evolutionary and experimental approach of the new botany, as well as for the movement for agricultural research stations in the United States.[7] MacDougal had then joined the Botanical Division of the U.S. Department of Agriculture (USDA) to conduct expeditions through Idaho and Arizona, before taking the job in Minnesota. MacDougal loved the outdoors, but from the start his interests were not primarily in exploration and collection. He came to plant physiology after eagerly working through experiments in Arthur's laboratory manual. He was fascinated by the phenomena of plant movement—the twisting and curling of vines and roots, leaves that track the sun, and flowers that close up at night. This subject had become a point of controversy after Charles Darwin and his son Francis published *The Power of Movement in Plants* in 1880. Plant physiologists, particularly Julius Sachs in Germany, questioned their "country-house experiments."[8] During his graduate work in Arthur's lab, MacDougal had jumped into the fray, using the laboratory techniques of the

German experimental physiologists to study the tendrils of *Passiflora*, plants that the Darwins had also worked with.[9] MacDougal now sought leave to pursue these studies further at the best facilities Europe had to offer, a common practice for American botanists and zoologists at the time. He left Minnesota for Germany, Holland, and England in 1895, touring the laboratories of influential experimentalists, including Sachs, his former student Wilhelm Pfeffer, and Hermann Vöchting.

While these men differed in their views of the Darwins' work, MacDougal soon found that all of them kept *Mimosa pudica* among their laboratory flora. After experimenting with the plant in Leipzig and Tübingen, MacDougal became excited about its potential.[10] *Mimosa* is a curious plant: if touched or shaken, it dramatically folds up its leaflets and droops, lending it a variety of evocative common names, including sensitive plant, shameweed, and *morí-viví* ("I died, I lived"). *Mimosa* first aroused scientific curiosity as early as the seventeenth century, becoming fodder for debates about mechanical philosophy after being imported into Europe from the New World tropics. Was *Mimosa*'s behavior a purely mechanical reaction, or did it require sensibility or even consciousness? Its rapid movements blurred conventional divisions between plants and animals, a problem reinvigorated by evolutionary theory and the Darwins' *The Power of Movement in Plants*. Moreover, because most other plants' movements were "so slowly made as to be incapable of detection except by repeated or long continued observation," *Mimosa*'s rapid reactions impressed MacDougal as a tool for investigating the general phenomena of plant irritability and movement.[11]

MacDougal not only encountered this new experimental organism, however; he also came into contact with a group of botanists whose interest in such physiological processes emphasized their relationship to the plant's natural environment. A group of European plant physiologists had enthusiastically taken up Darwinian natural selection. Among them were Andreas Schimper, Ernst Stahl, Georg Volkens, and Gottlieb Haberlandt, and they began to approach plant structures and physiological processes explicitly as evolutionary adaptations.[12] They reacted against what they saw as their teachers' narrow focus on the study of internal structures and functions of organisms in the laboratory. Such an approach ignored the external conditions to which plants had adapted. For them, the question was not just how, for example, *Mimosa* moved but why. What had it evolved to respond to in its native environment?

The study of plant-environment relationships was not itself new. Through the nineteenth century, botanists had classified vegetation and its correspondence to physiographic regions.[13] Known as plant geography, or *pflanzengeographie,* the approach was exemplified by Alexander von Humboldt's correlations of vegetation along latitudes and elevations of equal mean annual temperature. MacDougal had himself taken part in this tradition during his expeditions in the U.S. West mapping "life zones."[14] Nevertheless, the Darwinian botanists MacDougal encountered on his European travels doubted that correlation and mapping alone had the capacity to explain the adaptation of organisms to the environment. It could reveal global patterns of distribution, but the *causes* of distribution, they believed, could be found only at the level of the physiological adaptations of individual organisms.

Seeking to correct the laboratory-based perspective of the previous generation, while adding experimental rigor to plant geography, they sought to understand evolutionary adaptation by uniting physiology with the study of organisms in their natural environments. The result was the beginning of the science of ecology—at first called physiological or ecological plant geography, drawing on Ernst Haeckel's previously little-used term *oecologie.*[15] The first book to explicate this approach was Danish botanist Eugenius Warming's *Plantesamfund: Grundtræk af den Økologiske Plantegeografi,* published in 1895 and translated into German the following year—just in time to make a splash during MacDougal's trip.[16] (Schimper subsequently took a similar approach, and its English translation also inspired the first generation of U.S. ecologists.)[17]

The tropics played a critical role in the development of this new field of research. Warming and Schimper had both conducted significant fieldwork in Brazil, Venezuela, and the West Indies between the 1860s and 1890s.[18] The tropical site that had done the most to convince European botanists of the potential of tropical research, however—and that clearly made the strongest impression on MacDougal—was the laboratory at the Government Botanical Garden ('s Lands-Plantentuin), Buitenzorg, Java, Dutch East Indies.[19] MacDougal saw Buitenzorg as "a sort of Mecca to the botanists of the world."[20] Although a Dutch colonial institution, it welcomed a cosmopolitan community of researchers, among them the supporters of the new field of ecology. Originally established as a garden on the estate of the governor's palace in 1817, it had had a joint identity as a pleasure ground and an introduction garden for economic plants. Dutch botanist Melchior Treub, its director since 1880, expanded the laboratory facilities, transforming it into the most advanced site for basic physiological research available in the tropics.[21]

Many referred to Buitenzorg simply as "Treub's laboratory," but it was actually a complex for a wide variety of botanical work; there were not only gardens and experimental plots but also a herbarium, library, museum, and studio for photography and engraving. Substations on the seashore and in the forested mountains at Tjibodas gave the new ecological plant geographers access to a variety of environmental conditions and myriad adaptations and species. Tjibodas in particular seems to have shaped early ecologists' impressions of tropical rainforests. Buitenzorg was thus more than a garden or a laboratory; it was a research station. Accommodating a wide variety of scientific work, including access to living tropical organisms under natural conditions, it was an outgrowth of a broader movement for biological field stations that had begun in Europe in the 1870s.[22] It was, however, unique in its location in the tropics and emphasis on plant science. Although agricultural research continued to be Buitenzorg's primary focus—and the justification for the Dutch colonial administration's support of the institution—the international community increasingly recognized it as a place for "pure" physiological and ecological plant research. In the late nineteenth century, U.S. scientists seeking institutional reform frequently used pure science as a rallying cry. The most famous example is the physicist Henry Rowland's "A Plea for Pure Science," which pressed for remodeling U.S. universities as research institutions on the German model.[23] What MacDougal heard and read of Buitenzorg captured his imagination as a model for a new kind of botanical research institution. On his return to the United States, he reported to the readers of *Popular Science Monthly* that "almost all branches of purely scientific and applied botany may be pursued to advantage within it." Among botanical gardens "organized chiefly for research" and "with regard to advantages of geographical position and botanical possibilities, the garden at Buitenzorg in Java occupies a foremost position."[24]

What were the "botanical possibilities" of Buitenzorg's location in the tropics? Through the 1890s, the new ecologists would call for laboratories in a wide range of environments. In particular, under extreme conditions—in the dry desert, the cold arctic, the warm and wet tropics—plants' adaptations to the physical factors of light, temperature, moisture, air, and soil would be "strongly marked and easily understood," as the Chicago ecologist Henry Cowles later explained in promoting Schimper's and Warming's work to U.S. readers.[25] Schimper even called for a "counterpart to Buitenzorg" in the arctic.[26] Nevertheless, tropical environments were of special concern from a botanical point of view. While cold and aridity produced interesting physiological adaptations for survival in harsh conditions, they were limiting factors that resulted in a sparse vegetation with a stunted appearance. In contrast, the tropics were marked by extreme abundance. Unlike the meager vegetation of desert, arctic, or alpine regions, tropical foliage was notoriously lush and profuse: gigantic leaves and flowers; towering, dense jungles; vegetation that blanketed the landscape—the luxuriance of plant growth that was possible in the moist tropics captured the imagination of botanists as it did other travelers from the North. Plentiful moisture and constant warmth and sunlight were just those conditions most favorable to plant development. Without the seasonal challenge of winter, plant growth was unconstrained. Haberlandt (a devotee of Buitenzorg whose descriptions of *Mimosa* MacDougal read with interest), for example, explained that because the "external conditions for growth and nourishment are favorable throughout the entire year . . . the plant world can develop and grow with a freedom which our native flora is denied."[27] In the tropics, plants could reach their full physiological potential.

Beyond the sheer mass of plant life in the tropics, botanists were fascinated by the intricate and sometimes puzzling adaptations—like *Mimosa*'s movements—that could be found there. In contrast to the uniformity of northern forests, tropical trees presented in close proximity a bewildering array of shapes and textures in their leaves, trunks, and bark. To a general observer, such structures might evoke an ornate and whimsical beauty, but from a Darwinian point of view they cried out for explanation. In the desert it might be easy to infer that cacti retained water as an adaptation to an arid climate, but the purpose of the structures and behavior of tropical plants was not always so clear. Did the buttresses of some tree's trunks serve the same purposes as those of the Gothic cathedrals they so resembled?[28] Lacking the seasons of the temperate zones, what signals induced flowers to bloom or trees to shed their leaves when they did? Were elongate leaves with pointed tips an adaptation to excessive moisture? Why were certain adaptations—climbing, parasitism, the production of flowers on trunks—common in the tropics but rare elsewhere? The tropics offered a multitude of plant adaptations for the physiologist to study but also raised the question of why so many different forms could be found under seemingly identical environmental conditions.

Setting aside the bewildering variety of structures and physiological adaptations that plants exhibited in the tropics, the numerical diversity of tropical floras themselves was a striking fact—one that MacDougal would comment upon. As Alfred Russel Wallace had noted in *Tropical Nature,* tropical climates appeared to be able to support many more species than those in the temperate zones. Most taxonomic groups of plants contained a conspicuously greater number of tropical than temperate species. This pattern was especially noticeable among the cryptogams or "lower plants"—those reproducing without flowers or seeds, including ferns, mosses, liverworts, lichens, algae, and fungi.[29] Cryptogams were attracting increasing attention among botanists because they appeared to be evolutionarily primitive and because they had been little studied in the past. (Lacking flowers, identification without the examination of living specimens under a microscope was difficult.) The diversity or "richness" of tropical floras was nowhere more on display, however, than in what Warming, Schimper, and the other European ecologists called the *rainforest*.[30] None of them invented the long-standing word, but they gave it a new technical ecological meaning: a forest type determined by high annual rainfall. In contrast to the other common alternative, *Urwald*, meaning primeval forest or jungle, *rainforest* emphasized the role of physical environmental factors in determining the structure and physiognomy of a plant community. The outstanding characteristic of the tropical rainforest was its quantity of species; Warming called tropical rainforests the "richest of all types of communities."[31] There, "hundreds of species of trees [grow] together in such profuse variety that the eye can scarcely see at one time two individuals of the same species."[32] Seeking later to "[illustrate] this curious tropical richness of species by numbers,"[33] Warming noted the nearly 3,000 species he had collected in a 150-square-kilometer patch of forest and savanna in Lagoa Santa, Brazil, in the 1860s. He estimated that the whole of his native Denmark harbored half as many species. Compared to the tropics, the flora of Europe or the United States seemed poor indeed.

Many of the qualities that marked the tropics as exotic and strange to other travelers likewise fascinated botanists, but it is important to note how they began to suggest a reversal of certain elements of the usual discourse about the tropics. The tropics might be foreign to botanists, but to the plant kingdom itself it appeared to be home. The luxuriance, intricate adaptations, and diversity of plant life in the tropics indicated this was the most hospitable climate for plants' physiological development—the most "normal." Perhaps it was temperate, not tropical, plants that were aberrant and unusual? According to Haberlandt, for example, recognizing that "the long winter rest has impressed its stamp on all the plants in our native flora" would help overcome the potential bias of attributing as universal to plants "adaptations that relate only to the peculiarities of the European climate."[34] The pattern of increasing species diversity also pointed to the tropics as the plant kingdom's evolutionary homeland. The persistence of the most ancient forms of plant life alongside the most highly developed and extravagantly complex adaptations also seemed to demonstrate plants' evolution there over long ages. Whether primarily due to favorable climatic conditions or a result of the absence of prehistoric glaciation (as Wallace and Warming both argued), the tropics were seen as an evolutionary Eden for plants.

Such arguments convinced MacDougal of the significance of the tropics to the study of plants. They were born out by *Mimosa,* the new experimental plant that he brought into his Minnesota laboratory on return from Europe. Plant sensitivity was one of the variety of adaptations that was curiously common among tropical plants. The temperate zone contained a handful of unrelated species of rapidly moving plants, "but in the tropics the number is enormously multiplied," MacDougal noted. Genera harboring sensitive plant species seemed to be particularly "at home in countries near the equator," exhibiting a southward increase in species numbers.[35] What concerned MacDougal most, however, was the luxuriant, healthy growth of *Mimosa* in the tropics compared to what botanists could attain in Europe or the United States. The plant had to be coddled in the laboratory, but it surprised and delighted traveling botanists by carpeting fields as weeds in many parts of the tropical world.[36] In northern greenhouses *Mimosa* grew tall and spindly, but in sunny tropical climates it was a prostrate, woody ground cover. MacDougal recognized that *Mimosa*'s sensitivity—not just to touch but, rather, to environmental factors—caused changes in form and physiology that complicated attempts to infer explanations of its adaptations.

MacDougal's frustration at botanists' lack of attention to *Mimosa*'s natural habitat sparked his desire for a tropical laboratory. The plant had received "no attention except that given it in the north temperate zone, in warm houses, and for the greater part under highly artificial conditions." In this state, it had become the subject of "many highly ingenious speculations." MacDougal noted botanists' suggestions that *Mimosa*'s behavior was a defensive adaptation against hailstorms or grazing—without regard to whether these conditions prevailed where *Mimosa* had evolved.[37] These conditions could not be reproduced in a northern laboratory. There was neither the technical capability nor the basic knowledge of the plant's habitat. "The work upon the subject under conditions necessarily artificial has been so futile in real results" that he became convinced that "a satisfactory solution . . . may be accomplished only by researches prosecuted in the tropics, in the habitat of plants which have acquired a high degree of sensitiveness."[38]

He quickly leaped, however, from his own research problem of plant sensitivity to a broader issue that could galvanize the entire community of U.S. botanists: the need for a tropical laboratory. He took his proposal both to *Popular Science Monthly* and to Coulter, who had considerable power to organize U.S. botanists through the *Botanical Gazette.* "In order to come in contact with the problems that are now pressing," he wrote in what Coulter printed as the November 1896 editorial, "the botanist must establish his laboratory in the midst of the normal conditions, and no condition is so vitally needed by botanical research, and so overwhelmingly lacking, as that furnished by the tropics."[39] By applying the editorial banner, Coulter contrived to throw his considerable professional weight behind the project—"we can push the thing through," he promised. He instructed MacDougal to follow up with an open letter supporting the tropical laboratory (which he had himself proposed!) to give the issue momentum.[40] MacDougal obliged, arguing that a tropical laboratory was among "the most pressing needs of American botany, and . . . the most important means of advancement and preservation of the integrity of the science in general."[41] Explaining more fully the importance of such an institution in *Popular Science Monthly*, he echoed the refrain common among the European ecologists that study in the tropics could dispel biases borne of narrow familiarity with temperate zone plants: "Before a permanent foundation for the science can be laid, research along all lines must be extended to include the most highly developed forms, in the primitive habitat of the plant kingdom, in the tropics. The principles of the relations of plants and their relations to the animal kingdom may only be attained by the study of undisturbed communities of plants in the natural groupings resultant from the struggle for existence. Here are to be found such rapidity of growth and metabolism that the adaptive possibilities of the organism reach their highest expression."[42] To comprehend the full diversity of plant life, to assess normal plant growth, and to attain botanical knowledge that was truly general, U.S. botanists must study plants where they lived and grew. Most grew in the tropics.

The Tropics "at Our Very Door"

MacDougal built the foundations of his call for an American tropical laboratory on the emerging principles of ecology, but his motivations were not purely intellectual. He, along with the botanists who soon joined in support of the project, voiced considerable professional and nationalist ambitions. The response to MacDougal and Coulter's proposal reveals the contours of the network of U.S. science in the tropics at the end of the nineteenth century, before the transformations of 1898. It also maps out the territory of the tropics in the imaginations of U.S. scientists, a perspective colored both by their perceptions of European colonial science and by broader American ideas about the Caribbean and its relationship to the United States. U.S. botanists tended to ignore Buitenzorg's colonial economic foundations in appropriating it as a model, but their endeavor was no less colonial in its scientific aspirations.[43]

MacDougal's laboratory proposal was at its heart a claim for land. "To an American botanist," MacDougal noted, "a visit to Buitenzorg is equal to a trip around the globe."[44] The journey required a month in each direction and cost well over a thousand dollars, putting it far out of reach of the average university botanist. (The only U.S. scientist then to have worked in its laboratory was the soon-to-be-prominent plant explorer David Fairchild. Financed by an eccentric Chicago millionaire, he was the exception to prove the rule.)[45] "The American botanist needs to make no such journey," MacDougal maintained: "At the distance of a week's travel from almost every important laboratory there lies a tropical region whose teeming flora is but little known, even to taxonomists. At our very door there lies a vast *terra incognita*, with excessively luxuriant vegetation, and inviting endless research."[46] U.S. botanists need only turn south to reach "their own" corner of this scientific paradise. "A study of the map," argued MacDougal, "will show that the conditions to be met favor either the eastern coast of Mexico or the islands near the Caribbean Sea."[47] Why go to Java when they could establish an outpost of tropical science just off U.S. shores?

MacDougal, following European ecologists, had been persuaded by a geographical and biological argument about the necessity of performing tropical research to achieve a general understanding of plant life. In campaigning for a site in the circum-Caribbean, however, he and the scientists who joined him in this movement added another layer of geographical discourse, one of proximity and possession. As the Cuban War of Independence escalated in 1896, MacDougal's characterization of the Caribbean as a "neighbor," "at our very door," echoed language that historian Luis Pérez has identified among U.S. citizens clambering for military intervention. In part, such emphasis on proximity reflects significant reductions in travel time between U.S. and Caribbean ports, as steamship lines expanded to accommodate the growing fruit trade in the late nineteenth century. This language also carried an unmistakable political valence, however. Geographic determinism charged turn-of-the-century American expansionist discourse, casting the Caribbean as inviting colonial possession simply by its nearness.[48] This rhetoric functioned just as well for a scientific foray into tropical research.

By the logic of nineteenth-century geopolitics, the closeness of the Caribbean made colonization not only a right but also an obligation—a duty to the world to "spread civilization."[49] Likewise, developing an institution for tropical botany in the Caribbean would fulfill American responsibilities to science and to the international community of scientists. With this region at their door, MacDougal argued in the 1896 editorial, "the duty seems to be laid upon American botanists to establish and maintain an international laboratory in the American tropics."[50] Indeed, his repeated reference to the "American" tropics itself elided geographic distinctions, rhetorically bringing the entire Americas into the presumptive domain of U.S. science. At the same time, this language of duty made the internationalist norms of scientific culture compatible with U.S. scientists' growing nationalist ambitions. At the end of the nineteenth century, U.S. scientists increasingly sought to play a greater part—even a leading part—in the international scientific community. Establishing a foothold in tropical botany was thus also a play for scientific prestige. MacDougal promised that a tropical laboratory "would go far toward making America the scene of the greatest development of the biology of one of the two great groups of living organisms."[51] He hoped it could become a means to raise the stature of American science, a sentiment mirroring how some in the United States saw the capture of Spain's colonies as a route to world-class power.

Response to MacDougal's call for an American tropical laboratory was overwhelming and positive. (Coulter perhaps need not have engineered the editorial page exchange.) In private correspondence, further open letters in the *Botanical Gazette,* and notices in other journals, botanists from across the United States voiced their support for the creation of such an institution. The question now became where it should be located. Coulter and MacDougal secured the participation of William Gilson Farlow of Harvard and Douglas Houghton Campbell of Stanford, forming with them a Tropical Laboratory Commission to tour potential sites throughout the Caribbean.[52] With Campbell and Farlow, the project now had the support of two senior and extremely influential botanists, representing major East Coast and West Coast universities, as well as MacDougal and Coulter in the Midwest. Abroad, Schimper declined MacDougal's invitation to join the group but recommended they consider the island of Trinidad as a laboratory site. Francis Darwin and Karl Goebel also offered cordial support for the plan—Coulter and MacDougal hoped to secure their involvement as British and German representatives to the commission and thereby raise the plan's international profile.[53] Cooperation with U.S. zoologists seemed possible as well. Johns Hopkins University botanist James Ellis Humphrey shared MacDougal's avid interest in ecology and even hosted a seminar on Warming's work that winter.[54] He thought "it would be an unfortunate mistake to make such an establishment as is proposed exclusively botanical or zoological." The "cooperation of both biological groups" would give it "added strength," and "the very great mutual advantage of the association must be self evident."[55] Humphrey and a group of Johns Hopkins University biologists had spent three seasons working at impromptu "laboratories" set up at Jamaican hotels at Port Antonio and Port Henderson, but their choice of location had had as much to do with available accommodations as with the proximity of marine life.[56] Humphrey was eager to establish a permanent biological station. Offering his assistance, he suggested that the two groups meet in Port Antonio during the commission's tour that summer.

As the commission planned its Caribbean itinerary, the broader scientific community weighed in on the proposed laboratory. Those who had collected or visited botanical gardens in the tropics debated the virtues of competing locations and enumerated the qualities they desired in an ideal research station. Coulter also solicited reports on existing Caribbean institutions, as well as on Fairchild's experience at Buitenzorg.[57] "Botanical opinion seem[ed] united" on several points about the natural requirements of a suitable site.[58] The general agreement was that the laboratory should be "as near as possible to a primitive forest"; forests stood as the ideal plant community in the tropics, and, presumably, only an "undisturbed" condition would reveal ecological relationships in a natural state.[59] It is clear that they were not interested in establishing an agricultural experiment station, despite the frequent references to Buitenzorg, with its strong colonial economic imperative. Indeed, Coulter clarified that in using "Buitenzorg as an illustration of what was intended" for the "laboratory in the American tropics," he referred only to "such *scientific* work as has brought that station into *botanical* notice." Evoking the pure science ideal, Coulter explained that the pertinent components of Buitenzorg to be modeled were its "facilities for *research*" in a tropical locality, not its broader work on "economic problems."[60] In that respect, they seem to have had Buitenzorg's Tjibodas substation more in mind. Many correspondents recommended a site from which both mountains and seashore would be accessible—reflecting ecologists' interest in the relations of plants and environment at different altitudes. MacDougal also inquired specifically about the prevalence of sensitive plants in different locations. Beyond this, however, there was little discussion of environmental characteristics. That the site be "tropical" was the essential prerequisite.

The human and medical geography of potential sites was more controversial. Despite the enticing portrait painted by MacDougal, those who wrote to promote a favored site had to defend it against perceptions of the Caribbean as a dangerous region—a place of tropical disease, civil unrest, and foreign races and cultures. In the years before the sanitation efforts of the U.S. Panama Canal project, images of the tropics as the "white-man's-grave" or a "green hell" remained vivid. Yellow fever and other ills were "no laughing matter to the new-comer," cautioned USDA plant pathologist Walter Swingle, especially in the Caribbean port towns that otherwise seemed attractive for their accessibility.[61] Another USDA botanist warned MacDougal, "There is liable to be some loss of life no matter where the laboratory is located South of 27°."[62] To assuage such worries, some suggested a station in the highlands, at Orizaba, Mexico, or Cinchona, Jamaica. In favor of Jamaica, Humphrey argued, "As compared with many other parts of the tropics, the climate of Jamaica is exceptionally healthful, and it is remarkably free from poisonous animals."[63] For researchers intending to visit for an extended period, the healthfulness of the site was ranked at least as important as the quality of its flora, fauna, and scientific facilities.

Threats of violence seemed to lurk in the region as well. Cuba's ongoing War of Independence loomed over the discussion. Several correspondents recommended the island for its natural advantages of diverse species and climate, but all expressed similar sentiments: "Cuba is at present out of the question."[64] Respondents weighed the risks of commission members being mistaken for filibusterers and the relative trustworthiness of Mexicans and Cubans.[65] Most correspondents intimated some level of fear of unrest or discomfort with alien, nonwhite societies. Only one respondent, however, voiced opposition to any site that was not "within our own borders." Jared Smith, a USDA agrostologist (grass specialist) who favored a subtropical station in Florida, warned of the "restrictions and vexatious rules" that would be encountered working anywhere in Latin America or the Caribbean. Against Jamaica he fretted about the island's racial composition and the danger of a "sudden uprising of this black horde." To him any foreign country would present "strange customs" and "trouble with people who are more or less jealous of America and everything and everybody American."[66] Although Smith was the lone dissenter in MacDougal's correspondence, his sentiments may reflect elements of the broader botanical community who simply chose not to take part in the discussion at all. At the same time, it was in his self-interest to favor a site within the jurisdiction of his agency.

To many of MacDougal's respondents from universities and botanical gardens across the country, however, the British West Indies appeared to be the safe alternative to their stereotyped images of backward and turbulent Spanish-speaking countries. Of Jamaica, Humphrey wrote, "The island is a British colony, which means that life and property are secure, the roads fine, the language English."[67] To Humphrey, the British colonies offered a haven of "civilization" in the American tropics. In Jamaica, Dominica, or Trinidad, they would find the comfort of a more familiar culture, social order, and scientific community. "Jamaica is the most civilized has the most stable government and the best start in science of the West Indies,"[68] opined Swingle. In Jamaica or another British colony, U.S. botanists would have "the sympathy and aid of fellow workers."[69] Jamaica was the paragon of colonial botany, with several nodes of Kew Garden's vast, botanical network—the gardens at Hope, Castleton, Cinchona, and Bath.[70] As at Buitenzorg, Goebel recommended that "the laboratory should be placed near a botanical garden," where botanists could examine a catalogued collection of living plants, including those from other tropical regions.[71] A good herbarium and library to consult on taxonomic matters remained necessities even for those with a physiological focus, and Jamaica also had the advantage of a history of significant taxonomic publications—from the classic seventeenth-century work of Hans Sloane to the more recent contributions of August Grisebach. As Farlow put it, having the resources of "existing botanical gardens to fall back on" would be a major advantage.[72]

The way that supporters of the tropical laboratory movement heaped praise on Jamaica's scientific infrastructure, however, tended to obscure similarly long-standing institutions for botany, zoology, agriculture, and medicine in Cuba, Mexico, and other Latin American countries. U.S. biologists tended to rigidly equate civilization and the scientific spirit with Anglo-Saxon race and culture. Humphrey, for example, wrote revealingly after an earlier expedition to Jamaica,

> We are apt to think, when speaking of American botany and botanists, only of those of the United States and Canada, assuming that our southern neighbors, both continental and insular, have not yet reached that stage of civilization that encourages the cultivation of the sciences. And so far as those regions are concerned which have felt the influence chiefly of Latin civilization, this is measurably true. But some of the neighboring islands have been under Anglo-Saxon rule for two centuries or more, and have felt different influences. Not, indeed, that their people, as a class, have been much affected by contact with their rulers, but in the British islands the mother country has especially fostered botanical study from an early time, and British residents have carried with them the scientific impulse.[73]

In calling on Anglo-Saxonism to support scientific expansion, botanists like Humphrey drew on an emergent strand of political discourse that emphasized a shared U.S.-British imperial destiny.[74] They also set a precedent of giving recognition to English-speaking colleagues and predecessors while disregarding the established work of Latin American scientists.

As U.S. scientists weighed the benefits and drawbacks of various sites, questions of health, safety, infrastructure, and culture came to the fore. For all of their specialized concerns about tropical plant communities and climates, their imagined cultural geography of the Spanish and British Caribbean mirrored that of the broader U.S. public. Even as interventionists increasingly clamored to enter Cuba's independence war, however, the experiences and prejudices of the botanical community tipped the balance away from the Spanish Caribbean. The supporters of the tropical laboratory proposal emulated the ecological approaches pioneered by German- and Danish-speaking scientists at a Dutch institution, but their search for a site in the circum-Caribbean drew them into the British sphere of influence.

The "Loveliest Spot in the British Empire"

MacDougal, Campbell, and Humphrey converged on Jamaica in May 1897. The Tropical Laboratory Commission had planned to tour southern Florida, eastern Mexico, Trinidad, Dominica, and Jamaica as a group, but administrative duties forced Farlow and Coulter to back out unexpectedly, and they planned to make a follow-up inspection in the winter.[75] The group had to scale back their plans, but this was a small setback. Privately, MacDougal had confided that he was already convinced that Jamaica presented the "greatest number of advantages for the Laboratory." Although the commission was all but decided on Jamaica, to unite the botanical community MacDougal felt they should "preserve an impartial attitude" until they made their report.[76]

Arriving together, MacDougal and Campbell were gratified by their first sight "of cocoa palms, breadfruit trees, bananas, and other evidences of the tropics."[77] They had to go no farther than the roadsides to find "abundant and interesting material for the plant collector"; MacDougal found *Mimosa* growing as a "common weed." Excited by the island's exotic, "luxuriant vegetation," the men were also relieved by the familiar comforts that greeted them.[78] Port Antonio was the center of the West Indian banana trade in the 1890s and the seedbed of the emerging Caribbean tourist industry.[79] The steamships of the Boston Fruit Company (amalgamated to form United Fruit in 1899) ferried bananas and tourists between Boston and Port Antonio twice a week in the summer and from other U.S. ports almost every day of the week.[80] MacDougal and Campbell found not only hotels with "comfortable beds and spacious rooms" but also fellow countrymen.[81] They traveled more like pioneers of the packaged vacation than as followers in the footsteps of Wallace and Henry Walter Bates.

Joining Humphrey, they made their way south to Kingston. The rail journey gave them vantage on the relation of topography to vegetation.[82] At Kingston they had arranged to meet William Fawcett, director of the colony's Department of Public Gardens and Plantations and their enthusiastic guide. He accompanied the party to the nearby experimental plantations at Hope Gardens, his department's headquarters, as well as north to Castleton Gardens, with its "remarkable collection" of palms, cycads, and orchids.[83] Presented at a meeting of the Institute of Jamaica, a local scientific and literary society, MacDougal's sensitive plant research also found a favorable reception.[84] Fawcett also assisted the group in obtaining free passes to ride Jamaica's rails, and they explored the rest of the island, including the venerable Bath Gardens, which a century earlier had been the destination of Captain Bligh's infamous breadfruit plants.[85]

Fawcett's eagerness to assist the Americans can be explained not only in terms of Anglophone scientific camaraderie but also in the context of the pressure Jamaica's gardens were under at this time. The 1880s and 1890s were a period of economic retrenchment for the colony; members of the colony's legislative council had begun to see botanical institutions as "expensive luxuries" better placed in private hands.[86] The closure and sale of some of the Department of Public Gardens landholdings had been under discussion for the better part of a decade. Fawcett, who had recently helped secure a lease by the Boston Fruit Company of some land at Castleton Gardens, evidently saw the U.S. group as an asset in his battle against the wave of budget slashing.[87] If he could inspire the group to lease a portion of his department's lands, he would not only gain needed income but also showcase the continued value of his institutions. MacDougal's tropical laboratory scheme could not have come at a better time for Fawcett.

Humphrey, the only member of the party with experience in Jamaica, had previously enjoyed Fawcett's hospitality but had not yet been able to visit one of his department's treasures: the "Hill Garden" or Cinchona.[88] He had been told it was "the loveliest spot in the British empire."[89] Guided by Fawcett, the commission members rode by carriage the first nine miles from Kingston to Gordon Town. The final thirteen miles to Cinchona took them up into the Blue Mountains by pony on a narrow bridle path.[90] The site was difficult to reach, but it was in fact because of its montane location that the Department of Public Gardens had made Cinchona its headquarters until the recent budget

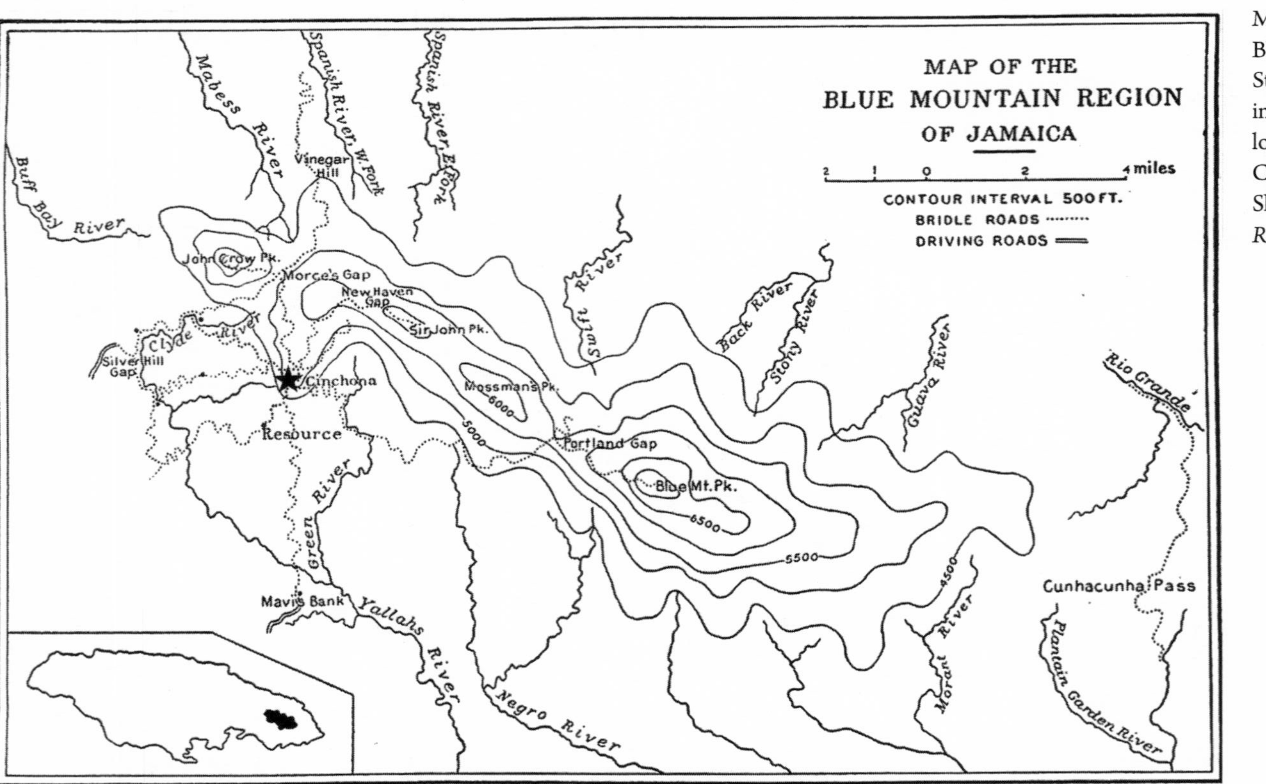

Map of Jamaica's Blue Mountains. Star added to indicate the location of the Cinchona station. Shreve, *Montane Rain-Forest*, plate 2.

cuts. The site was named for its plantation of cinchona, the iconic "plant of empire" whose bark is the source of the antimalarial drug quinine.[91] Native to the Andes, cinchona trees thrive in cooler tropical highlands. As a consequence, Cinchona was chosen in 1868 for its "temperate" climate—botanists envisioned a Humboldtian climatological and botanical correspondence between altitude and latitude. By the 1890s, Jamaica had been pushed out of the cinchona market by competition with Dutch Java, but botanists also tended a "European garden" at Cinchona, introducing peas, potatoes, cabbage, strawberries, and other familiar vegetables and fruits craved by settlers. Throughout the British Empire, moreover, whites saw "hill stations" like Cinchona as respites from tropical dangers.[92] A montane site could be expected to be free from the fever and disease that haunted swampy lowlands. Moreover, hill stations were highly racialized as white spaces, offering escape from black populations in the urban centers.[93] High in the Blue Mountains, Cinchona offered (literal) transcendence from what many whites perceived as undesirable aspects of Jamaica's tropicality.

Where the British valued it as a comparatively temperate site, however, Cinchona offered everything the Americans wished of an ideal tropical site. In particular, MacDougal and the other commission members marveled at its huge variety of species. The same microclimates that made the area particularly useful for experimental horticulture also encouraged local species diversity. The steep mountains had the effect of creating substantial variations in local moisture, light penetration, and temperature—not only gradations with altitude but also contrasts between the windward and leeward slopes. These produced a multitude of habitats, each harboring species with their own unique adaptations. Although expecting tropical abundance, Campbell still found himself astounded by the "extraordinarily rich" flora.[94] Humphrey remarked on the "rich and varied flora . . . within a few miles one may pass from the sea level to the summit of Blue Mountain peak, 7360 feet high."[95] To attempt to put the sheer numbers of species into perspective MacDougal coined a numerical and spatial comparison: "The equatorial zone is the home of an enormous number of species. The island of Jamaica, with an area of about 6,000 square miles, about that of Connecticut, furnishes approximately four-fifths as many species of flowering plants as are to be found in the United States east of the Mississippi river."[96] In coming decades, such juxtapositions of tropical countries and U.S. states would grow into a common trope illustrating tropical diversity. To help the tropical laboratory cause, Fawcett prepared a talk for the upcoming meeting of the Botanical Society of America, in which he, too, placed particular emphasis on Jamaica's large number of species.

The station and garden at Cinchona, 1903. The buildings behind the flowering pampas grass in the background served as botanical laboratories for U.S. visitors. Smithsonian Institution Archives, no. SIA2016-009530.

"Jamaica is a paradise for the botanist," he wrote, "whether he specializes in algae, fungi, mosses, ferns, or flowering plants." He cited 450 species of ferns and 2,180 flowering plants as inducement for specialists of all taxonomic groups.[97]

Not only did the commission members think they could entice a wide variety of U.S. botanical specialists by the abundance of almost any desired group of plants, but such diversity and its relation to environment were also significant objects of the new ecological perspective that MacDougal hoped to promote. The commission admired Jamaica's rich vegetation and the botanical collections of Fawcett's Department in Jamaica, but privately Humphrey opined, "Very little scientific work is being done there."[98] For Humphrey and MacDougal *scientific work* meant not economic plant introduction, as they saw at Hope, but new research in the areas of physiology and ecology. Unlike Buitenzorg, Jamaica's gardens did not yet house a modern laboratory. In their eyes, from a purely botanical perspective, little was being done to exploit the colony's vast botanical resources. The field seemed wide open for an American "invasion."

The group returned home in time to attend the August meeting of the Botanical Society of America in Toronto, where they reported on what they had seen and hoped to marshal the support of a wide variety of botanical institutions.[99] Although the follow-up trip by Farlow and Coulter remained, Jamaica seemed a foregone conclusion. The U.S. fruit industry's inroads into the colony not only paved the way, developing the transportation and tourist infrastructure that funneled Americans into Port Antonio, but even established the precedent of a lease of land from the Department of Public Gardens. Cinchona in particular rose to the top of the list. Intercolonial competition had stripped the garden of its original purpose. Once an experimental plantation of a prized tropical commodity, it was now vulnerable to budget cuts, and Fawcett was desperate to find ways to keep the site viable. Although the commission kept up the pretense of a rational search for an ideal tropical locale, economic and political conditions had already channeled them toward Jamaica.

A Scramble for the Caribbean

Even as MacDougal, Campbell, and Coulter reported on their commission's progress, however, tragedy struck. News arrived that Humphrey, who had so confidently proclaimed the "exceptionally healthful" climate of Jamaica, had fallen ill and died in Port Antonio on August 17.[100] Upon return to Boston, Franklin Story Conant, another member of the Johns Hopkins party, perished just shy of his twenty-seventh birthday.[101] The deaths were a difficult blow. Only thirty-six, Humphrey had a promising career ahead. He was well loved as a lecturer and was fast gaining a reputation in experimental botany. Conant was a recently minted PhD and a Johns Hopkins Fellow with great prospects ahead. The deaths of the two young men shocked the scientific community. Further yellow fever outbreaks led to port quarantines, and fear of the illness was amplified as it spread into the U.S. South.[102] The renewed disease anxieties and loss of Humphrey's advocacy slowed action on the tropical laboratory project to a snail's pace.[103]

At the same time, the U.S. public became increasingly preoccupied with the situation in Cuba. To many, the prospect of an independent Cuba suggested the stoppage of trade, the further irruption of disease, and general anarchy.[104] In April 1898, interventionist voices won out and the United States entered the war. The Spanish-Cuban-American War was over by August, but the consequences of the changed political landscape took much longer to resolve. As broader U.S. publics wrestled with the prospect of governing a tropical empire, U.S. scientists also faced an altered set of circumstances. In the Caribbean, Spain's cession brought to the United States not only Cuba but also Puerto Rico, which had until then barely entered U.S. botanists' discussions of a tropical laboratory. On paper at least, the Treaty of Paris also gave the United States charge of the Pacific territories of Guam and the Philippines—though the bloody Philippine-American War would last until 1902. In June 1898, the United States also annexed Hawaii, five years after a coup by U.S. business elites. For botanists, as for the rest of the nation, however, the acquisition of this overseas empire fueled uncertainty as well as excitement. Coulter suggested to Farlow that it was "a good thing that we did not undertake to settle in the West Indies until after the war."[105] "Cuba and Porto Rico seem open to us now," he explained to MacDougal, but "the whole thing is so indefinite at present that I cannot say anything positive."[106] He had "had Porto Rico prominently in mind ever since we secured it," but the new territories' relationships with the United States remained unsettled. "It may be that Cuba would be even more available," Coulter opined, "but its fate does not seem to be so well fixed as that of Porto Rico."[107]

Far from merely opening new possible research sites, the aftermath of 1898 revealed fractures in the U.S. scientific community along institutional lines. With the acquisition of formal territories came an influx of Washington-based, state-sponsored science; the USDA established experiment stations in Hawaii and Puerto Rico in 1900, and the Smithsonian Institution sent out brigades of botanical explorers.[108] Compared to the European colonial powers, however, the U.S. federal government gave minimal support to science. The nation's major scientific institutions were largely private and spread across multiple metropolitan centers, each of which began to pursue its own interests in tropical research. MacDougal and Coulter had originally envisioned an "American tropical laboratory" supported through the collaboration of a unified national community of "American botanists," but as opportunities for research throughout the circum-Caribbean widened, U.S. scientists had fewer incentives to collaborate to get a foot in the door.

The New York Botanical Garden (NYBG), led by Nathaniel Lord Britton, emerged as a particularly powerful player in tropical botany. Britton had worked self-consciously to build a scientific empire. He envisioned his garden as an American Kew and, situated in the nation's financial center, Britton had the connections to wealthy philanthropists that allowed him to realize this ambition. Britton, seen to be declaring a "botanical Monroe doctrine," directed his institution's efforts on the U.S. West and the Caribbean, setting out to amass a collection of Western Hemisphere plants that could surpass those of Europe.[109] Britton dispersed collectors to Cuba, to Haiti, and throughout the British West Indies, but like Coulter, his attention focused particularly on Puerto Rico. With funds from Cornelius Vanderbilt and local assistance from the U.S. naval station, Britton began sending private collectors to the island in 1898, just three months after the war's end.[110] After himself visiting in 1906, Britton became instrumental in organizing a multidisciplinary scientific survey of the island through the New York Academy of Science. It would operate from 1914 through 1934 and become a powerful shaper of Puerto Rican science.[111]

Coulter expected that Britton would "jump at" Puerto Rico but felt that the NYBG did not yet have a total scientific monopoly on the island. "Britton is fortunate in having a Vanderbilt to back him," he told MacDougal, but "they are going at it purely from a taxonomic standpoint, to which they are, as far as I am concerned, entirely welcome. I would wish the tropical laboratory to concern itself chiefly with physiology, morphology, and ecology."[112] Britton's unilateral actions threatened to undermine plans to support a tropical laboratory through the collaboration of U.S. scientific institutions, but the fact that Britton seemed more interested in collecting expeditions than in place-based ecological or physiological research suggested at first that the efforts could coexist. "Congratulations on your annexation of Porto Rico," MacDougal conveyed to Britton playfully. "I hope you will not devastate the native flora to such an extent as to render it unfit for a tropical lab."[113]

The tropical laboratory project had another source of competition that MacDougal and Coulter were less aware of, however. At Harvard, Farlow had met some difficulty uniting members of the university's disparate botanical institutions in support of the tropical laboratory. Encountering a lack of cooperation from George Lincoln Goodale, director of Harvard's Botanic Garden and Botanical Museum, Farlow had had intimations that Goodale "intend[ed] to act individually."[114] In fact, Goodale's interest in tropical botany had become cemented after his 1890–91 tour of the East Indies, which included a visit to the Buitenzorg gardens.[115] Although Harvard remained firmly committed to the ideal of pure science, Goodale had a deep interest in economic botany and sought to promote connections between academic botanists and horticulturalists. Working near the heart of the U.S. fruit industry in Boston also gave Goodale access to patrons. The 1898 war was barely over when the sugar baron Edwin F. Atkins met with Goodale. Interested in disease-resistant sugar cane, Atkins donated $2,500 for Goodale's student E. Mead Wilcox to visit the East Indies (including Buitenzorg) and report on cane breeding techniques.[116] By 1899, Goodale, with the assistance of Oakes Ames, then assistant director of the Botanic Garden, had persuaded Atkins to establish a garden and sugar experiment station on his plantation at Soledad, in Cienfuegos, Cuba. Although the institution's ties with Harvard remained informal, Atkins dubbed it the "Harvard Botanic Station for Tropical Research and Sugar-Cane Investigation."[117] It would grow into a premier station for tropical research and education, but the initial focus of the Harvard station was far more economic in orientation than what MacDougal and Coulter imagined for their tropical laboratory. Nevertheless, the direction of Harvard faculty attention toward Soledad diverted an important set of potential allies.

Indeed, as competition heightened, MacDougal himself was scooped up into Britton's expanding botanical empire. Eager for a more prestigious position, MacDougal was hired as NYBG's director of laboratories in 1899. He brought the tropical laboratory project along with him into the domain of the NYBG. Coulter had in fact had a chance to use his department's endowment to acquire a station for the University of Chicago. He could have done so as early as 1897, but Coulter felt strongly that the station should be a cooperative effort, not controlled by a single organization. He had explained to Farlow,

> Of the $1,000,000 gift for biology, under which I am here, there is a provision for the establishment of a station or stations at a distance from the university. . . . Under this provision I could go to the tropics & establish a station for the Univ[ersity] of Chicago. But I don't want to do this. In my notion it is essential that this station be free from the domination of any institution or society. I don't want a trop[ical] station for this Univ[ersity], but I do want this Univ[ersity] to gain with others in establishing one that we can all feel free to use.[118]

Envisioning a station run cooperatively by U.S. botanical institutions, likely through subscription, Coulter missed a chance to found the tropical station through the auspices of his own university.

Britton, in contrast, saw the potential of private support to finally get the project off the ground. As an adviser to the newly formed Carnegie Institution, he attempted to obtain funding for "a botanical station, laboratory, and garden" in the West Indies in 1902.[119] Under the Carnegie plan, the tropical station would be part of a much larger project to study the relationship between vegetation and the environment throughout the United States and the "American Tropics."[120] The Carnegie Institution did fund the project's proposal for the Desert Laboratory at Tucson. (There, from 1905 on, MacDougal would follow the funding, shifting his attention to administration and desert research and away from the tropics.) The tropical portion of the proposal did not find support, in part because the head of the institution's executive committee was opposed to involvement in projects outside U.S. territory.[121]

Nevertheless, the proposal reveals how the transformations of 1898 shifted botanists' rhetoric in justification of tropical research. The proposal replicated much of MacDougal's language from the original *Botanical Gazette* editorial, reminding readers of the neglect of physiological research on plants in the tropics, where "the more important physiological and morphological types are abundant, exhibiting a range of adaptation and a luxuriance of development unknown in the temperate regions" and arguing that it had "been made evident many times that generalizations reached by such limited and circumscribed methods are unsafe and misleading." A new justification emerged, however. Unlike MacDougal's original proposal, the version revised for the Carnegie Institution emphasized the "scientific *and economic* importance" of a tropical station: "The establishment of a botanical station of the intended scope and functions would not only afford opportunities for the furtherance of *research in all of the strictly technical aspects of the science*, but the results obtained would include much of *economic importance* at a time when it seems necessary for all tropical American countries to improve their methods, or modify the character of their agricultural operations."[122]

The ecological and physiological orientation of the original plan remained but now sat alongside a general claim for future economic applications in tropical agriculture. It, too, used Buitenzorg as a reference point, noting that the Dutch maintained a "research laboratory in pure botany in connection with its great plantations and collections."[123] In this case, however, the institution became a model of the proper relationship of "pure" research and "economic" results in colonial science. Combined with the larger proposal, the plan set forth a vision of tropical botany as an "accessible, rich, and comparatively unworked field" from which commercial applications would naturally flow; "the uses of tropical plants are multifarious, and many new applications of them and their products in the arts and manufactures would be likely to ensue." The proposal asserted that "a revolution in the methods of tropical agriculture is a probable, almost inevitable, result of American influence."[124] It suggested that the proposed projects of plant taxonomy, physiology, and ecology would play an obvious role in this revolution but provided no detailed explanation of how. Such claims vanished when Carnegie funding failed to materialize, but this was among the first in a line of future efforts by

U.S. scientists to fund pure or basic tropical research on the basis of promised economic outcomes.

Britton's proposal had made no mention of Jamaica, but he was frequently in touch with Fawcett about plans for securing Cinchona as the station's site. Despite the uncertain funding, when the Jamaican legislature finally decided to relocate the remaining work taking place at Cinchona and open the grounds for lease in 1903, Britton was ready. Intending to find a means of support later, he sent MacDougal to Jamaica in July to secure the site.[125]

A Tropical Education

Of the several hundred acres of the original cinchona plantation, the NYBG's ten-year lease brought it the use of a modest residence (Bellevue House), two twelve- by thirty-foot buildings to act as laboratories, two greenhouses, a stable, servants' quarters, and ten acres of surrounding grounds.[126] The station came stocked with some basic supplies, but visitors had to bring along their own microscopes and any other specialized equipment. It was far from the massive research complex, manned by hundreds of staff scientists and servants, that one could find at Buitenzorg. Considering the station's ecological orientation and tropical, colonial locality, if not its scale, botanists' frequent references to the station in Java are understandable. The station afforded U.S. botanists an opportunity to begin to extend physiological and ecological research into the tropics. Moreover, as MacDougal had suggested, botanists began to identify flaws in more hasty generalizations as they gained a deeper familiarity with this tropical site.

It was just this potential to broaden botanical horizons that Cinchona's supporters touted as the station's most valuable role for U.S. visitors. The NYBG framed a "visit to Cinchona [as] an important part of the liberal education of a botanist," and although the number of visitors was never very high—it averaged about half a dozen a year—the station attracted a wider range of U.S. scientists than might otherwise have attempted tropical work.[127] Fear of tropical fevers began to fade along with the memory of the deaths of Humphrey and Conant; U.S. mosquito control methods were widely publicized as a success in Cuba and were beginning to be applied in Panama as the United States broke ground for the new isthmian canal in 1904. The station offered comfortable accommodations, even if the ride up the switch-backed mountain trail left travelers jostled (or soaked by the frequent rains). Dense forest lay close at hand, but the rooftops of Kingston could still be viewed in the distance on a clear day. Cinchona permitted U.S. scientists to gain experience with tropical organisms and native plant communities without large expense or a strenuous expedition. Several high school teachers made the trip to Cinchona. Male researchers often brought their families—MacDougal traveled there with his wife and daughter, and the renowned Puerto Rican–American bird illustrator Louis Agassiz Fuertes even saw fit to make a visit to Cinchona part of his honeymoon in 1904.[128] Several accounts of visits by women botanists (nearly a third of visitors were women) were clearly calculated to create a domestic and welcoming image for the station.

Rather than recounting heroic feats and tropical dangers overcome, such narratives promised safety and highlighted the rounding effects of wider travel. As Vassar biology instructor Winifred Robinson explained, a visit to Cinchona "gives the American student the advantage of life in a foreign country in addition to what he gains botanically."[129] "To have personal knowledge of banana plantations, sugar-cane and coffee, of Sapadillos, chochoes, and akee, the experience of tropical living, together with the scientific knowledge which is gained make a trip to the Cinchona Laboratory a delightful part of a botanist's education," she wrote.[130] U.S. scientific travelers to Cinchona delighted in becoming intimately acquainted with the exotic. Visitors could return home having collected not only dried specimens and data but also tales and photographs of picturesque coffee plantations, unusual fruits, and barefoot "negro market women" carrying baskets atop their heads.[131]

They could be sure of a successful visit and a safe return because in fact they came "not altogether as strangers in a strange land," explained Mary Brackett, a Wadleigh High School botany teacher and an editor of the popular journal *Plant World*.[132] There to greet and guide them was the "gracious" Director Fawcett, as well as the superintendent of Hope Gardens, William Harris, and his wife, Mrs. Harris, who sometimes attended to visitors' household needs at Cinchona.[133] The botanical department loaned U.S. visitors reference books and herbarium specimens. It also ensured that Afro-Jamaican servants and assistants were there to handle domestic chores and carry supplies and specimens. Robinson advised that "the collector finds it necessary to take . . . a 'boy' (a negro anywhere from fifteen to fifty years of age) to carry lunch and drinking water, to cut a trail if necessary with the huge cutlass which he always has in a leather sling over his shoulder."[134] Servants also aided scientists working at the Marine Biological Laboratory at Woods Hole, Massachusetts, or at the Stazione Zoologica in Naples, Italy, but to U.S. visitors the novelty of the colonial labor regime was a real difference—one striking enough for frequent comment.

Jamaican labor not only allowed U.S. visitors to work at ease, however; it also enabled them to become familiar with the bewilderingly diverse flora of

An unidentified Jamaican man serves for scale in Forrest Shreve's photograph of rainforest vegetation. Shreve, *Montane Rain-Forest*, plate 17.

the area. In contrast to manual labor, the intellectual labor of local people often remains invisible in traveling scientists' writing. The Afro-Jamaican plant collector David E. Watt's relationship with visitors nevertheless provides a glimpse of how science at the station depended on the interchange of knowledge with resident assistants.[135] Employed for years as caretaker of Cinchona's gardens, Watt also acted as "guide and counsellor" for visitors making longer excursions into the mountains.[136] The impression he made certainly suggests that Watt had a talent for managing botanists as well as for identifying plants. Watt's relationship with botanists afforded him a relatively privileged local status; for his labor, he received rent-free access to land at nearby Resource, with permission "to take tenants . . . and to cultivate the property for his own benefit."[137] In turn, visitors' recollections show that they relied on him not only to collect specimens but also to identify species and vouch for long-term observations of phenological events, such as the timing of flowering.[138] His botanical skills are still celebrated in the name *Phoradendron wattii*, a mistletoe species that he could proudly "spot a mile off."[139] He gained his encyclopedic knowledge of the Blue Mountain flora not only through long residence in the area but also through the experience of working with both British and U.S. botanists over two decades.[140] In this way, Watt's role in mediating botanical knowledge and maintaining continuity at the station was significant. At Cinchona, U.S. visitors found not only a base from which to accomplish botanical work in close proximity to a tropical forest but also an existing Afro-Jamaican and British network of local knowledge and scientific practice that helped them make sense of an otherwise foreign land.

The cosmopolitan atmosphere of a visit to Cinchona extended to the range of botanical research that could be accomplished there. The early twentieth-century laboratory movement has usually been associated with the growth of sharp divisions between experimentalist and natural historical approaches, but it is notable that such boundaries were hardly enforced at Cinchona. The station had multiple uses. With its modest laboratory benches, herbarium, greenhouses, and botanical garden, as well as the surrounding forest, the site accommodated a wide array of types of research. Although originally conceived as a laboratory for plant physiology and ecology, Cinchona was also used by Britton as a way station for propagating tropical plants for the NYBG. In part because of its tenuous financial circumstances, Cinchona's promoters sought to attract as many visitors as possible, of whatever scientific stripe. A handful of zoologists also visited, using it primarily as a base for collecting insects or birds. The site was particularly ideal, though, as Yale botanist Alexander W. Evans explained, for projects requiring "careful work in the field followed by careful study in the laboratory and herbarium, and the facilities offered at Cinchona and Hope Gardens are probably unequalled anywhere else in the tropics except at the botanical garden of Buitenzorg."[141] MacDougal had originally envisioned the benefits such boundary-crossing work could provide for physiology and ecology in the tropics, with researchers making measurements and conducting experiments indoors and out. Evans, however, was a taxonomist. Since identifying cryptogams required microscopic study, he too needed to combine work in multiple scientific spaces.

Significantly, the diversity of plant life around Cinchona was itself seen as having a broadening effect on U.S. botanists. For plant physiologists who recognized a need to incorporate comparative methods into their laboratory work, the site offered a source for acquiring a wider range of biological material than was available at home. The cell physiologist Clifford H. Farr explained, "Much light is frequently thrown on perplexing cytological problems by the study of related genera and species; and thus to one whose investigations have been confined to those species growing in the temperate zones, Cinchona furnishes splendid opportunity for the extension of his work

Research at Cinchona combined field collection with experimentation in the laboratory. The ecologist Forrest Shreve integrated outdoor and indoor measurements and observations to understand the relationship between climate and plant physiology. In this instance, Shreve juxtaposed photographs of both types of work in a single plate. The top image shows small-leaved specimens collected at the highest altitudes; the bottom image displays an atmometer and plants used in a transpiration experiment. Shreve, *Montane Rain-Forest,* plate 22.

to such allied tropical species."[142] Botanists also studied embryology, seed development, and the ecology of ferns, lichens, bryophytes, and parasitic plants—groups whose diversity was far greater surrounding the tropical laboratory than near the institutions where these botanists usually worked. Cinchona also enabled the comparison of plants growing in tropical and temperate environments; MacDougal himself tested the heritability of environmentally induced variations by arranging to have the same species cultivated at Cinchona, in New York, and in Arizona. The more intensive study enabled at the station could also expand understanding of the true diversity of the plant kingdom for taxonomists accustomed to northern floras. Although botanical study of Jamaica went back two centuries, Evans noted, "the student of taxonomic problems soon becomes impressed by the imperfection of our knowledge of tropical species and by the difficulties of interpreting the older records regarding them. It will, in fact, be a very long time, unless the number of workers becomes much greater, before our knowledge can even approach completeness."[143] Evans was right. In 1897 botanists recognized about 2,630 Jamaican plant species; today more than 4,000 are known.[144]

Even experienced tropical travelers encountered the unknown in the forest around Cinchona. Fuertes, who incorporated his Cinchona notes into a "comparative study" of tropical and temperate bird song, reflected that "the principal sensation one gets in the tropical forests is the mystery of the unknown voices. Many of these remain forever mysteries unless one stays long and seeks diligently."[145] Fuertes regretted that he could not get to know the possessors of all of these unknown voices during his "hasty survey," but his comments point to a significant quality of research stations that goes well beyond equipment or facilities: the ability to stay or return repeatedly to a site over extended periods.[146] By enabling long-term research, Cinchona permitted U.S. visitors a greater familiarity with the ecology of a tropical site, a familiarity that, as MacDougal predicted, began to reveal flaws in generalizations that had been based solely on research in temperate regions.

Duncan Johnson's eighteen-year study of the "natural experiment" of revegetation after a landslide near Cinchona is an important case in point. In 1909, unusually heavy rains—18.3 inches in twenty-four hours at Cinchona—caused disastrous flooding and landslides.[147] From the lookout at the station, "great gray scars" were visible where acres of coffee plantations and native vegetation had been washed away.[148] A Johns Hopkins botany professor who had been Humphrey's first and only graduate student, Johnson had visited the station in 1903 and 1906.[149] After the landslide, he was struck by the dramatic change in the landscape and saw it as an opportunity to observe the process of plant succession. He thought such a study "ought to prove as interesting, from many aspects, as the island of Krakatoa," where Buitenzorg's Melchior Treub had famously initiated studies of plant colonization following the 1883 volcanic eruption.[150]

Returning to the site of the slide in 1919 and 1926, Johnson believed his repeated observations threw doubt on prevailing assumptions about plant succession in the tropics. In the first place, revegetation occurred much more slowly than he had expected given the favorable environmental conditions for plant life found on the tropical site. In the denuded valley "there are present right through the year all of the climatic factors, such as moisture, heat, and light, that are needed for the production of a rich vegetation"; he was "surprised" to find little new growth after the first decade.[151] In the second place, the order of plant succession also defied expectations. Whereas Warming had suggested that lichens and mosses would be the initial colonizers of fresh ground, Johnson found shrubby perennials returning first. Documenting the species and groups of plants that he observed appearing over the decades, he argued "that in this tropical valley the whole order of immigration and establishment described by WARMING and by TANSLEY and CHIPP has been almost completely reversed. Whether this inverted order of establishment is generally characteristic of tropical regions must be determined by further observation."[152] Of interest is not only his concern to take care in making universal statements about tropical forest succession but also the truly long-term scope of the investigation he envisioned. Johnson expected that recovery to the state of vegetation he had first observed "will probably take centuries," explaining that this period "indicates the length of time through which . . . the reinvasion must be studied."[153] Johnson not only was one of Cinchona's most dedicated repeat visitors but also routinely brought Johns Hopkins graduate students along with him. Perhaps he envisioned that his students would ultimately settle the question.

The work of Johnson's student Forrest Shreve, in fact, did the most to widely demonstrate the potential of place-based ecological research to disrupt assumptions about tropical plants. Shreve first visited the station with Johnson in 1903 and again in 1905 as an assistant at the NYBG.[154] These early tropical encounters left him wishing he could stay longer; Shreve felt he had had "just about enough experience" to identify interesting problems but "not enough time afterwards to go at them."[155] At first he imagined "making a study of the Physiological Plant-Geography of Jamaica as a whole," but he now began to see the advantage of focusing on the area around Cinchona so that he could study in depth the "whole history of transpiration, photosynthesis and growth in the most hygrophilous plants of the rain-forest."[156] The idea pleased Mac-

Dougal, who promised him several months in Jamaica if he would agree to leave his job at the Women's College of Baltimore (now Goucher College) and come work for him at the Desert Laboratory in Arizona. Shreve returned to Jamaica in 1909 with his new wife, Edith Bellamy Shreve, a scientist in her own right who would also later publish on physiological ecology. Arriving at Cinchona, they found themselves "in free possession of this ancient landmark." Conditions were primitive. The laboratory remained little more than a room with a few rough-hewn wooden tables. Finding it necessary to make improvements, Shreve drew on comparisons with contemporary ventures in "taming" tropical nature. He likened his efforts to raise the station's sanitation standards to running "a young Canal Zone . . . with carbolic acid, white-wash and spade." He soon found himself at home in both the tropical environment and colonial social order. "By the aid of a sinewy mule, three black people and my wife's tactful management the domestic ills that flesh is heir to have been reduced to the minimum," he reported to MacDougal.[157]

Following in his mentors' footsteps, Shreve set out to study the forests surrounding Cinchona from an ecological perspective. To do so, he scrupulously applied Schimper's categories and methods. He amassed rainfall, humidity, and air and soil temperature data—he had the benefit of the decades of historical climate records collected by the British at the station. Cinchona's facilities allowed Shreve to combine laboratory plant physiology with forest observation: he measured the loss of water (transpiration) throughout the day in eight rainforest species, rigging up a darkroom and moist chamber in the lab to measure the effect of light and humidity on this process. In the field, he examined the physical conditions and vegetation that prevailed in the mountains' various microclimates—the windward and leeward slopes, ridges, and ravines and the high forest canopy. He saw no monotonous rainforest but instead classified several distinct habitats, each with its own assemblage of plants.

Shreve's eleven total months on Jamaica's mountainsides and in the laboratory led him to challenge a variety of common assumptions about rainforest vegetation. Although botanists generally presumed that plants in moist climates transpired more than those in deserts, Shreve found the reverse: transpiration is essentially proportional to evaporation. Carefully attending to phenological changes, Shreve also noticed unexpected seasonal periodicity in certain rainforest species—evidence, he hypothesized, of "inherited differences of physiological constitution" stemming from north temperate origins.[158] Moreover, he disputed the supposition that the elongate leaf tip morphology of many tropical plants was necessarily an adaptation for shedding water.[159] Most significant, like Johnson, he also found reason to question the applicability of temperate models of plant succession and climax communities to the tropical rainforest. Rather than an orderly linear procession, Shreve perceived a mosaic of environmental conditions, each harboring its own suite of plants. "No one of the forest types . . . may be looked upon as possessing a closer adjustment to its own complex of physical conditions than does any of the others," Shreve wrote. Given the diversity of conditions, he saw no possibility that one forest type could come to dominate the area. "In other words," he argued, "there is no means by which it might be possible to fix upon any one of the five types as representing the so-called 'climax' forest of the Jamaican montane region."[160] In his pioneering study, *A Montane Rain-Forest: A Contribution to the Physiological Plant Geography of Jamaica,* Shreve argued that, rather than any specific association of plant species, the vegetation of the Blue Mountains was characterized by diversity.[161]

For Shreve, this diversity was a product of the topography of the montane forest, which produced a patchwork of physical conditions, not one monotonously moist and warm climate. The adaptation of different species to these extremes of moisture, light, and warmth necessarily produced diverse plant assemblages. In the mode of physiological ecology, he conceived of plant diversity not in terms of species numbers alone but, rather, with regard to the diversity of adaptive forms. He wrote, "There is no type of vegetation in which may be found a wider diversity of life forms than exist side by side or one above the other in a tropical montane rain-forest. Together with the structural diversities, discoverable in the field or at the microscope, are diversities of physiological behavior, discoverable by observation or experiment."[162] The mixture of techniques enabled at Cinchona had allowed him to more fully appreciate this diversity and, he believed, to recognize its environmental causes.

Shreve's observations also led him to suggest caution in assuming the uniform hospitability of tropical environments for plants. The extremes of moisture found in the montane rainforest, he argued, could present as harsh an environment as the aridity of the desert. "The optimum conditions for the operation of all essential plant processes," he suggested, would instead be found in "the regions of the earth which present intermediate conditions between those of the desert and the reeking montane rain-forest." Concluding his monograph with an uncharacteristically broad assertion, Shreve maintained, "It is, indeed, in such intermediate regions—tropical lowlands and moist temperate regions—that the most luxuriant vegetation of the earth may be found, and it is also in such regions that the maximum origination of new plant structures and new species has taken place."[163] Like Warming and the other early European ecologists, he assumed that "optimal" environmental conditions would be the most productive of new adaptations and species. He indicated, however, that such favorable "intermediate conditions" could be found in both the tropical and temperate zones. In this claim, and in his findings about seasonality and transpiration, Shreve painted a portrait of a tropical vegetation not wholly Other but sharing unexpected characteristics with that of his own temperate home.

Shreve never elaborated on these conclusions. Like MacDougal, his new job at the Desert Laboratory shifted his attention from the rainforest to the desert in the decades that followed. *A Montane Rain-Forest* remained, however, an important step toward a more refined understanding of tropical forests. One reviewer wrote, "At almost every point one realises that the author's work marks a new era in the investigation of tropical vegetation, clearing away various false conceptions and generalizations drawn by earlier writers from too general and casual observations."[164] University of Chicago ecologist George Fuller likewise stated that it "necessitat[ed] a readjustment of many generalizations resting upon less definite evidence."[165] In only one decade, the station at Cinchona had begun to enable researchers to interrogate deeper dimensions of change over time and heterogeneity across space in tropical environments. If U.S. scientists' preconceptions about tropical people remained little changed through their experiences at this borrowed colonial outpost, gaining familiarity with the forest around Cinchona did begin to break down long-held assumptions about tropical ecology.

A Place for "Pure Botany"?

"This sort of opportunity for studying tropical plants where they must be studied—in their tropical surroundings—has seldom been offered to American investigators until within the last decade," Duncan Johnson wrote in 1914 as the NYBG lease of Cinchona approached its end. As Johnson explained, "Aside from the investigations accomplished, many botanists have here had their first opportunity of perceiving the intimate dependence of certain types of extreme specialization in plant structure on the accentuation of definite climatic or edaphic conditions. In other words, many have here first seen ferns and seed plants living as epiphytes, and have first appreciated the extreme diversity of habit and complexity of composition attained by the plant life of a primeval forest under tropical conditions."[166] For U.S. botanists at the turn of the twentieth century, the establishment of a tropical laboratory promised first-hand encounters with plant life at its most diverse and complex. As MacDougal argued forcefully a decade earlier, universal knowledge of living plants could not be produced in northern laboratories alone. Botanists must work where plants lived in nature. Above all, they must work in the tropics, the presumed evolutionary home of the plant kingdom and the region where plant life attained its greatest development.

For U.S. botanists, tropical research seemed to offer not only more solid foundations for scientific generalizations about plant life but also a path toward recognition on the international stage of science. U.S. botanists looked toward the European scientific community, which appeared to be advancing rapidly by exploiting the botanical resources of its colonial institutions. While "colonial science" has been criticized as an analytical category for suggesting a "distinct phenomenon separate from science proper"—perhaps even a phenomenon productive of "pathological distortions" in the service of a state—for U.S. botanists in this period, science in the colonies of the European empires held a powerful mystique and served as an ideal upon which to map their own scientific ambitions.[167] They drew heavily on the imperialist discourse that surrounded them but had little interest in becoming tools for any state. Even before the United States itself obtained tropical territories, U.S. botanists began to organize to acquire a tropical outpost of their own. To do so, they drew on existing colonial infrastructure, labor structures, and transimperial discourses of racial and scientific solidarity. Ultimately, borrowing from German and Danish scientific approaches and British and Dutch institutions, they experimented with their own version of colonial science at Cinchona, Jamaica.[168]

This strategy appeared highly successful, securing U.S. botanists their desired tropical foothold. Within a decade, Cinchona began to show the fruits of a place-based approach to tropical research. The economic and political geography surrounding the station, however, had radically transformed since MacDougal's initial appeal for an "American tropical laboratory." At a broad scale, U.S. tropical science exploded in the years between the Spanish-American War and the opening of the Panama Canal. This period, however, was in some ways as disruptive as it was productive for the nascent community of U.S. ecologists working in the tropics. Johnson lamented that Cinchona remained "the only station in the western tropics for the study of pure botany." He explained, "In the last decade, it is true, several other botanical laboratories, all of them primarily for experimental-station work, have been organized in the American tropics."[169] Among these were Harvard's new station at Soledad, Cuba, as well as the USDA stations at Mayaguez, Puerto Rico, and Miami, Florida. In Mexico and Dutch Guiana (Suriname) new government- and corporate-sponsored agricultural experiment stations had also appeared. Nevertheless, Johnson fretted, "at none of these has much botanical research thus far been carried on, except the economic agricultural work of the regular staff members."[170] If it was now much easier for U.S. scientists to access Caribbean sites, it was still not clear whether there would continue to be a place for "pure botany" within the institutional landscape that was taking shape.

Maintaining a station for basic ecological research in the tropics proved difficult. Failing to secure private sponsorship for Cinchona, Britton had fallen back on Coulter's original scheme of voluntary subscriptions. The NYBG made several unsuccessful attempts to solidify the cooperation of U.S. botanists in the maintenance and improvement of the station. In 1907, for example, after petitioning forty-five universities, gardens, and museums, only five of the necessary twenty institutions would donate $100 to install a resident investigator.[171] Operating on a shoestring budget, without permanent staff and with little dedicated equipment, Cinchona could still offer vital access to a tropical forest, but clearly it could not fulfill more lofty aspirations of "furnish[ing] a second Buitenzorg on this side of the world."[172] By the end of the ten-year lease, Britton decided that the trickle of mostly non-NYBG researchers could not justify renewal under the auspices of his institution. British botanists—from the start somewhat rankled that U.S. scientists had "invaded" and "annexed" what should have been British scientific territory—took over the site through the British Association for the Advancement of Science.[173] With war looming in Europe, this arrangement would be only temporary. Largely through Johnson's efforts, U.S. scientists would again rent the station under the aegis of the Smithsonian Institution between 1916 and 1921.[174] It would remain a site of intermittent scientific observation and tourism thereafter, but not an active research station.[175]

In Cinchona's origins within the Kew network, in the tropicality of the aesthetic experiences recounted by visitors, in the transportation infrastructure that brought them there and the labor regime that supported their work, science at Cinchona was deeply marked by imperialism—both the imperialism of the British state and that of expanding U.S. agribusiness. Fundamentally imperial, too, were U.S. botanists' ambitions to establish institutions for tropical research in "their own" hemisphere. Nevertheless, research at the station was overwhelmingly "pure" in orientation—it did little to serve either U.S. or British colonial interests in the exploitation of tropical plant resources. Maintaining a pure science ideal in the context of the early twentieth-century Caribbean was a challenge that even the promising station at Cinchona could not overcome. As the century progressed, the stations with staying power would be those that could marshal the support of U.S. government agencies and corporations in support of a broad program of tropical biology that linked economic and basic scientific concerns.

CHAPTER TWO

Making Biology Tropical

We are on the eve of a great increase and diffusion of public appreciation of the value of biology. This is due in large degree to our increasing dependence upon, and our greater familiarity with, the tropics.
—Thomas Barbour, 1924

In the decade following the opening of the Panama Canal, U.S. botanists and zoologists began to make much wider and more forceful territorial claims on the tropics. Increasingly, U.S. scientists argued not only that experience in the tropics would broaden their understanding of living things but also that a broad-based program of tropical research was integral to U.S. national interests in the Caribbean and tropical world. Underpinning such claims was the emergence of a new type of station for basic biological research in the tropics. While the station at Cinchona, Jamaica, still operated intermittently, it continued to be oriented toward plant ecology and rooted in the infrastructure and research models of tropical botanical gardens. Between 1916 and 1923, U.S. zoologists began to establish their own tropical stations. Emphasizing the study of wild animals as well as plants in the tropics, they rallied under the banner of "biology."

This chapter follows the most important builders of biological stations in the tropics during the early twentieth century: Thomas Barbour and William Beebe. No two individuals did more to shape the emergence of this new type of institution. In British Guiana, New York Zoological Society (NYZS) curator of birds William Beebe founded the society's Tropical Research Station at Kalacoon in 1916 (moved afterward to Kartabo). At the same time, Thomas Barbour, then associate curator of reptiles and amphibians and later director of Harvard's Museum of Comparative Zoology (MCZ), began to expand Harvard's station at Soledad, Cuba, into a full-fledged biological research station. By 1923, he would also help found a new station at Barro Colorado Island in the Panama Canal Zone, which today remains the oldest continuously operating tropical biological station in the Western Hemisphere. These stations would set the modern standard for place-based terrestrial biological research in the tropics.

Like the botanists at Cinchona, U.S. zoologists were increasingly interested in studying tropical organisms within the context of their natural habitats. Rather than coming primarily from a physiological background, however, the supporters of these new stations in the circum-Caribbean tended to work within the natural history tradition, based at museums and zoos, as well as in university departments of zoology or biology. Many were particularly interested in problems of animal distribution, both at a regional biogeographic scale and, increasingly, at a local ecological level. Zoologists began to perceive a need for field stations to gain a more intensive understanding of the geographical distribution of species in tropical countries than collecting expeditions could provide on their own. Stations offered opportunities to collect at a finer geographic scale—documenting the presence of even very rare and endemic species in the surrounding area—as well as allowing more in-depth studies of the behavior, life cycles, and ecological interactions of living tropical animals.

The circum-Caribbean had changed significantly since U.S. botanists first called for an "American tropical laboratory," however. Although its opening in August 1914 was overshadowed by the irruption of World War I, the Panama Canal rapidly and profoundly reshaped the region's economic and transportation geography. With this strategic advantage and Europe's attention focused inward, U.S. financial institutions increasingly applied pressure on regional governments to shift national debt out of European control and into their own hands. In this era of "Dollar Diplomacy," however, Western Hemisphere warfare was far from solely economic in nature. Following the U.S.-backed secession of Panama from Colombia in 1903—which opened the way for the canal project—and President Theodore Roosevelt's 1904 corollary to the Monroe Doctrine, the U.S. military carried out a series of campaigns. During the period of World War I alone, U.S. forces intervened in Mexico's revolution (1914, 1916–17) and occupied Nicaragua (1912–33), Haiti (1915–34), the Dominican Republic (1916–24), and, for the third time, Cuba (1917–22). The tentacles of the United Fruit Company, nicknamed *el pulpo* ("the octopus"), also spread into new landholdings, building railways and steamship lines and acquiring increasing political sway.[1]

Biologists might still choose to engage in research that had little direct connection with such worldly affairs, but they were nevertheless entangled in the ideologies and infrastructures that mediated engagement throughout the circum-Caribbean. It is no coincidence that the new tropical stations, although placing increased emphasis on the study of "primeval" tropical nature, were founded on land owned by U.S. agricultural or mining companies or were administered by the government of the Panama Canal Zone, an entity under the supervision of the U.S. secretary of war. In an era of increasing U.S. military and corporate hegemony, U.S. biologists who wished to work in the tropics could not—even if they wanted to—remain insulated from regional politics and economics. This scientific community was faced with making decisions about how to position tropical research to maintain professional and intellectual autonomy while securing access to land and funding.

Barbour and Beebe not only founded stations but also forged a new vision of tropical biology and its place in world affairs. Both Barbour and Beebe were captivated by tropical nature from a young age. As professional zoologists, both men had their specialties—Beebe in ornithology, Barbour in herpetology—but they shared wide-ranging interests that encompassed all of natural history. Both were comfortable in elite metropolitan circles and well connected to the titans of U.S. industry and government. Each was expert at turning these connections to support their station-building enterprises. Despite these similarities, they cut quite different figures in their contributions to tropical biology. (They also cut different figures physically: Barbour approached six feet six inches tall and nearly three hundred pounds, while Beebe's thinness became a hallmark of press caricatures.) Examining Barbour's and Beebe's fieldwork illuminates the intellectual and professional concerns of the emerging field, as well as its ties to U.S. power and patronage. At the same time, their efforts to promote tropical biology reveal tensions over both the place of the tropics within biology and the place of biology within U.S.-Caribbean relations.

From the East to the West Indies: Thomas Barbour's Biogeography

Like many museum-based zoologists, Barbour was fascinated by zoogeography, the study of how geological and evolutionary change combined to create patterns in the distribution of animals throughout the world. (The broader term *biogeography* is more common today, but in Barbour's day the study of animal and plant geography was not yet well integrated.) Although he would describe his own research as a "matter of great theoretical interest with no practical application of any sort whatsoever," his most lasting contributions would nevertheless be in the consolidation of commercial, government, and academic support for institutions for biology in the tropics.[2] Tracing Barbour's travels from Cambridge to the Caribbean (via tropical Asia) reveals the key linkages between expeditionary fieldwork and commercial and imperial networks in the tropics during the early twentieth century. It also demonstrates how an itinerant collector like Barbour came to be a major supporter of station-based tropical biology.

Barbour owed his love of the tropics in equal parts to ill health and his family's great wealth and connections. After suffering an attack of typhoid fever in 1898, at the age of fourteen, his parents sent him away from the inclem-

Thomas Barbour at Soledad, Cuba, 1941. Barbour used his Model T Ford to access the Cuban countryside. Photograph by Dr. E. G. Stillman. © President and Fellows of Harvard College. Arnold Arboretum Archives.

ent Northeast to convalesce with his grandmother at her winter home in central Florida. An avid outdoors-woman and devotee of Thoreau, she encouraged his fledgling interest in natural history and gave him a taste for the tropics. On a visit to Miami they joined her friend Henry M. Flagler, the Florida real estate developer and cofounder of Standard Oil, on a land-buying trip to the Bahamas. While Flagler laid foundations for tourism, Barbour collected lizards and snakes. He later wrote, "I got in Nassau my first glimpse of the tropics—an iron which entered so deeply into my soul that it is still completely embedded."[3] Cementing these feelings, when Barbour was fifteen the father of a school friend—who also happened to be Henry Fairfield Osborn Jr., the president of the NYZS—recommended he read the tropical narratives of

Alfred Russel Wallace, Thomas Belt, Henry Walter Bates, and William Henry Hudson. Wallace's *Malay Archipelago* made an especially strong impression. Reading and rereading the book, Barbour set his heart on following in Wallace's footsteps.

As the son of a successful Irish-American textile manufacturer, Barbour was free to chase his vocation at Harvard's museum, as well as to indulge in extensive travel. In October 1906, freshly graduated and newly wed, Barbour and his wife, Rosamond Pierce, embarked on a tour of India, Burma, and the Dutch East Indies to retrace Wallace's travels. Barbour's Harvard connections put a well-established interimperial scientific network at his service everywhere he went—learning to navigate such networks would be essential to his future endeavors. When the Barbours arrived in Singapore, a note from Harvard botanist George Lincoln Goodale established contact with the director of the botanic gardens there, who recommended an experienced Chinese collector and translator, Ah Woo, to accompany them on the rest of their voyage. In Java, with a letter of introduction from MCZ director Alexander Agassiz to Melchior Treub, Barbour made himself at home at Buitenzorg's zoological museum. He used Buitenzorg as a base for collecting excursions, including a visit to the montane forest station at Tjibodas.[4] Treub also helped him to hire additional skilled Javanese collectors. News traveled quickly, and in many places where the party landed throughout the archipelago locals came in droves to sell specimens. In fact, at Ternate, Barbour was greeted by Ali "Wallace," the now elderly Malay man who had assisted Alfred Russel Wallace almost fifty years earlier.[5]

Barbour was not motivated simply by admiration for the codiscoverer of natural selection or a romantic interest in tropical travel, however. He hoped to make significant theoretical contributions to the understanding of animal dispersal and evolution based on his observations of animal distributions. With a few exceptions, biogeography has been treated by historians of biology as primarily a science of the nineteenth century—one culminating in the theories of Darwin and Wallace, but stagnating thereafter in a quagmire of sinking continents and imaginary land bridges until the rise of plate tectonics in the 1960s. In fact, biogeography remained an important and hotly debated area of research in natural history during the early twentieth century.[6] Even as laboratory studies in genetics and experimental evolution gained higher profiles, many zoologists continued to believe that exploration and field collection could reveal patterns of species distribution crucial for understanding the role of isolation, dispersal, and environment in the processes of evolution. (And indeed, public interest in such expeditions never abated.)[7] Nowhere was this truer than in the tropics, where so many species remained to be discovered and mapped. Barbour's expedition amassed over a thousand specimens to ship back to the MCZ. The meticulously recorded locality data that accompanied each specimen served as the basis, upon his return in August 1907, of his research for his doctoral thesis, later published as "A Contribution to the Zoögeography of the East Indian Islands."[8]

In his thesis, Barbour built on Wallace's idea that the islands had gained their faunas through continental connections during the last ice age. He disputed Wallace's assertion, however, of a single sharp boundary line dividing the characteristic fauna of Asia and Australia, corresponding with the deep Lombok strait.[9] In recent years, zoogeographers had in fact been debating the precise placement of Wallace's line through the complex archipelago. While Wallace had emphasized bird and mammal distribution, Barbour looked at a wider array of animal groups and characterized instead a "transition zone," mediated by a more complex history of geological and sea-level change that had connected and separated landmasses from one another over time. He also used his observations to make some tentative conclusions about island biogeography in general. Foreshadowing Robert H. MacArthur and Edward O. Wilson's 1967 *Theory of Island Biogeography* (see chapter 4), he noted, for example, that the "species population of an island has a very direct relation with the surface area of that island, other things [i.e., climate] being equal."[10] Thus, care should be taken in interpreting the species composition of smaller islands. Some species might be absent, not for a geologically significant reason, but simply because the island was too small to support its population. Significantly, he also argued that, by paying attention to the nature and habits of different groups of animals, one could distinguish species intolerant of transport over saltwater (such as delicate frogs or freshwater fish) from hardy "waif" species that readily disperse. Based on such evidence, he argued that most islands were populated through past continental connections rather than by fortuitous "flotsam and jetsam" dispersal over sea. His broader message was that careful fieldwork—in terms of both more intensive collecting and attention to animals' life habits—could provide biogeographic knowledge that would enable scientists to reconstruct the geological history of past land connections.

Barbour hoped to apply the same field methods and insights in the Caribbean. As a tropical island chain situated between two continents, the West Indian islands seemed a near-perfect analogy with the Malay Archipelago. Barbour could test the generality of Wallace's and other zoogeographer's claims about animal dispersal and evolution on islands by comparing observations in both regions. In fact, Wallace himself had speculated that the flora and fauna of the West Indies had once extended across a single, now submerged landmass spanning from Yucatan to Cuba, and perhaps even as far south as Venezuela.[11] Many U.S. naturalists, however, rejected the idea of past North American connections with southerly neighbors. Those who saw land bridges as overly speculative favored the idea of oversea dispersal. Barbour knew that if he could contribute to this debate he could make a real mark on his profession.[12] In this way, the Caribbean figured for Barbour much as it had for the supporters of an "American tropical laboratory": an open stage for making new contributions to a field dominated by Europeans working in the East Indies.

Barbour's first extensive travels through the Caribbean came in 1908 and 1909. Having some early experience with the Spanish language, Barbour was asked (despite his young age) to represent Harvard at the First Pan-American Scientific Congress in Santiago, Chile. The event was meant to foster closer intellectual and commercial ties among the nations of the Americas.[13] He used the return voyage as an opportunity to collect. At the Pan-American Scientific Congress he had met fellow delegate Colonel William Crawford Gorgas, the famed manager of the U.S. mosquito control programs in Cuba and Panama. Gorgas provided Barbour a base for a month of collecting in Panama at the Board of Health Laboratory of the Isthmian Canal Commission. Afterward, Barbour traveled overland across Panama, where the water reserves for the canal were just beginning to rise. He then traced a common steamer route to stop in Jamaica and Cuba on his way back to Boston.[14]

These islands were of particular importance for Barbour's biogeographical research because of their sharp faunal differences. Barbour could assume that, as relatively large islands, their faunas were effectively complete. Jamaica's species, however, shared far more in common with those of Hispaniola than with Cuba, even though these islands were equidistant from Jamaica. Cementing the analogy with Wallace's line (or the transition zone in Southeast Asia) was the existence of the 3,000-fathom Bartlett Deep fracturing the seabed between the islands.[15] To tease out the history of the islands' disconnections and reconnections, Barbour had access to decades of Caribbean collections at the MCZ, where after completing his PhD in 1910 he would be appointed associate curator of reptiles and amphibians. Nevertheless, he felt that he needed higher-quality and much more intensively collected Caribbean specimens.

Why weren't existing collections enough? There were two primary reasons. First, Barbour believed that closer attention to freshly caught and better-preserved specimens would help document "distinct island races"—that is, species and varieties that had evolved through isolation from allies on neighboring islands.[16] Older specimens did not always preserve the subtle variations of color and form that might distinguish island species. If naturalists used less well-preserved material and were unfamiliar with living forms, they might mistakenly lump together taxa from different islands that had in fact begun to speciate. For example, lizards from the genus *Ameiva*, the "jungle-runners," are found throughout tropical South and Central America, but, Barbour pointed out, nearly every island in the Caribbean harbors its own unique species.[17] Distinguishing among closely related island species was important not only for understanding the significance of isolation in the process of evolution but also to reconstruct the region's geological history. Presumably, Barbour explained, a population constantly replenished from the mainland by "raft[s] bearing individuals from the parent stock would by preventing breeding in, at least in some cases, prevent speciation by isolation taking place."[18] Conversely, if island species were common (and he believed they were), this would suggest a common origin across a lost landmass, as Wallace had suggested.

Second, although the West Indies were geographically much smaller and less complex than the Malay Archipelago, knowledge of the faunas of individual islands was still incomplete. Reptiles and amphibians were less well collected than birds, for example. Human-introduced species, moreover, complicated the task of attaining complete faunal lists for each island. The mongoose, imported in 1872 to control cane rats on sugar estates, was at the time notorious for decimating Jamaica's native fauna.[19] Barbour and other naturalists believed they were in a race against time to document the occurrence of species before they disappeared.[20] Rare, endemic reptiles and amphibians were among the most threatened and also the most significant for Barbour's biogeographical work.

Barbour considered reptiles and amphibians essential for understanding the geological history of land connections, both because he believed many were intolerant of oversea dispersal and, significantly, because they represented evolutionarily more ancient groups. Mammals and especially birds might be more prestigious and easy to collect, but their origins dated to the Mesozoic era, 245–65 million years ago. The distribution of more "primitive" groups of animals—reptiles, amphibians, and invertebrates—could, Barbour believed, reveal far more ancient geographical relationships.[21] *Peripatus* was a key example. A "living fossil," this genus of velvet worms appeared unchanged from fossils of the approximately 570-million-year-old ancestor of worms and arthropods. They were also delicate creatures. Barbour had difficulty keeping the specimens he collected from the forest floor in Jamaica alive for even a short period; the animals were far too sensitive for oceanic transport, and yet they could be found on islands and mainlands across the tropical world.[22] Such a distribution pointed to very ancient land connections indeed.[23] Barbour's central research questions had more to do with the biogeography of islands than of the tropics per se. Yet he argued that the groups of animals most crucial to understanding island biogeography were tropical, in particular the "primitive" herpetofauna and invertebrates of the tropics.

Indeed, this methodological argument was at the crux of Barbour's 1916 dispute with William Diller Matthew, paleontologist of the American Museum of Natural History.[24] Matthew argued in his influential *Climate and Evolution* that the north temperate zone was the center of animal evolution.[25] In sharp contrast to the views of the plant ecologists discussed in chapter 1, Matthew held that the challenging environment of the North produced stronger species, which dispersed south in successive waves, pushing weaker species to extinction. In Matthew's view, tropical islands—which he also believed were populated wholly through oversea dispersal—would be the last refuge of weak and less adaptable species. The Caribbean held a peripheral position in Matthew's biogeographic scheme and in his overall argument. For Barbour, however, the region was a central proving ground for universal biogeographical theories.

Barbour was willing to accept a northern origin for mammals—the group from which Matthew drew most of his evidence. Other groups, however, told a different story. Like most naturalists, Barbour envisioned areas of high species numbers as centers of evolutionary development. Looking at the distribution of reptile and amphibian groups, he found it hard to believe that northern species were not the "depauperate offshoots of the elaborate southern stock."[26] He concentrated his attack on Matthew's views on island biogeography, criticizing his lack of familiarity with environmental conditions and the full diversity of species inhabiting tropical islands. While Matthew cited the "poverty" of island faunas as evidence for fortuitous oversea dispersal, Barbour argued Matthew had "not realized the enormous sum total of different species which go to make up the fauna of such islands as Cuba and Haiti [Hispaniola]. Such a vast number of species would require squadrons of rafts."[27] Not just numbers but the nature of these species made them improbable immigrants. Calling on the authority of his own field experience, he wrote, "I only ask the reader to tramp West Indian shores" to judge the likeliness that amphibians, freshwater mollusks, or *Peripatus* could survive a sea voyage by raft and then crawl ashore.[28]

The biogeographical questions that interested Barbour operated at a regional and even global scale, but the incompleteness of scientific knowledge of the local faunas of the Caribbean drove him toward more intensive fieldwork. Collections made during a few brief expeditions would not suffice. From 1909 until the end of his life, Barbour would travel to the Caribbean every year, usually from February to May.[29] He could escape Massachusetts's hated winters ("I firmly believe that hell is full of snow," he once wrote), while probing into the prehistory of the Caribbean.[30] Although he remained an inveterate island hopper, no place came closer to becoming a second home than Cuba.

Seeing "the Possibilities of the Place": Soledad, Cuba

Cuba's biogeographic significance brought Barbour to the island, but he was drawn to the station at Soledad by a powerful set of economic and institutional circumstances. When he first visited it during his 1909 collecting expedition, the site was known officially as the Harvard Experiment Station. It was located on the Colonia Limones, part of E. Atkins & Company's Soledad estate (today Pepito Tey) just east of the port city of Cienfuegos. Most visitors would simply call it Soledad, conflating the station with the sugar estate that made its operation possible.

E. Atkins & Company was one of the largest plantation owners in Cuba. The Soledad estate was just one of the Boston-based company's many landholdings acquired as Cubans defaulted on debts during the depression that followed Cuba's Ten Years' War with Spain (1868–78).[31] Harvard's connection with the company went back to 1899, when George Lincoln Goodale and Oakes Ames began efforts to establish a private experiment station there (see chapter 1). Edwin F. Atkins, a perennial modernizer, was eager to implement scientific management and the latest technologies throughout his operations.[32] He had sought out the Harvard botanists' expertise in an effort to improve the productivity and disease resistance of Cuban cane—chiefly the variety Cristalina, which planters believed had "degenerated" over time through the practice of propagating cane asexually.[33] Cubans also engaged in their own significant cane improvement efforts.[34] Whereas the U.S. Department of Agriculture (USDA) could establish stations in Puerto Rico and Hawaii, however, the 1898 Teller Amendment (which appeased U.S. sugar beet farmers by promising that Cuba would not be annexed) prevented this in Cuba. The establishment of such a station fell to private hands.[35]

Having succeeded Goodale as director of the Harvard Botanic Garden in August 1909, Ames, an increasingly close friend of Barbour, was now in charge of the station from Harvard. The plant breeder Robert M. Gray became the station's local superintendent, remaining on site year-round engaged in efforts to produce sugar cane from seed and hybridize new varieties for Atkins. The institution was already beginning to grow well beyond its initial purpose, with the addition of a tropical botanical garden. Knowing the slow cane experiments would be a "long uphill pull," Ames worked "to render the Garden a thing of beauty in the meanwhile."[36] The Atkins family supported this gradual expansion over the years, valuing the aesthetics of the station's botanical garden and using it to entertain and impress guests. Like most botanical gardens, however, it served dual aesthetic and economic roles. Ames had started his career as an orchid specialist and taxonomist but had also developed an uncommonly broad view of economic botany. His views were so unusual that his students sometimes dismissed his Harvard course on the subject as "Uneconomic Botany"; they were perplexed when he "scarcely mentioned wheat" but spent "an entire month on arrow poisons."[37] In fact, Ames was a pioneer of ethnobotany and what would by the end of the century be called bioprospecting. He was interested less in the development of existing staple and cash crops than in the scientific recognition of plants with wholly unexploited chemical properties—plants known and used by people in corners of the tropical world but that were not yet commercialized as medicines, insecticides, or industrial materials. For such purposes, a broad collection of living tropical plants was essential for study and comparison.

Thus, during annual visits, Ames worked to expand the station's botanical collections. Entering into exchange relationships with other gardens and agricultural stations in Cuba, throughout the circum-Caribbean, and across the tropical world, Ames was able to acquire a wide variety of economic and ornamental plants—everything from mango and rubber trees to roses and gigantic water lilies.[38] The grounds devoted to experimental crops also gradually expanded. Privately, Ames dared to hope that Soledad might one day stand alongside Rio de Janeiro, Peradeniya (in Sri Lanka, then British Ceylon), and Buitenzorg as the site of one of the world's great tropical botanical gardens.[39]

However innovative, an agricultural experiment station and botanical garden might seem to have little to offer Barbour as a zoologist. He was no stranger to the possibilities of stations for zoological research, of course. Zoologists had access to a variety of well-established European and U.S. stations—most famous were the Naples Statione Zoologica and the Woods Hole Marine Biological Laboratory, established in the 1870s and 1880s. Following these models, most zoological stations continued to be located at the seaside within the United States and Europe.[40] A few groups, however, like the Johns Hopkins Marine Laboratory, had experimented with seasonal "laboratories" in the tropics, temporarily based in hotels (as in Jamaica; see

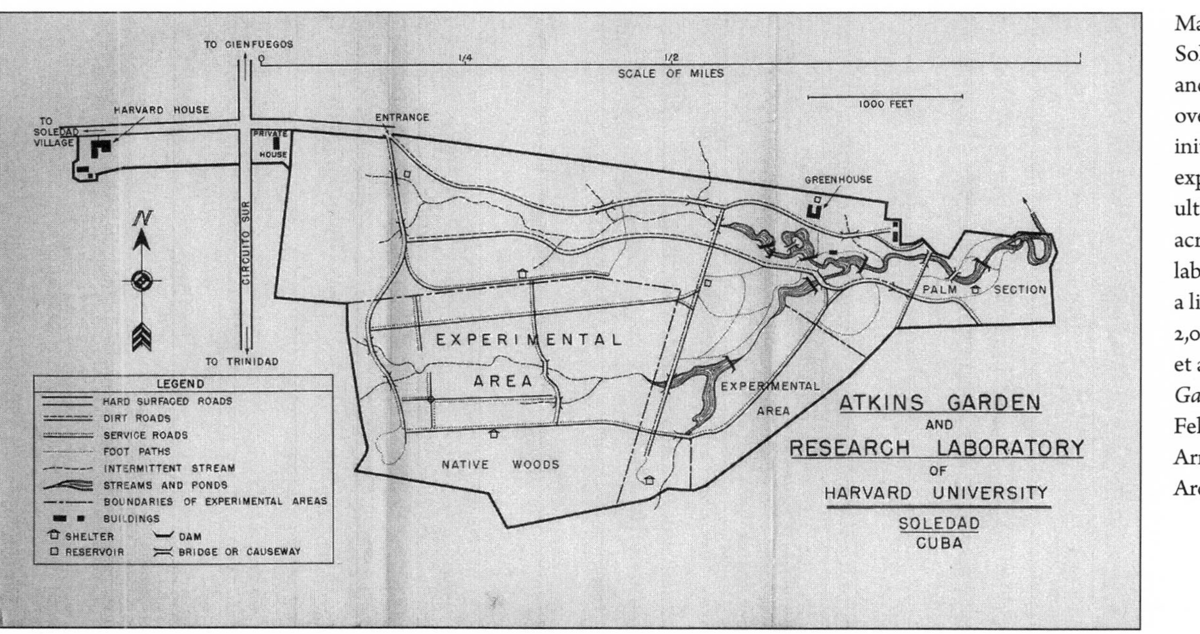

Map of Harvard's station at Soledad, 1954. The station and its grounds would grow over the years, from an initial 11 acres for cane experimentation to an ultimate extent of 221.6 acres, including a laboratory, dormitories, and a living collection of over 2,000 species. Clement et al., *Guide to the Atkins Garden*, © President and Fellows of Harvard College. Arnold Arboretum Archives.

chapter 1). As an undergraduate, Barbour had even been present in Bermuda in 1903 when his future graduate adviser, Edward Laurens Mark, helped found the Bermuda Biological Station for Research, a subtropical marine station made possible by collaboration among faculty at Harvard, New York University, and the Bermuda Natural History Society (and initially based out of rooms rented at the Hotel Frascati).[41] In fact, several of Mark's many other graduate students also went on to found and promote marine biological stations: Alfred G. Mayer directed (1904–22) the Carnegie Institution's station at Dry Tortugas, Florida; William E. Ritter was the founding director (1903–24) of the station that would become the Scripps Institution in California; and Charles Kofoid was involved with Scripps and the Illinois River research station and in promoting travel to European biological stations.[42] Zoological stations ranged widely in size and emphasis. Some, like the Marine Biological Laboratory and Statione Zoologica, focused on experimentation and microscopical work in a laboratory, using the nearby ocean primarily as a source of biological material. Others, like the Illinois station or Wimereux in France, encouraged research on the interactions of animals and their aquatic environments and played a key role in fostering the emergence of animal ecology.[43]

The station at Soledad, however, appeared poorly equipped for either approach. Apart from the modest workspace and microscope Gray used to inspect for cane inflorescence, no general laboratory was available. Moreover, the station was situated on agricultural land, which to early ecologists seemed an undesirable environment in which to study animals under natural conditions. The station's focus, under Ames, was on diversifying Cuban crops and assembling a broad tropical botanical collection, not the investigation of the natural diversity of the surrounding landscape. Station-based research itself might even seem to be fundamentally at odds with Barbour's research goals. After all, his primary purpose in Cuba was biogeographical. He was not primarily interested in studying the fauna of one site deeply. Rather, he wanted to obtain a comprehensive and geographically extensive inventory of the whole island's reptiles and amphibians.

Nevertheless—as Barbour's earlier travels should suggest—stations in fact played an essential role in facilitating the more intensive but still geographically extensive form of collection that his work required. Historian Robert Kohler has called this mode of collecting "survey science"—a type of expeditionary science aimed not at brief reconnaissance and exploration along a linear path of travel but at complete coverage of a broad area.[44] Temporary field stations played a key role in this mode of collecting; the U.S. Biological Survey, for example, set up a succession of temporary camps in the western states for days or weeks while scouring the surrounding countryside for species. Barbour was part of no official surveying organization, but like many naturalists he also utilized ad hoc stations during his travels. These might simply be hotels or private residences, but the choice of station in the tropics was in any case strongly conditioned by existing networks of science and empire.[45] Colonial botanical gardens (including Buitenzorg) often served this role for Barbour, as had Gorgas's Board of Health Laboratory in the Panama Canal Zone. In Cuba, Soledad, with its existing Harvard ties, became his preferred home base.

Even without facilities tailored especially for zoologists, Soledad offered several advantages. In many other parts of the Caribbean, Barbour gathered most of his specimens by operating local markets for specimens. Stationing himself at a hotel or docked yacht, he encouraged local residents, especially children, to bring him lizards and snakes in exchange for a few cents. The technique enabled him to amass large series of specimens within a short time, but it often missed rare or nondescript species and could leave misleading impressions of species' relative abundance.[46] With Soledad as a reliable and permanent base, however, Barbour could gather more first-hand knowledge of Cuba's fauna. (Never mind that he also found the market approach less successful in Cuba than in Jamaica or Haiti, where poverty sometimes induced hundreds of residents to participate.)[47]

As a headquarters, Soledad offered not only a space to gather supplies and prepare specimens but also a place to build local social connections. Barbour's friendship with the Atkins family grew as he and his wife stayed with them at their Soledad residence annually over the passing years. Gatherings at the plantation also enabled him to socialize with both U.S. and Cuban elites. Among these were members of the well-established Cuban scientific community, including the country's most eminent naturalist, Carlos de la Torre y Huerta. An expert on mollusks, La Torre had been a student of Felipe Poey. He was a major proponent of Darwinian evolution and a leading figure in Cuba's Academia de Ciencias and the University of Havana. He became Barbour's valued friend and collecting partner during his annual sojourns on the island.[48] La Torre collaborated with many visiting U.S. naturalists, but Barbour's facility with Spanish and seasonal residence helped foster an exceptionally close relationship.

From Soledad—and with guidance from La Torre and other Cuban naturalists—Barbour could reach several prime collecting localities. Although the estate and the surrounding countryside were dominated by a sugar monocrop, in the early twentieth century these lands were still crisscrossed by living fences—hedgerows where a surprising diversity of plants and animals could be found.[49] Lands less amenable to sugar agriculture, however, harbored an even wider array of rare and endemic species of high biogeographic interest. Unknown species remained to be discovered in habitats reachable on a day trip from Soledad—in the serpentine savannas, among coastal mangroves, and on limestone *mogotes*. In one nearby remnant grove, standing on rocky limestone soil "like an island in a sea of cane fields," Barbour even encountered the rare *Phyllobates limbatus*, at a quarter inch long the smallest frog then known. It was a species whose type specimens had deteriorated and whose range was in doubt (and Barbour was quite sure it could not withstand "flotsam and jetsam" dispersal from Central America, where its closest relatives lived).[50] The caves riddling the limestone just east of Soledad particularly sparked Barbour's interests. They harbored blind, colorless fish and shrimp that had apparently evolved in place. Cuba's caves were also the source of the only available fossils of the island's past life—important corroborating evidence of past geological connections.[51] By horseback or automobile—Barbour imported a Ford Model T, which he kept at Soledad for the purpose—one could also reach the forests of the Sierra Trinidad just twenty miles southeast (now part of the Topes de Collantes nature reserve). About sixty miles to the west lay the Caribbean's largest wetlands, the Ciénaga de Zapata, which is today a UNESCO biosphere reserve. Barbour collected a variety of species even on the grounds of Soledad's garden, but the station's location was important primarily because it was central to a variety of undeveloped areas—places seen today as havens of Cuba's biodiversity.

Thus, operating from Soledad facilitated Barbour's work on the first comprehensive list of Cuban reptiles and amphibians, his "Herpetology of Cuba."[52] It was only after the U.S. entry into World War I in April 1917, however, that Barbour began to involve himself in the development and management of Harvard's station. The war exacerbated circumstances that had led to the so-called sugar intervention (1917–22): U.S. planters, fearing the destruction of crops by liberal forces who disputed the 1916 elections, lobbied for the deployment of U.S. troops. When Cuba joined the United States in declaring war on Germany, the possibility that Germany would aid the rebels loomed on the horizon, precipitating this third U.S. occupation of Cuba.[53] Barbour served as an agent of the State Department during this period, living in Cuba for most of 1917–19. Although the feared German assistance to the insurgency never came, he reported on the country's general political and sanitary conditions and other information of U.S. interest.[54] His proficiency in Spanish and wide array of personal contacts suited him to the work. He had access to the circles of Cuban and U.S. planters through Atkins (who himself had been a key informant on Cuban affairs to the U.S. government during the Spanish-American War).[55] In Barbour's case as in so many others, scholarly work provided excellent cover; collectors' activities might seem eccentric to the layman, but their movements were less likely to be questioned. Ironically, it was during his time acting as a U.S. informant that he began to feel "almost as Cuban as North American." In his 1945 memoir of his time in Cuba, Barbour describes finally getting "the 'feel' of the island," moving beyond his fascination with Cuba's wildlife and scenery to a deeper appreciation of its people, food, and culture.[56]

Spending most weekends as a guest at Soledad through this period, Barbour began to see "the possibilities of the place."[57] He saw the Harvard station's potential to become a more broadly based tropical biological station. It was already in a location attractive to naturalists like himself, who were interested in surveying Cuba's native flora and fauna. Its expanding botanical gardens and arboretum were useful to taxonomists and horticulturalists wishing to compare living specimens of tropical plants. What it needed to attract a larger number and wider array of biologists was a laboratory and a more formal scientific administration. The first step toward these was an endowment.

Barbour, with his gregarious personality and bellowing voice, got along better with Atkins than did Ames, who, shy and restrained, found "begging for gifts" to be a "distressing duty."[58] During his stays, Barbour took Atkins on afternoon drives to the garden and often discussed Cuban politics and the role of the university in the agricultural problems of the tropics.[59] In close proximity during the war, Barbour was better able to bend the ear of the aging sugar baron than was his friend Ames, who was increasingly burdened by administrative duties back at Harvard. Significantly, by 1919, in the wake of the sugar boom that accompanied World War I as competition from European beet sugar evaporated, Atkins was also more receptive to the idea of putting the station on firmer financial ground. That year, too, Harvard researchers helped show that the threat of cane mosaic disease could be mitigated by horticultural methods.[60] The station had begun to prove its worth, and Atkins was also considering his own legacy. In December 1919, he made his first donation, $71,395, to Harvard to establish the Atkins Fund for Tropical Research in Economic Botany.[61] Although this endowment specifically supported research on tropical agriculture and horticulture, it laid the foundation for the station to increase its capacity for a wide variety of both economic and basic biological work.[62] Barbour sailed home to Cambridge in 1919, but his plans for the Soledad station had only begun to be set in motion.

The interior of the "Harvard House" laboratory, circa 1930. It was built in 1924 with funds from the Atkins family. Photo 37.410, Collection of the Massachusetts Historical Society.

A "Wholly New Type of Biological Work": William Beebe in British Guiana

Even as Thomas Barbour began to reimagine Soledad as a biological station, however, William Beebe was already at work building his own. While Barbour spent his war wooing Atkins in hopes of expanding Harvard's tropical outpost, Beebe set out for British Guiana having already secured substantial backing. Through the NYZS's pool of railroad and industrial philanthropists, he had $6,000 for the first six months and a handful of dedicated staff members devoted to the project. In March 1916, he established the NYZS's first "Tropical Research Station" at Kalacoon in the colony's Bartica District.[63]

It is a sign of the strength of Beebe's social connections that, just three days after the station opened, none other than former president Theodore Roosevelt appeared as its guest. Roosevelt was not only a hero among U.S. advocates of tropical empire but also a noted amateur naturalist and personal friend of Beebe. Stopping to visit while on a tour through the West Indies, Roosevelt was quite taken with the station and gave it his enthusiastic and public stamp of approval. Rather than ranging across the region for specimens

William Beebe in British Guiana, 1917. © Wildlife Conservation Society. Reproduced by permission of the WCS Archives.

like Barbour, Beebe intended to delve deep. As Roosevelt explained to the readers of *Scribner's,* in the tropics, "here and there desultory roaming or more systematic and extended exploration will still yield zoological results of prime consequence. But what is now especially needed is restricted intensive observation in carefully selected tropical stations."[64] The emphasis of Beebe's station was on the study of *living* tropical wildlife; this, Roosevelt argued in an introduction to the station's first major publication, "marks the beginning of a wholly new type of biological work, capable of literally illimitable expansion. It provides for intensive study, in the open field, of the teeming animal life of the tropics."[65] The vast diversity of tropical life promised rewards for a more sedentary form of field science, one focused on the endless puzzles of the lives of animals in the tropics.

How had Beebe come to see the need for this new type of tropical research? Like Barbour, he had been drawn into natural history by reading

William Beebe (far end of table) dines with (counterclockwise) Edith Roosevelt, Donald Carter, George Inness Hartley, Sir Walter Egerton (governor general of British Guiana), Rachel Hartley, Theodore Roosevelt, Anna Heyward Taylor, and the Withers family at Kalacoon, 1916. © Wildlife Conservation Society. Reproduced by permission of the WCS Archives.

Wallace and had also fallen in love with warm climates during a trip to Florida for his health.[66] He did not follow a formal academic path, however. In 1899, at the age of twenty-two, he dropped out of Columbia before completing his degree to accept an invitation from Henry Fairfield Osborn Jr. to work at the NYZS. He collected for the zoo throughout tropical Asia, as well as in Mexico, Trinidad, Venezuela, and British Guiana. On these trips, he, like Barbour, often made use of temporary stations—using what he called "improvised laborator[ies]" at campsites, hotels, gardens, and plantation residences as bases for excursions, specimen preparation, and as the loci of specimen markets.[67]

As an employee of the NYZS, Beebe primarily aimed to capture living animals for display, not amass the prepared specimens that interested Barbour. In fact, Beebe had become increasingly disillusioned with the collecting practices of his fellow naturalists. He joined the growing number of amateur and professional ornithologists who urged that more attention be devoted to the study of living birds. This movement was in part a backlash against the practice of collecting very large series of specimens (hundreds of the same species), which had emerged during the late nineteenth century as ornithologists sought to understand how populations varied across wide geographic areas.[68] Although taxonomists and biogeographers like Barbour maintained the importance of building more comprehensive and intensive geographical collections, Beebe stood among those who saw such activities as mere stamp collecting by naturalists more intent on naming new subspecies than observing living creatures. His interest in the observation of living animals aligned him with the emerging community of animal ecologists. Beebe was among the first members of the Ecological Society of America, founded in 1915, and he maintained that "the collecting of thousands of skins will be of no service" to the fundamental problems in ornithology. Only the "careful and minute tabulation of . . . ecological conditions" could provide solutions.[69]

On his tropical expeditions, Beebe became more and more aware of the potential of prolonged scientific study in place. He took the opportunity to observe the behaviors of the captive animals he collected. He marveled at the slow reactions of a sloth, or the voice of a male curassow—at least until his note-taking was disrupted by the need to ship the animals back to the Bronx Zoo for public display.[70] On his first trip through British Guiana, while gazing upward into the tantalizing forest canopy, he imagined erecting an "observing station in the tree-tops" to discover the secrets of the forest life above.[71] It was in Belém, in the state of Pará, Brazil, in 1915, however, that he began to fully recognize the opportunities of place-based research in a tropical forest. Sent to establish a new supply line for the zoo, Beebe had only a short stay in the city, at the mouth of the Amazon River. He felt that the best use of his time would be to concentrate his efforts on a very limited project in a forest in the suburbs, called Utinga.[72]

Since the area's birds had already been "collected assiduously and offer little chance of novelty to the transient ornithologist," Beebe struck upon the idea of observing the bird life of a single tree over the course of just one week.[73] On the trail near a wild cinnamon tree, *Canela do Mato*, whose ripe berries had attracted a raucous mixed flock of birds, he set up a canvas steamer chair (the site was only a five-minute walk from the end of the Belém tram line). There he sat for two to six hours each day, watching through binoculars the bird life that came and went. He shot comparatively few specimens but took notes on all the activity he could observe—foraging habits, courting, and vocalizations. In the immediate vicinity of the tree, he identified no less than ninety-seven bird species. By the final day, he felt loath to "turn my back on all this wildness, this jungle seething with profound truths."[74] In awe of the "unparalleled insurgence of such a variety of organisms as can occur only in the tropics," he considered how to convey a sense of the diversity he had encountered:

> Now that there remained only a brief space of time I tried to conceive of some last thing I could do to re-emphasize this important phase of tropical life. . . . I brought from my camping stores an empty war bag, and carefully scraped together a few handfuls of leaves, sticks, moss, earth and mold of all sorts. From directly under the Canella do Matto, I gathered four square feet of jungle debris, filled my bag and shouldered it. Then I said adieu to my trail and my tree, a sorrowful leave taking . . . for the bonds which bind me to a place or a person are not easily broken.[75]

In his article "Fauna of Four Square Feet of Forest Debris," along with a popularized version in the *Atlantic Monthly,* Beebe recounted the tiny world of worms, insects, and arachnids discovered within the pile.[76] Essentially, Beebe had loosely adapted the quadrat—a method of randomly sampling the species in a given area, which had been in use by plant ecologists since 1897.[77] During their steamship voyage north, he and an assistant sifted through the material with forceps and a magnifying lens. Beebe recorded the relative abundance of individuals of different animal groups. "The same tropical law," he noted, "which holds good in regard to plants and the larger creatures of the sunlighted world overhead applies here. I found numbers of different species, but very few collections of individuals of the same kind."[78] He was not able to obtain precise identifications of every genus or species in the sample, but Harvard's ant specialist William Morton Wheeler would identify two previously unknown genera among the specimens saved.[79] By the time the ship docked, Beebe had recorded "over *five hundred* creatures. . . . At least twice as many remained."[80]

As the ship's steward disposed of the remaining detritus, Beebe extrapolated its meaning. In his imagination, he multiplied the sample to "a square mile of jungle floor" and then to "the three thousand straight miles of continuous jungle which had lain westward up the course of the Amazon. . . . My mind faltered before the vision of the unnamable numerals of this uncharted census, of the insurgence of life which this thought embraced."[81] Although he never referenced Eugenius Warming's work, the course of his thought was similar to Warming's 1899 meditation on Lagoa Santa, whose store of species dwarfed that of Denmark (see chapter 1).[82] Comparing his Brazilian haul with four square feet later collected in New York, Beebe found temperate life only half as abundant. In another sample from the tundra of Labrador, he found a mere twenty-seven organisms.[83] He felt awe at the species richness of the tropics, but sheer numbers were never his only interest. For Beebe, the idea of diversity—the "insurgence of life"—was inseparable from the activities and interactions of the living animals that he encountered. His experiment collecting four square feet of forest floor left him lamenting "how little knowledge we have of the life histories" of tropical organisms.[84] He longed to settle in one place and observe—to find his own "square mile of jungle," from which there would be no "sorrowful leave taking."

As early as his 1909 expedition to British Guiana, Beebe had seen the potential of the colony as the site of the more intensive, prolonged observations that had begun to take form in his mind. In British Guiana not only would he have "a continent to draw upon"—he knew that a mainland site would have more species than an island—but there were also social advantages.[85] As a British colony, Guiana attracted Beebe for reasons similar to those that continued to make Jamaica and Bermuda attractive to U.S. scientists in the early twentieth century. He need not learn Spanish. Health conditions seemed to be of no serious concern. Georgetown, Guiana, was in regular contact with New York via cable, as well as by the steamers that island hopped through the West Indies. Beebe had already begun to develop contacts who happily invited him to use their land as the base of his excursions—not only did the Georgetown Botanical Garden welcome his party, but so did the U.S. millionaire Henry Gaylord Wilshire, who owned several mines in the colony, and the Englishman George Withers, manager of a rubber plantation and himself an amateur naturalist.[86]

Despite these prospects, some zoo administrators were initially hesitant about the cost of establishing a station, as well as about encouraging Beebe's extended absences from duties in the Bronx.[87] The onset of World War I, however, transformed the economics of the matter in unexpected ways. The zoo depended on live animal markets in tropical Asia and Africa for its popular tropical exhibits. It accessed these largely through middlemen in Germany and Britain. As Beebe explained, even the zoo's South American species "came by the roundabout way of Hamburg" rather than the direct route. The war's "throttling of the live bird trade" necessitated the development of "new arteries of supply."[88] The kind of tropical research station he had in mind could become an important source of supply for the zoo's public exhibitions, as well as boosting the NYZS's scientific prestige. He got approval in 1915.[89]

The innovation of Beebe's station was its emphasis on the observation of tropical animals in place, in the "wild."[90] The zoo worker Beebe yearned to see tropical animals displaying their full range of natural behaviors, and a wild life could be lived only in a wilderness.[91] Lying in the interstices between the

Spanish and Portuguese empires, the Guianas had been known since the seventeenth century as the "Wild Coast," a region of wetlands and dense forest that resisted environmental and political control.[92] For Beebe, no tropical environment embodied wilderness more than the "jungle."[93] In British Guiana, he could access it. Unlike Soledad, which offered a base for collecting excursions to a variety of undeveloped habitat types, Beebe wanted his station "to be near primeval jungle."[94] Accepting Withers's offer of free use of his property, "Kalacoon House," a "great house in the heart of the Guiana wilderness," Beebe established the NYZS's new Tropical Research Station.[95] The site lay near Bartica, at the confluence of the Cuyuni, Mazaruni, and Essequibo Rivers. Bartica was called the "gateway to the interior."[96] As frontier country, it was no wonder that it appealed to Roosevelt, as well as Beebe. About a five-hour journey from the capital and port city of Georgetown, it was the last stop for miners seeking gold and diamonds in the lands to the south—the entrance to a vast region of forest continuous with that of the Amazon.[97] Beebe's imagination was captured by this land, the legendary home "of the ill-fated search for El Dorado by Sir Walter Raleigh."[98] The station's home lay directly along this borderland, "at the very edge of the jungle."[99]

While plant ecologists might make fine distinctions among different types of tropical forests, Beebe distinguished chiefly between the area of clearing and second growth immediately surrounding the station, and the "real jungle" just beyond.[100] Not until the 1950s would Beebe embrace the term *rainforest*.[101] In 1916, that word still had the air of rarified jargon to Beebe's writerly ear. *Jungle* conveyed less the specificity of a particular plant community than a fecund tropicality, a stage full of creatures whose lives and habits awaited his discovery. In both his scientific and popular writings, he drew on the romantic sense of the jungle as an untamed and unexplored wild place, but he downplayed connotations of danger. At the same time, *wilderness* did not necessarily mean an unpeopled land—the presence of Afro-Guyanese and indigenous people only heightened his sense of the country as one set apart and unspoiled by modernity. The "wild voices" of black boatmen were for him part of his aesthetic experience of exotic, picturesque tropical scenes.[102] He recognized that Akawaio Indians "had passed up and down" the waterway "for centuries," but in his mind they "had left no trace" on the land.[103] Often calling the station a "wilderness laboratory" or "jungle laboratory," Beebe relished the juxtaposition of the "primitive" lands and peoples outside with the station's status as a space of modern science.[104]

In fact, he explained, the station was "delicately balanced between the absolute primitive wilderness and [the] comforts of civilization."[105] Kalacoon received "ice, fresh vegetables and mail" three times a week and maintained telegraph contact with Georgetown.[106] Beebe was a crusader for the idea of a healthy tropics that posed no especial risk to whites; the station became part of a mission "to show that scientists from north temperate regions could accomplish keen, intensive, protracted scientific work in tropical jungles without injury to health or detriment to the facility of mental activity."[107] (The comforts of the station indeed impressed Roosevelt, still bruised from his disastrous Amazon expedition two years earlier.)[108] As at Cinchona in Jamaica, the comforts of "civilization" were also the leisurely comforts of British colonial life. At Kalacoon, a black servant who Beebe referred to only as Sam, his "boy Friday," cared for the group's domestic needs, as well as scientific chores, like animal skinning and tree climbing.[109] Beebe clearly enjoyed playing "sahib"—a term and identity he had picked up during his fieldwork in India and kept even when not interacting with Indo-Guyanese "coolies."[110] His group likewise appropriated the British colonial ritual of evening cocktails and attended teas and parties at local plantations and the British Colony House club.[111]

Kalacoon became both "a home and a laboratory to study the wild things about us—birds, animals, and insects; not to collect them primarily, but to photograph, sketch, and watch them day after day."[112] As miners trekked past the station on their way into the colony's interior, Beebe settled down to mine, in his own way, the natural resources of the forest at Kalacoon. Beebe and his party focused most of their efforts on an area within two miles of Kalacoon, a "patch of jungle of about the size of Central Park in New York City."[113] Paul G. Howes focused on the area's bees and wasps. George Inness Hartley studied the piranha, electric eels, and other exotic fish from the Mazaruni River, as well as bringing a wide variety of bird eggs into the laboratory for embryological study. They were assisted by collector Donald Carter and recruited indigenous Akawaio hunters to secure additional specimens (and game for dinner). The artists—Beebe called them "girls"—Rachel Hartley (sister to George Inness Hartley) and Anna Heyward Taylor painted the group's finds.[114]

Beebe was particularly fascinated by the hoatzin, a bird emblematic of the seemingly primordial nature of the tropical forest. In their juvenile form, the birds sport claws upon their wings, making them appear, as Beebe noted, much like the fossil *Archaeopteryx*.[115] To sight hoatzins in the wild, Beebe wrote, "permitted a glimpse of what must have been common in the millions of years ago."[116] He observed their habits and, with George Inness Hartley, analyzed the development of the wing of the primitive bird. He hoped to be the first to keep them alive in captivity. Like Barbour's sought-after *Peripatus*, hoatzins promised to reveal primal evolutionary secrets that could be discovered only in the tropics.

Mary Blair Beebe (William Beebe's first wife) and an unidentified woman in the field in British Guiana, circa 1909. © Wildlife Conservation Society. Reproduced by permission of the WCS Archives.

Even this fascinating topic could not keep Beebe from becoming distracted by the sheer variety of life and unsolved scientific problems around him. Although the study of living organisms—overwhelmingly of plants—was also the purpose of the Cinchona station, the general approach at Kalacoon was quite different. Ecological botanists studied living plants largely by describing their associations in the landscape and, significantly, by measuring them—correlating carefully measured physiological processes with the physical factors of light, moisture, temperature, and soil. Beebe's ecological interests had more to do with the behavioral interactions of animals with their environment and with one another. His work also carried a much more immersive quality. He intended the station to facilitate the "direct personal investigation" of tropical life.[117] In encountering wild animals he hoped not only to produce sound science but also to share with others his vision of tropical wildlife "as actors and companions rather than as species and varieties." At times this scientific calling took on a spiritual element as he allowed himself to become enveloped in this "great green wonderland." "The peace of the jungle is beyond all telling," Beebe rhapsodized in the opening to his 1918 popularization of the station's work, *Jungle Peace*.[118] If primitive nature was a balm to men's souls, a war-torn world was in need of stronger stuff. Beebe found his preferred refuge in tropical nature: "When nerves have cried for a time 'enough,'" he wrote, "I turn with all desire to the jungles of the tropics."[119] For Beebe, the tropics offered not only innumerable scientific secrets but also a sanctuary from a modern world at war.

In the midst of his reveries on the tropical wilderness, it might be easy to forget that, precisely like Harvard's station at Soledad and the Cinchona station in Jamaica, Beebe's Tropical Research Station was also situated on a plantation. Kalacoon House sat on a disused corner of the Hills estate of the Bartica Rubber Company and had been intended for use as an overseer's residence and rubber drying house. Its vacancy owed in part to the company's difficulty finding investors following a major financial scandal. (The company's fiscal agent, Sterling Debenture, had committed mail fraud by selling stocks on false claims about the discovery of hundreds of thousands of wild rubber trees already growing at Hills, a scam that played on fantasies of the unexploited richness of tropical nature.)[120] Beebe was far quicker to acknowledge the availability of civilization's comforts at Kalacoon than to note the ever-present hand of commerce. Nevertheless, much like the gold seekers, planters, and financiers, he was himself a participant in an extractive industry—albeit one of comparatively minute scale. For all his emphasis on in-place observation, the Kalacoon crew shipped back "an exceptionally rich harvest" to the major scientific institutions of New York.[121] Crates of preserved specimens went to the American Museum of Natural History. A veritable menagerie flowed north for display at the Bronx Zoo and New York Aquarium. The biological richness of this tropical site promised an endless supply of scientific material, as well as deep questions for Beebe to ponder. After six months, the group returned to New York having completed a surprisingly wide array of work. Yet, Beebe wrote, "compared with the problems still to be solved and the researches of the future, our efforts seem like the scratch of a single dredge along the bottom of an unknown ocean."[122]

Beebe was keen to imagine the "jungle peace" at Kalacoon as a respite from the world of human affairs, but the site's direct linkage to the modern global economy was soon made shockingly plain. After the U.S. entry into World War I, Beebe took on volunteer duty training pilots in New York. During his leave time following a minor air crash, Beebe found an opportunity to make a

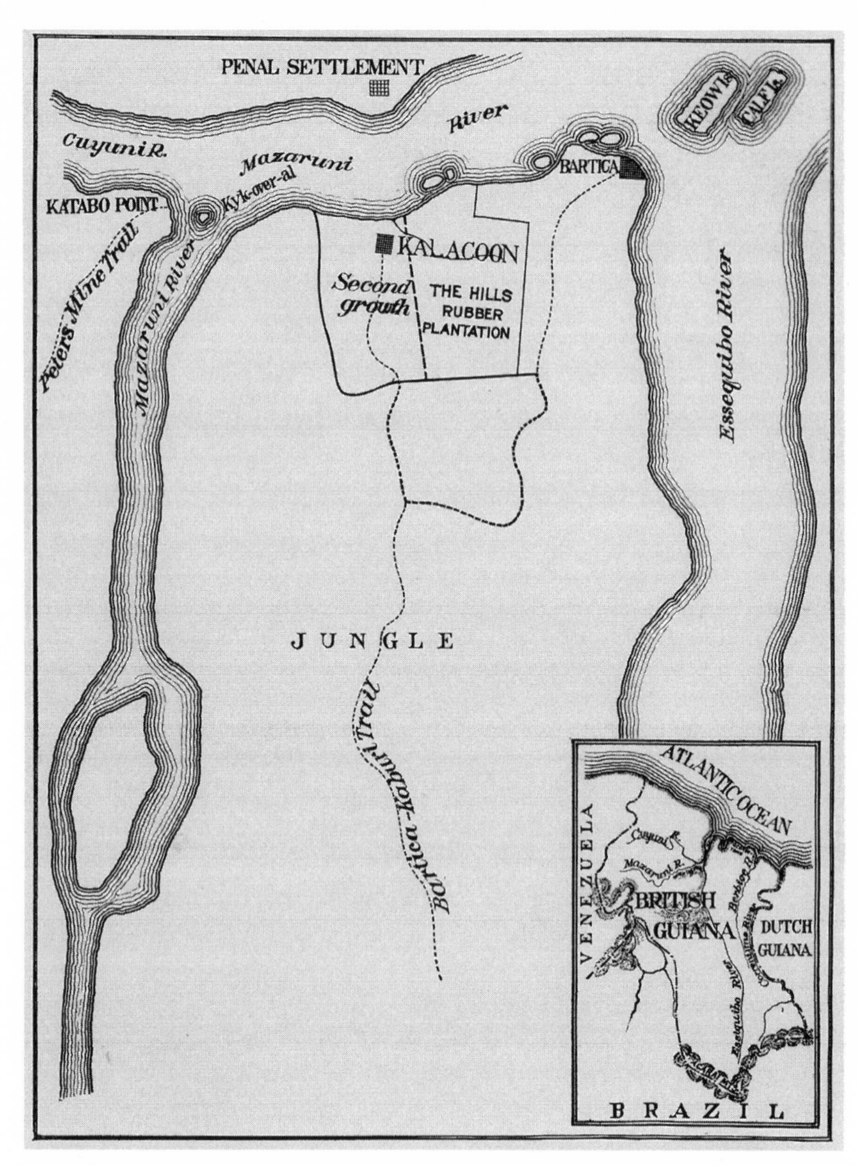

Map showing the location of the Kalacoon station relative to the "jungle" and the Hills Rubber Plantation. William Beebe would relocate the station to Kartabo (here labeled "Katabo Point") in 1919. Beebe, Hartley, and Howes, *Tropical Wild Life*.

short visit to British Guiana in August 1917. Arriving at Kalacoon, he was devastated to find that his beloved forest had been cleared.[123] It seemed that the devastation of war had touched even his beloved jungle. Wartime demand for rubber had induced Withers to expand his plantings all the way to Kalacoon's doorstep. For Beebe, the loss of the station was a blow far more severe than his airplane accident; he had envisioned it as the site of a lifetime of work.[124]

Barbour versus Beebe

The NYZS curtailed its work for the remainder of the war, but NYZS president Henry Fairfield Osborn Jr. stood behind Beebe's plans. The New York–based British Guiana Gold Concessions Company offered a new site at Kartabo Point, just a few miles up the Mazaruni River from Kalacoon, in 1919.[125] The station's annual budget grew to $10,000.[126] Although Beebe had felt an attachment to Kalacoon, he soon came to consider the new site even more ideal. The group quickly settled into the routine set in motion during the 1916 field season. At Kartabo, Beebe again ran a small fiefdom, with research assistants, hunters, servants, and artists under his direction. The Guiana government obliged, authorizing prisoners from the penal colony across the river to be pressed into service clearing trails around the station.[127]

Despite the wartime interruption and relocation, the Tropical Research Station attracted increasing attention, as publications based on the 1916, 1919, and subsequent seasons began to flow. Among these were the variety of articles for *Atlantic Monthly*, *Harper's*, and *House and Garden*, which were collected in Beebe's *Jungle Peace* and its sequel, *Edge of the Jungle*. Scholarly works included *Tropical Wild Life in British Guiana*—full of photographs and graced by Roosevelt's introduction.[128] In all, the station was clearly giving a return on the NYZS's investment. Among the 1919 haul for the zoo were a five-foot iguana, an armored catfish, a cock-of-the-rock, a jaguar, an ocelot, three capybaras, four golden agoutis, four anacondas, and a jabiru.[129] During the 1920 season, Beebe was ecstatic to capture hoatzins on film for the first time. He hoped his studies of their diet and ecology would enable the zoo to keep them alive for display (a feat that eludes zoos to this day).[130] In the meantime, the hoatzin movies made "a splendid advertisement" for the zoo.[131] Less than any particular prize, however, Beebe was most proud of the breadth and depth of research made possible by focusing on a circumscribed locality. The station also began to attract researchers from other U.S. institutions, including Harvard's William Morton Wheeler and the ecologist Alfred Emerson, then a Cornell graduate student.[132]

The scientific community's praise was not undiluted, however. Beebe's harshest critic, in fact, was Thomas Barbour. Despite their differences in field approaches, Barbour was enthusiastic about the idea of a station entirely devoted to research in a tropical forest. He and others, he wrote in *American Naturalist,* watched the development of the Tropical Research Station "with great pleasure and with high hopes" that it may "be thought of always in [the] future as bearing a relation to the tropic rain-forest in the same way that one subconsciously recalls the Naples Station when thinking of or discussing the fauna of the Mediterranean Sea."[133] Yet, while praising Beebe's "facile pen" and the "romantic style" of his essays, Barbour suggested "they sometimes have the defect of capitalizing supposed 'new discoveries' at a rather high advertising value when the history of our earlier knowledge has not been determined from the literature."[134] Key to Beebe's claims for the significance of his station was the way that in-place observation could uncover details of behavior, life history, and ecology that could not be disclosed by dead specimens alone. One example, lauded in particular by Roosevelt, was the roosting habits of the tinamou. Ornithologists had long recognized two forms of this ground bird, distinguished by a rough or smooth tarsus. Only Beebe, he claimed, had observed the former roosting in trees, thus illuminating the biological meaning of this anatomical characteristic; for at least "a hundred and six years, . . . during all this time ornithologists have accepted this character without thought or question."[135] Other naturalists had pointed out, politely, that these climbing behaviors were well known—indeed, Beebe's own account suggests the fact was no news to his Akawaio guide.[136] Elsewhere, Beebe also tended to downplay or ignore Dutch and British naturalists' existing descriptions of animal behavior in the colony.

Barbour's criticism, however, focused on the checklists of species produced by the station. Not mincing words, Barbour found Beebe's catalogue of reptiles and amphibians "phenomenally bad. . . . There is no evidence that specimens of many of the obscure species discussed have ever been preserved for examination by a herpetologist."[137] Finding no voucher material at the American Museum of Natural History, he was unable to verify observations of several species that he knew of only from as far away as Cuba, the Bahamas, or even Peru. The list also included a large number of names discarded as synonyms by contemporary taxonomists. Beebe had admitted the work was "preliminary" and that he had "made no attempt at a thorough search of literature . . . deeming this the special province of the literary systematist."[138] No doubt, as one equally at home in the field and museum, Barbour did not appreciate the sharp dichotomy Beebe repeatedly drew between the "modern field naturalist" and the mere "literary systematist."[139]

This spat was not merely over the names of a few toads. At stake was precisely what constituted proper practice in modern natural historical research. Barbour was in whole-hearted agreement with Beebe about the need for more knowledge of living tropical organisms. He believed, however, that an accurate species inventory was of fundamental importance, especially in a species-rich tropical forest:

> It is hard to review a paper of this sort without being personal, for personality becomes most inexplicably pushed into what at first sight is pure dry-as-dust. It is only fair to offer constructive criticism. . . . Beebe should first learn the value and importance—and the use—of an adequate library. He should have attached to his staff trained taxonomists who are also skillful collectors. These men should, taking the fauna group by group, make careful determinations so that the observers at the station may *know what they are working with*. Any reliance in the future on such a list as the one published—admitted to be necessary—yet "wholly foreign" to the station's aim—will be regarded by sincere naturalists with pity at the great opportunity lost and sorrow at the misuse of resources and energy.[140]

Beebe acknowledged that a catalogue of local species was a prerequisite for the more significant work of ecological observation, but Barbour felt he did not recognize the seriousness of the task. Inaccurate lists would be useless for biogeographers interested in placing Beebe's "jungle" into a broader regional context, and they would leave future naturalists unable to verify the station workers' observations. Beebe had wasted his efforts, Barbour contended, by attending more to popularization than to practices fundamental to rigorous fieldwork. Beebe's crew's observations would simply be meaningless if they did not know what species they were working with.

To Beebe, however, Barbour's interests were old-fashioned and pedantic. The future of biology lay in observing *life*—life histories, behavior, ecology. Endless bickering over dead specimens had left the teeming tropics a virtual unknown to modern science. The ponderous pace of the "museum people" was inexcusable—he complained of waiting four years with no identifications forthcoming from the American Museum before "butting into herpetology" on his own.[141] Time was too short and the field of research too large. Beebe passionately believed that naturalists needed to let go of their traditional emphasis on taxonomy and immerse themselves in living nature in new ways. This meant disciplined practices of observing a circumscribed unit of forest, as well as the use of deep description to reveal the interconnections of tropical life. The true significance of tropical research lay not in merely specifying innumerable species but in grasping life's full diversity.

Their squabble might appear as a matter of fieldworker versus sedentary systematist or, alternatively, of credentialed scientist versus popularizer. Fundamentally it concerned what role natural historical practices would play within the new field called "biology." The nature of fieldwork in the species-rich tropics had sharpened this broader debate to a fine point. How much time, money, and effort should be devoted to collection and taxonomy? How much to studies of life histories and behavior? How should such work be completed with rigor? The problem of how best to devote scarce resources for research in a little-studied but highly species-rich locale would continue to plague tropical biologists through the rest of the century. And how to secure these resources? Both depended on private patronage, but whereas Soledad grew into a biological station based on support initially granted for economic botany, Kalacoon and Kartabo's industrialist benefactors enjoyed the good press of Beebe's popularizations of "wild life." From both men's perspectives, the biological richness of the tropics and the shortage of funds for basic fieldwork abroad meant that researchers must make the most of their opportunities.[142] They agreed on the need for stations, but clearly, these sites could facilitate a wide variety of field practices. How should research be organized at a tropical institution?

As Barbour ended his review, sarcasm crept in. He mocked Beebe's exoticized and picturesque tales of local informants and collectors: "Be the urchin yellow, red, white or black," Barbour wished one had told Beebe, "'Oh, kind Sir! Do *keep the lizard* for your less happy colleagues at home.'"[143] Given the tenor and public nature of Barbour's "constructive criticism," Beebe not surprisingly dug in his heels ("let the poor man hang himself," he privately told William Hornaday, director of the NYZS).[144] He insisted that his identifications were correct and maintained, to boot, that Barbour's supposed Guianese *Ameiva* subspecies were nothing more than individual variants that could all be observed living within five hundred yards of Kalacoon.[145] Despite the icy interaction, they would cross paths again soon, confronted with an opportunity to combine their resources with those of other U.S. scientific institutions into an even more ambitious project of tropical research than either had yet attempted.

Organizing Research in Tropical America

In 1921, Beebe and Barbour found themselves serving together as members of a newly formed Institute for Research in Tropical America (IRTA), a consortium of U.S. research institutions sponsored by the National Research Council (NRC). Formed in 1916 as World War I created a pressing need for cooperation among government, industry, and research organizations to develop defense technology, the NRC expanded its focus after the war to the broader coordination of science in the national interest.[146] As U.S. economic interests grew in Latin America and the Caribbean, as well as the tropical Pacific, tropical science fell within this mission.[147] Barbour's and Beebe's central concerns may have been professional and intellectual, but IRTA's purpose was to bring biological expertise to the exploitation of economic resources in the tropics.

From the standpoint of the NRC, the efforts of the nation's scientific community in the tropics were proceeding in a piecemeal and uncoordinated fashion. IRTA brought together members representing nineteen major U.S. universities, botanical gardens, museums, and scientific societies, along with representatives of the USDA and Bureau of Fisheries. Beebe acted as representative of the NYZS and, temporarily, also the Ecological Society of America.[148] Barbour played a more substantial role in the organization, representing Harvard, serving on the institute's executive committee, and eventually becoming its chairman. With founding documents that characterized "tropical America" as "one of the few large areas of the world now awaiting development," IRTA had as its purpose to organize the U.S. scientific community to effect that development in the U.S. national interest.[149]

In broad outline, IRTA members agreed on four major lines of research of interest to biologists in the tropics, and these give insight into what the U.S. scientific community imagined "tropical research" to be. Three of these—tropical medicine, agriculture, and the development of "raw products"—had clear commercial and political ends. IRTA's plans framed these research areas explicitly in terms of their role in supporting both colonial resource extraction and white settler colonialism. Thus, echoing Oakes Ames, IRTA flagged not only the development of existing plant products such as "a great variety of woods" and rubber but also the need for systematic exploration and taxonomic study to identify new "oils, resins, drugs, dyes and numerous other products" for use in U.S. industry. Tropical agriculture was couched in terms of feeding the growing "white population of the world," which could no longer subsist only on the products of the temperate regions. Likewise, IRTA's members framed the goal of tropical medicine as one of making the tropics "a home for the white race."[150]

IRTA's members argued that research on the elimination of tropical diseases was best left to the Gorgas Memorial Institute in Panama and other agencies, but they reinforced an argument—made repeatedly by Beebe—that tropical fieldwork could be made safe for northern researchers. Equating scientific progress with the presence of white U.S. scientists, IRTA's documents highlighted the improvement of health conditions for whites as a necessary prerequisite for "a united effort to promote scientific research in the tropics." IRTA framed the advancement of the nation and "the race" as inseparable from the advancement of science. The number of U.S. biologists working in the region would not be increased "by isolated workers here and there," willing to risk their health and comfort. IRTA's goal was to "make the matter a business"—a routine, organized endeavor. The problem of promoting U.S. science in the tropics demanded "that we eliminate risk of disease and unusual hardships, that the worker be surrounded with adequate facilities, scientific atmosphere, social and bodily comforts, in short duplicate as nearly as may be the conditions under which the scientist works in Washington, New York, or St. Louis." To adequately study life in the tropics, IRTA members agreed, U.S. scientists must be able to live much as they did at home.[151]

In making tropical research "a business," IRTA was only to go so far. While the institute's members emphasized the benefits of scientific research to commercial interests in the tropics, this did not mean that science should simply act in service to industry. IRTA's plans included a fourth line of research, implicitly central to the others, that they called "pure science" or simply "scientific research."[152] Research directions, they believed, should be up to the judgment of the nation's best scientists, not narrowly directed toward immediate commercial or governmental needs. "In our broad plan for research," IRTA planning documents enjoined, "we must provide facilities for the scientists who wish to carry on investigations in what is often termed pure science, that is, investigations which may not be directed at any particular industrial problem"—curiosity-driven research like Barbour's on island biogeography or Beebe's on the fauna of his four square feet of forest debris.[153]

Rather than insisting on a strict barrier between applied and practical problems, however, IRTA members' view of pure research was more akin to what later generations would call basic research: fundamental scientific work that would serve as the foundations of future, perhaps unexpected, applications.[154] Their position was that innovations and the solutions to practical problems would naturally flow from undirected research, guided by the disciplinary concerns of professional scientists rather than by outside groups. Pointing to "the history of industrial research," IRTA plans claimed, "most epoch-making discoveries in science have been the result of scientific research in general rather than as the result of a specific demand."[155] Or, as one IRTA proposal stated more bluntly, "If the fundamental exploration and research are done, the practical matters can be trusted to take care of themselves."[156] Such a position was common in the years following World War I, as the U.S. federal government and corporations became involved in funding and directing scientific research at an unprecedented scale. Scientists were at once eager to take advantage of new funding sources and wary about the potential loss of intellectual freedom this would entail.

No one in the U.S. biological community expressed this unease better than Beebe and Barbour's mutual friend William Morton Wheeler, whom Barbour kept abreast of developments within IRTA. In his 1920 speech to the American Association for the Advancement of Science, Wheeler criticized the current "vogue of 'organization' "—as practiced above all by the NRC—as a bad "habit" picked up during the mobilization for war.[157] Businessmen and funding agencies keen for results might favor the encouragement of the most practical fields, but this would be a mistake, Wheeler argued. Pure and applied research—or, in his reframing, "discovery and invention"—were "most fruitful only in symbiosis, for the neglect of discovery must lead to impoverishment of the theoretical resources of the inventor."[158] Wheeler's speech was widely circulated; during their time with IRTA, Beebe and Barbour each corresponded about it with glee. Barbour told Albert Spear Hitchcock, the chair of IRTA, that Wheeler "states our point of view so adequately that there is not much use expanding."[159]

The other IRTA members agreed: "The most rapid progress will be made if facilities are offered to all biologists and for all kinds of biological work."[160] Above all, this meant the development of stations, institutions that could facilitate a wide range of research. While members also discussed IRTA's potential role in coordinating expeditions, the need to support tropical stations became the focus of their attention. Echoing the debates over an American tropical laboratory nearly two decades earlier, they weighed the benefits and drawbacks of different locations. Ultimately they agreed that "a station or laboratory should be so situated that the workers may be surrounded by healthful conditions, and normal social and bodily comforts. It . . . should be easily accessible to the larger centers of learning and commerce. . . . The station should be in or near the great tropical forests or jungles. It need not be emphasized that localities subject to political instability must be avoided."[161]

Johns Hopkins representative Duncan Johnson was particularly eager to secure permanent support for the Cinchona botanical station in Jamaica, whose rental by the Smithsonian would come to an end in 1921. Cooperation through IRTA might offer a way to keep the station afloat, perhaps through affiliation with other stations like Soledad or Kartabo. IRTA members discussed some of the necessary attributes of a potential new station but emphasized no particular research program. A well-equipped station could give support "to plant pathology and to medical zoology" alongside work in "general biology."[162] The purpose of a station was to attract U.S. researchers to the tropics, not to limit the type of research they would pursue.

The emphasis on building facilities rather than developing research projects in part reflected unease about relations with industry. Although all members agreed that biological science should play a role in "developing" the tropics, most were hesitant to form direct relationships with corporations or government agencies for fear of losing professional control over research directions. IRTA's constitution made specific provisions against "control passing into the hands of industrial concerns."[163] After much discussion, full membership was limited to representatives of scientific institutions. Corporations and commercial organizations could be represented only by "associate members."[164] In fact, despite the strident imperialism of most IRTA documents, some members worried specifically about the negative appearance of close relations with commercial interests:

> If the spirit of the Monroe Doctrine, which presupposes a community of interests among nations of the New World, is to survive, it must be based upon something more than political exigency. . . . Commercialism always has an element of exploitation about it and since Latin American nations are especially suspicious of stronger powers, something more than commerce is needed to maintain confidence and good feeling. . . . In spite of their backwardness in some respects, the Latin Americans have a very high appreciation for the value of science. . . . Scientific workers with their high ideals and disinterested motives might exert a powerful influence for good.[165]

From this point of view, only pure science, unsoiled by industrial interests, could promote hemispheric good will.

A few members felt quite strongly about the need to foster international scientific cooperation, whether because of idealistic Pan-Americanism or because of their practical experience working with Latin American scientists. Barbour, of course, worked closely with the Cuban scientific community, as well as the Pan-American Scientific Congresses, and in future years would take on Latin American graduate students. Likewise, Chairman Hitchcock and associate member Franklin Sumner Earle both had many years of experience working with colleagues and institutions in Cuba, Puerto Rico, and Panama, as well as throughout the circum-Caribbean.[166] They attempted to push IRTA toward a more active collaborative stance and "advise[d] very strongly against limiting the control of the Institute in such a way as to exclude Latin-American scientists from full membership." Such a move would "prevent any real cooperation of Latin-American institutions and . . . any implication of inequality would be deeply resented."[167] The possibility of opening IRTA to Latin American membership was discussed at length but indefinitely postponed.[168] A variety of proposals emerged that could have built a foundation of true U.S.–Latin American cooperation: programs encouraging the interchange of specimens with Latin American institutions, the international exchange of naturalists, fellowships, and the distribution of publications in both English and Spanish. Supported vigorously by only a minority of members, however, these proposals languished.

Proposals for international collaboration received a lukewarm reception, but members agreed on the need to free U.S. science from "dependence" on Europe in tropical matters. Their goal was "supremacy . . . for American biologists" in this field.[169] "Such information as we have," one proposal noted, "we take second hand from the British or the Germans. Encouragement of work in this field, therefore, would develop our own experts in special branches of biology in which we are now deficient and dependent."[170] Johnson argued that stations were needed to give U.S. researchers and students tropical experience; this was "the only way in which the ranks of tropical researchers can be increased at a rate commensurate with the increasing importance of tropical problems."[171] Barbour agreed: "It seems absolutely necessary that American biologists should have the facilities for work in the tropics which are offered to British workers at Peradynea [*sic*], to Dutch at Buitenzorg, and to French at Saigon."[172] Thus, the imperative to build tropical stations meant not only supporting pure science, or biology in its broadest sense, but also indicated a fundamentally imperial project—despite its members' ambivalence about corporate and government connections. At key moments, IRTA could have fostered a collaborative approach to research in the tropics. Instead, the group chose to exclude Latin Americans and emphasize competition with the European powers. Stations were an avenue for the scientific colonization of the tropical field by U.S. biologists.

Despite consensus at the highest levels of overall aims, IRTA's members had difficulty cooperating. IRTA was intended as "a center which will increase the interest and contact with the tropics for the benefit of the whole country," but members remained cautious about ceding any authority over their own institutions' projects.[173] Ultimately, nothing came of the idea of creating a network of stations throughout the region.[174] Well-established institutions like Soledad and Kartabo had little to gain from official association with IRTA. Beebe made no specific offers to cooperate. He was never strongly committed to the group, an understandable position not only given his personal friction with Barbour but also because his tropical enterprises were comparatively well supported through the NYZS. Likely motivated by IRTA discussions, however, Beebe moved to form his own new Department of Tropical Research within the NYZS in 1921.[175] He managed Kartabo and future tropical endeavors through this organization, with no connection to IRTA. Likewise, Soledad was increasingly Barbour's pet project, and he was uninterested in weakening his growing authority over it. He made sure to keep Soledad well clear of IRTA's plans, in spite of others' suggestions of the suitability of Soledad as part of a network of IRTA stations.[176] Nevertheless, IRTA could still be of some use to him.

As IRTA's "drab, bickering meetings" ground on, Barbour decided to take matters into his own hands.[177] With no concrete source of funding, he thought the group "best put all its chips on one number": a research station in Panama.[178] From his dealings with Atkins at Soledad, he believed that the best way to get support was to recruit powerful allies well connected to U.S. agricultural and medical interests. In 1922, Harvard's specialist in tropical medicine, Richard P. Strong, had just accepted the position of scientific director of the Gorgas Memorial Institute, located on Panama Bay.[179] It seemed likely that a modest marine station affiliated with the institute could be founded there—an offshoot for basic research similar to the biological laboratory he was planning at Soledad. Rather than working directly through IRTA, Barbour in effect hatched a coup. He convened unofficially with Strong, Wheeler, and their close mutual friend David Fairchild, the USDA tropical plant explorer. An advocate for tropical stations since his 1898 visit to Buitenzorg, Fairchild had close connections with private funders and had been calling for more extensive facilities for agricultural research in the Panama Canal Zone since a recent trip.[180] They arranged visits to Panama and began to set their plans in motion.[181]

Despite Barbour's low opinion of IRTA, it was during his involvement with this committee that he began to more clearly articulate a vision for the broader importance of basic biology in the tropics. Making an appeal for support from Harvard alumni and potential funders, Barbour explained how fundamental scientific questions about the origins of the tropics' great diversity and abundance of life were inseparable from vital problems of commercial and national interest: "We are on the eve of a great increase and diffusion of public appreciation of the value of biology. This is due in large degree to our increasing dependence upon, and our greater familiarity with, the tropics. Who that has camped in cathedral forests on the Equator has not realized that there is the mother liquor and that nature as we see it about Cambridge, for example, is but a weak dilution? South, the forces of nature work at their maximum intensity, for our benefit and for our injury as well." Because life reached its "maximum" in the tropics, the science of life became of central importance. For Barbour, biology meant the field in its broadest sense, a field rooted in the practices of natural history. The tropics made the "value of biology" indisputable; there problems of tropical agriculture and medicine required fundamental research in natural history to solve. Taxonomy and distribution studies were essential to the work of economic botany. The complexity of host-parasite relations in human and plant diseases, he explained, depended on research on the ecology and life histories of both animals and plants. Such research required stations in the tropics. "We can study these forces only where they are active," Barbour explained. "Personally this belief has become perhaps almost an obsession."[182]

To emphasize the inseparability of the problems of controlling tropical nature from fundamental biological questions, Barbour began for the first time to call this field "tropical biology." By endowing the Cuba station, he explained, "Mr. Edwin F. Atkins has shown his faith in the future development of tropical biology." Pairing this with a station in Panama would place Harvard at the center of the development of tropical biology, "and who does not believe that, when we make known what has been done with small resources, then greater assistance surely will be vouchsafed? *We have made a beginning.*"[183]

For the "Future Development of Tropical Biology"

By the early 1920s, tropical biology began to take shape. In its intellectual commitments and institutional foundations, this new field owed much to Thomas Barbour and William Beebe. Beebe's contributions were visionary in articulating the outlines and philosophy of a new kind of station for tropical research, as well as in popularizing this idea widely in books, magazine articles, and exhibits at the Bronx Zoo. Barbour, in contrast, worked largely behind the scenes. By arguing for natural history not as pure science but as research fundamental to addressing the biological complexity of managing tropical nature for agricultural and medical purposes, Barbour was able to establish and maintain institutions that would last beyond his lifetime—at Soledad and, as chapter 3 demonstrates, also in the Panama Canal Zone.

Both men saw research in the tropics as fundamental to biology. Biology would not be a universal science until it took into account the region where life's forces reached their "maximum intensity," where the greatest "insurgence of life" could be found. Barbour sought universal biological truths by comparing the East and West Indies, while Beebe saw his square mile of British Guianese forest as a synecdoche for "the jungle," the pinnacle of tropical nature. As a tropical region harboring large numbers of species—especially "primitive" species, not yet well documented by scientists—for both the Caribbean figured as a storehouse of evolutionary secrets. Both men also affirmed the need to strengthen rather than retreat from natural history in an era when the prestige of laboratory-based biology was skyrocketing.

At the same time, the differences in their approaches to fieldwork highlight tensions within the U.S. biological community about the role of natural historical practices. Both advocated attention to the life histories and habits of little-studied tropical organisms, and stations offered a means to accomplish this. Nevertheless, Barbour adhered to a long-standing tradition of biogeographical research, which Beebe disparaged in favor of a more radically ecological focus. These differences came to a head in their very public disagreement about how best to organize the study of tropical life: was a taxonomic inventory a fundamental necessity or merely a necessary evil? At stake was the utilization of scarce scientific resources in a vast and urgent field of research.

Certainly, scientists like Barbour and Beebe were rarities amid the army of economic botanists, entomologists, engineers, and medical doctors dispatched to Panama, Cuba, and throughout the expanding U.S. corporate enclaves during this era. Nevertheless, their stories illuminate the full range of allegiances held by academic biologists—professional and intellectual as well as national and financial. These loyalties coincided only at times. IRTA failed to unite U.S. biologists into a coherent program of tropical research in the U.S. national interest, not because of any objection to settler colonialism and resource exploitation but because members sought to maintain their own intellectual and professional autonomy. Yet, in part because IRTA provided a forum that forced biologists to justify their work in relation to problems of the national interest, biologists began to articulate "tropical research" or "tropical biology" as a broad field of basic biological research. While Beebe's Department of Tropical Research continued to operate Kartabo (now sometimes called the Laboratory of Tropical Biology), Barbour and his associates leveraged IRTA to begin their own efforts in the Panama Canal Zone.[184] Their new station, founded at Barro Colorado Island, would have the support of the U.S. military, the United Fruit Company, and other U.S. interests, but in its basic research orientation it would owe far more to Beebe's vision of a "jungle laboratory" than Barbour would ever admit.

CHAPTER THREE

Jungle Island

We may regard Barro Colorado as an essentially complete faunal unit—a little world in itself. Herein lies its inestimable value to the ecologist.
—Frank Chapman, 1929

Between 1907 and 1913, rising waters transformed a Panamanian hilltop called Barro Colorado into a six-square-mile island. Lying halfway between the Caribbean and Pacific, it became the biggest island in what was then the world's largest man-made reservoir, Gatun Lake, constructed to provide a waterway for the Panama Canal and storage for the vast quantities of water needed to operate its locks. Below Gatun Lake lies a submerged forest. U.S. engineers inundated 42,500 hectares of valley floor when they dammed the Chagres River to create the lake, leaving on Barro Colorado Island a small fragment of a forest that once stretched into the Chagres Valley below.[1]

In 1923, Barro Colorado Island (BCI) would begin another transformation, one less dramatic but in many ways no less significant. It would become a site of science. That winter, circumstances brought the Panama Canal Zone entomologist James Zetek together with a party of visiting scientists who had connections with the National Research Council's Institute for Research in Tropical America (IRTA). With their encouragement, Zetek secured BCI as a nature reserve and site for the IRTA's new research station. Administered by Thomas Barbour from Harvard and managed locally by Zetek, BCI drew hundreds of researchers during its first two decades. Its location within a U.S.-controlled territory and its proximity to the Panama Canal, a hub of world commerce, helped make BCI a major scientific destination. More than any other single institution, the BCI station became the nucleus of an emerging U.S. community of tropical biologists.

By the mid-1920s BCI was in fact the only tropical research station of its kind. Although Johns Hopkins faculty would still routinely take students into Jamaica's Blue Mountains for fieldwork, the Cinchona botanical station there ceased regular operations in 1921 with the lapse of the Smithsonian's lease.[2] William Beebe's station at Kartabo, British Guiana, was an early rival of BCI, but by 1926 Beebe would abandon the station in search of new adventures in deep-sea diving off the shores of Haiti and Bermuda.[3] Barbour himself helped build a biological laboratory at Harvard's station at Soledad, Cuba, as

Station buildings on Barro Colorado Island viewed from Gatun Lake, circa 1930. To the lower right is a banana plantation left by the island's former lessees. Photographer unknown. Digital reproduction provided by the Smithsonian Tropical Research Institute, file referer no. 34459.

well as at BCI in 1924. They were in a sense sister stations—Barbour administered both until his death in 1946—yet BCI's unique setting made the character of research there quite distinct. Horticulture remained central at Soledad; studying Cuba's native flora and fauna still usually meant ranging beyond the station's grounds. BCI offered U.S. visitors on-the-spot access to a tropical forest. Although for many the visit to BCI was one stop along a larger itinerary, scientists increasingly came for longer stretches of time and returned over the years. As the focal point of studies of the ecology, behavior, life history, and population dynamics of tropical species, this small island played an outsized role in U.S. biologists' imaginations.

BCI's forest was generally taken to be primeval, despite the island's recent formation. Far from being seen as a liability, for U.S. biologists the fact that BCI was an island offered material and conceptual advantages. Surrounded by the waters of Gatun Lake and designated a nature reserve under the authority of the U.S. Panama Canal Zone, BCI appeared to be protected from the environmental changes sweeping the canal's far shores. The station's insular

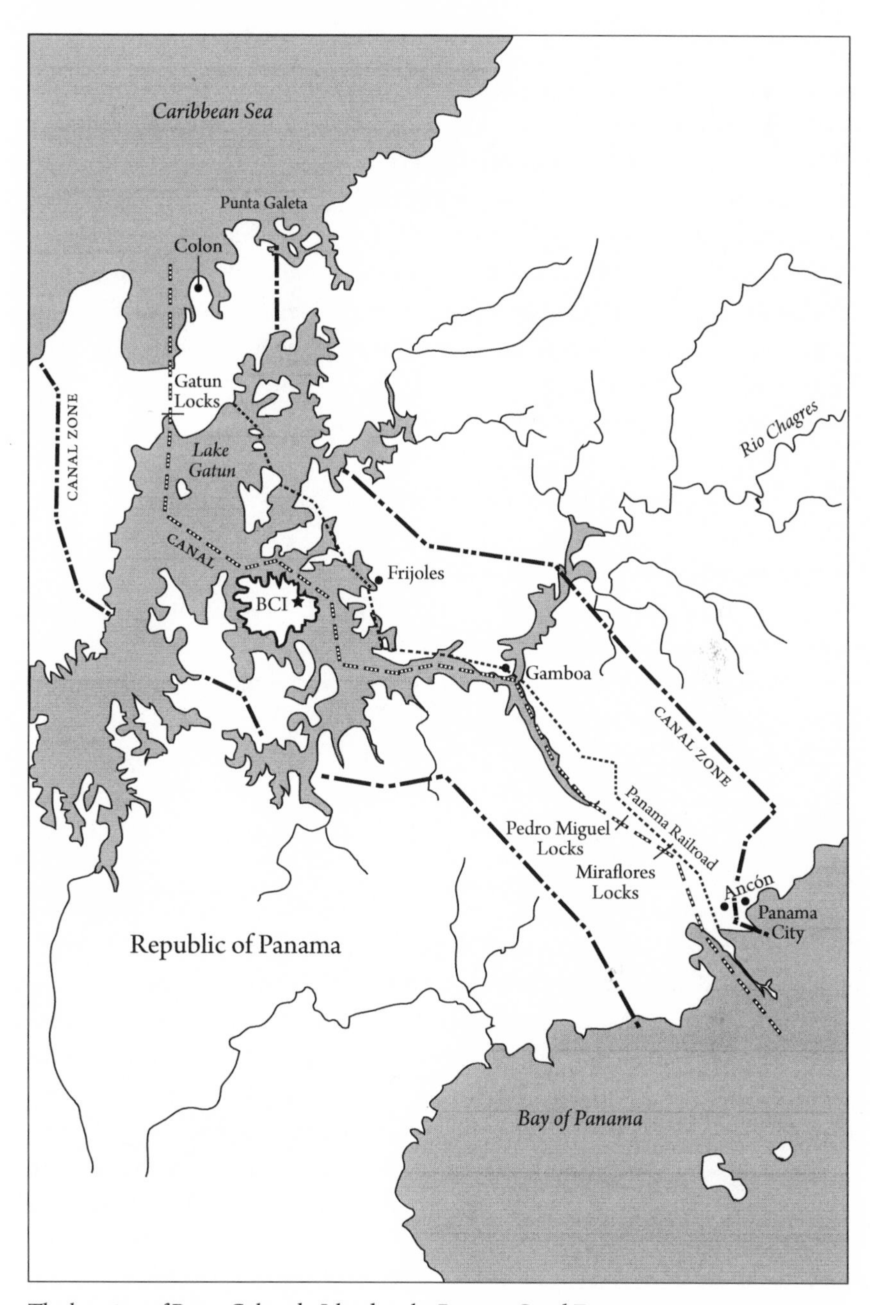

The location of Barro Colorado Island in the Panama Canal Zone.

political and physical geography enabled BCI's administrators and scientists to exercise unmatched control over space. BCI was managed with the primary mission of basic biological research. Unlike other tropical stations, it did not share land with agricultural enterprises, had little need to negotiate with foreign governments and institutions, and enjoyed the promise of continued land tenure into the future. Access and labor could be controlled almost as if the entire island itself were a laboratory. In fact, it fostered new place-based field practices. Its bounded, circumscribed environment became a powerful research tool as scientists began to study changes in the diversity of its ecological community over time. Seen as a "unit of jungle" and "a little world in itself," BCI became a microcosm of tropical life.[4]

A Man, a Plan

Unlike the stations at Soledad, Cinchona, or Kalacoon, BCI was conceived as a project of both nature preservation and scientific research. This was in large part due to James Zetek's role in establishing the station. A proud Canal "Zonian," Zetek had arrived in Panama in 1911 to work for the Isthmian Canal Commission in its mosquito eradication efforts. By the 1920s, he worked for the U.S. Department of Agriculture (USDA) Bureau of Entomology.[5] Much of his work now concerned the biology of the flies that plagued the territory's fruit crops, an economic resource second in importance only to the canal itself.

Although Zetek employed his expertise in service of the Canal Zone, he did not hold back his criticism of some of its policies.[6] He was particularly disturbed by the effects of a land lease program for former canal workers implemented in 1921. As leaseholders cleared land along the shores of Gatun Lake to cultivate subsistence crops and bananas for export, Zetek reported with alarm to his superiors, "The jungle is being cleared on all sides."[7] Destitute former canal workers were allowed up to fifty hectares. "Already some 1800 leases have been let out," Zetek warned, "and the result is a large amount of clearing which destroys the original tropical growth."[8] He never commented on the much larger environmental destruction caused by the construction of the Panama Canal itself, reserving his ire for the largely West Indian lessees and Canal Zone bureaucrats who initiated the program. While the administration offered the leases out of an interest in mitigating civic unrest and maintaining a labor reservoir for potential future projects, Zetek saw the policy as shortsighted.[9] He wanted "to see an official condemnation of the general practice followed here of clearing the land and burning everything."[10] Without conservation measures, he worried that rapid deforestation would soon leave nothing "of the original flora and very little of the original fauna."[11]

Zetek's concern for nature preservation stemmed in part from his professional background. As an undergraduate at the University of Illinois, he had studied under some of the most influential U.S. ecologists, including Henry Gleason, Charles Adams, and Stephen Forbes, who established the first river station at Havana, Illinois, and whose foundational 1887 paper "The Lake as Microcosm" emphasized holism in the study of ecological communities.[12] Zetek was also a member of the Ecological Society of America (ESA)—the only charter member based in the Canal Zone.[13] In 1917, the ESA had formed a Committee for the Preservation of Natural Conditions, chaired by the pioneering animal ecologist and first ESA president Victor Shelford. It approached advocacy for nature preservation as an applied science and a duty for professional ecologists.[14] In contrast to mainstream conservation movements, the goal of Shelford's committee was "preservation for science," not for future resource use, aesthetic, or recreational purposes.[15] As "a branch of biological science which obtains its inspiration in the natural order of original habitats," Shelford argued, ecology "must depend upon the preservation of natural areas for the solution of many problems."[16] The committee sought to identify an array of sites that could serve as baselines for understanding nature in a state unmodified by human actions, places where scientists could observe ecological processes as they unfolded over long periods of time. Seeking representatives of a full range of ecological communities, the committee ultimately declared hemispheric ambitions. Zetek was the committee's contact for information about Panama as they prepared *The Naturalist's Guide to the Americas.*[17]

In February 1923, then, Zetek was already familiar with the idea of preserving a small, typical example of Panamanian nature for scientific research when William Morton Wheeler, Richard Strong, and Charles V. Piper met with him in Panama. Both Barbour and IRTA chairman Albert Spear Hitchcock had asked them to look into conditions for the establishment of the IRTA's research station during their visit. Strong and Wheeler had come to help launch the laboratory of the recently formed Gorgas Memorial Institute of Tropical and Preventive Medicine. Piper was a former Harvard fellow working, like Hitchcock and Zetek, for the USDA. Zetek often hosted scientists passing through the Canal Zone (he had met both Barbour and the plant explorer David Fairchild during their previous trips). During their two-month visit, he shared his space in the USDA's entomological laboratory in Ancón, the Canal Zone's administrative center. They discussed the plan to site an IRTA marine station alongside the Gorgas Memorial Institute, but Zetek thought he knew just the place. He took them on an excursion to BCI.[18]

The island, about twenty-five miles from Panama City, was known to local land agents as the largest and least settled of the islands in Gatun Lake.[19] It was no stranger to scientific study, however. In fact, Hitchcock had visited the area as part of the 1910–12 Smithsonian Biological Survey of the Panama Canal Zone, a delayed effort to survey "the natural conditions of the fauna and flora of the land-bridge between North and South America" before their composition was altered by the construction and flooding of the canal.[20] The region was of major significance to naturalists and biogeographers as both the meeting place of the North and South American floras and faunas and as a barrier between the marine communities of the Caribbean and Pacific. Locally, BCI itself was understood to have a biogeographical significance of a somewhat different kind. Zonian sportsmen believed that the island had acted as a kind of Noah's Ark where the animals of the Chagres Valley had concentrated, escaping Gatun's floodwaters for higher ground. Increasingly naturalizing the artificial island, they proudly claimed it as the "greatest natural game preserve the hemisphere affords."[21] Zetek also held this view, describing the island as having been enriched by in-migration from the entire Chagres Valley, making it a truly "representative sample of the natural fauna and flora of the Canal Zone."[22]

With its "rich assembly" of species, BCI seemed an ideal site for a scientific nature reserve.[23] Wheeler identified thirty species of ants in his first hour on the island, reportedly exclaiming, "There is a greater variety of trees, plant, and insect life on this island than is found on any other equal area in Central or South America."[24] Wheeler had worked at Beebe's station at Kartabo. He had been impressed by the place (even if he gossiped disapprovingly about the "women element" there).[25] He quickly saw BCI's potential to host a similar institution—a promising prospect especially if taken under the wing of a colleague like Barbour. Zetek's scheme quickly merged with the IRTA plans. Urged forward by Wheeler and identifying himself "as a member of the Committee on Preservation of Natural Conditions" of ESA, Zetek petitioned Canal Zone governor Jay J. Morrow to preserve BCI as a "good sample of our virgin flora and fauna."[26] Governor Morrow promptly assented. On 17 April 1923 he designated BCI a natural park, guaranteeing that no new leases would be issued on the island and banning hunting "except for strictly scientific purposes."[27] A caveat remained that the Canal Zone could revoke the island's reserve status in case of military necessity. Nevertheless, founded as an official site of conservation as well as of science, BCI set off on a much firmer basis of land tenure than either Cinchona or Beebe's stations ever enjoyed. Canal Zone authorities were not liable to support BCI financially, so it was also an effortless act for the administration. Zetek promised Morrow that BCI "will be the center of biological research in the American Tropics."[28] The island was worth more to Governor Morrow for the political cachet of visibly supporting a progressive scientific and conservationist cause than for any immediate economic or strategic purposes.

IRTA now had a site for its station, not in Panama City, but on an island nature reserve about twenty-five miles away. Until his death in 1946, Barbour would direct BCI through annual visits and by correspondence from Cambridge, Massachusetts, while Zetek served as the station's manager in Panama until his retirement in 1956. As an artificial island of recent origin, BCI might easily have been seen as too exceptional, too modified by human activity to serve as a center for ecological study. Could knowledge produced there apply beyond the island's shores? The station's founders overcame this epistemological hurdle by casting BCI less as a by-product of the canal's construction than as a fortuitously preserved "sample" of a wild Panama rapidly being lost to deforestation.

The "Naturalists' Paradise"?

Writing for popular audiences, a host of journalists and scientists alike seized on the idea of BCI as a fragment of authentic tropical nature standing at the crossroads of the world. In *Jungle Island*, Warder Allee and Marjorie Allee's popular account of their 1924 season of fieldwork, the ecologist and the children's book author described BCI as "set aside" after the canal's construction so that scientists could "come and find how animals *really* grow and behave in a tropical forest."[29] For the travel writer David Edward Starry, too, BCI figured as a "laboratory" for nature's ongoing evolutionary experiment: "My cabin was on the borderline between civilization and the jungle. . . . My back and side windows looked out on an area that held its own against the advances of man—a veritable laboratory of primeval life. From my front window I could see ships from all parts of the world passing through Gatun Lake . . . pushing their blackened funnels above the green of the forests."[30] The busy signs of commerce and modern life visible from BCI's shores only heightened the awareness of primitive conditions on the island. Frank Chapman, an eminent American Museum of Natural History ornithologist and one of BCI's most loyal visitors and popularizers, commented on the island's "strange setting," surrounded by "water through which pass ships of the Seven Seas." Juxtaposing island life with the evidence of civilized society he could spy through binoculars, he recalled one day when "several Coatis, a band of Collared Peccaries, a family of Howling monkeys, and the Duke and Duchess of York passed my door . . . the last-named were on H. M. S. Renown."[31] "Civilization stops out there in the channel where the ocean steamers go by and nature begins here at the boat landing," extolled the visiting nature writer John C. Van Dyke. For him, the island was a "primitive wilderness where nature still her customs holds," a sanctuary for nature's "status quo."[32]

BCI appeared to supporters as an accessible piece of timeless, primeval tropical nature. It was celebrated as a refuge for both scientists and wildlife. Barbour and his close friend David Fairchild called it their "naturalists' paradise," suggesting not only the tropical island's variety of life and undeveloped state but also the freedom researchers could have there to pursue natural history investigations of all kinds.[33] BCI advertised an apparently "wild, virgin jungle," yet it could be reached in just a little over two hours from either side of the isthmus by rail and a short motorboat or cayuco ride across Gatun Lake.[34] Places offering such a "combination of wildness and accessibility," historian Robert Kohler argues, were fundamental to the increase in intensive studies of wildlife by the early twentieth century, as patterns of settlement and transportation put pockets of "relatively undisturbed nature" within scientists' reach.[35] BCI's visitors certainly perceived the island in this way. Yet, how wild was this artificial island, and to whom was it accessible?

In fact, the forest showed ample evidence of long-standing human occupation. A walk through the woods revealed feral cacao trees and an abandoned iron dump car from the abortive French canal effort, repurposed into a furnace by bootleggers.[36] Despite the often exaggerated publicity about BCI's wildness, botanists soon recognized about half of BCI's forest as second growth dating to the French canal period.[37] The ecologist Warder C. Allee acknowledged that the rest of the forest also might at some point have been "cut over in the four centuries during which the isthmus has been occupied by white people"—a statement ignoring land use by indigenous people and other Panamanians.[38] What mattered to naturalists was that BCI's forest could meaningfully be studied as representative of a broader tropical type. In the words of Paul C. Standley, the Smithsonian botanist who completed BCI's first flora, the island had "every aspect of the typical virgin forest occupying the humid lowlands of Central America." Little might be "really virgin forest, but the woods have been so long undisturbed that one will hardly recognize the fact," he reasoned.[39] Naturalists expected that, excluded from

Donato Carillo, Thomas Barbour, and an unidentified man on Barro Colorado Island (left to right), 1920s. Photographer unknown. Digital reproduction provided by the Smithsonian Tropical Research Institute, file referer no. 33733.

future subsistence farming and hunting, BCI would return to the essentially natural conditions that they wished to study.

Barbour and Zetek's first act of management was thus to remove the island's existing population of settlers. Although rarely mentioned in the station's publicity, several Panamanians had in fact begun to make BCI their home since the start of the 1921 land lease program. Cultivating about twelve of BCI's fifteen hundred hectares, they grew subsistence crops like yuca and corn, as well as bananas to sell to middlemen for export by United Fruit.[40] The island's comparatively sparse and recent settlement was the primary quality that had recommended it to Zetek. In establishing the station, however, he and Barbour took steps to make BCI an even more distinctly "natural" space. Barbour bought out the four men's leases himself.[41] The decision to remove BCI's Panamanian residents followed in the footsteps of the depopulation program that had emptied the Canal Zone—and not just those areas slated for flooding—of nearly 40,000 people during the construction period.[42] By transforming a landscape dotted with Panamanian towns into an unbroken jungle, historian Marixa Lasso argues, the policy enabled Canal Zone officials to depict their project as one of technological triumph over nature rather than a colonial struggle with the republican citizens of Panama.[43] Likewise, depopulating BCI produced a landscape that could meet U.S. visitors' expectations and epistemological needs for natural conditions in the tropics. The move also stood in accordance with long-standing practice at U.S. nature parks. Adhering to an ideology of wilderness, U.S. park managers saw subsistence and market activities as incompatible with the preservation of land in a natural state—largely for the recreation of urban tourists.[44] BCI was intended as a space for science, not tourism, but this fact demanded even more stringent controls over space. The removal and exclusion of Panamanians and their land use permitted scientists to understand BCI as a site restored to natural conditions. Zetek, advertising the station in the ESA's *Naturalist's Guide*, called it "practically undisturbed" and "practically virgin."[45]

Having depopulated the island, Barbour and Zetek next sought to recolonize it with scientists. BCI's location alongside the Panama Canal certainly helped make the station accessible to U.S. visitors; no other point in the tropics was in such frequent transportation contact with U.S. coastal cities.[46] Transportation geography alone did not guarantee the station's success, however. How Barbour and Zetek were able to attract and accommodate a population of visiting researchers—and *who* comprised that population—was largely structured by the station's position within a complex web of political, social, and economic ties. BCI's place in the Canal Zone made it dependent on the U.S. federal government and territorial administration in a way quite distinct from comparable stations in Jamaica, British Guiana, or Cuba, let alone within a U.S. state. As administrators, then, Barbour and Zetek's ability to develop and maintain BCI as a scientific facility drew heavily on their political and social connections in both the Canal Zone and Washington, DC.

Unlike Soledad, BCI had no primary corporate or university sponsor. Barbour's role was thus quite different when he traveled to Soledad and BCI in 1924 to oversee the construction of laboratory buildings at both stations. In Cuba, the Atkins family's generous funding supported the Harvard station's biological laboratory and visitors' quarters, known as "Harvard House."[47] At BCI, in contrast, a building was scrounged from Canal Company surplus and the bill footed by Barbour himself, along with Fairchild and a few of his wealthy friends.[48] BCI nominally operated on the "table system," like the Marine Biological Laboratory at Woods Hole or the Naples Stazione Zoologica, whereby scientific institutions paid a subscription for the use of lab space. Individual visitors also paid daily fees.[49] As at Cinchona, this cooperative effort provided barely a shoestring budget. Instead, BCI survived largely because of its relationships with the U.S. government and corporations. IRTA was increasingly a "paper organization," essentially synonymous with the BCI station itself.[50] Nevertheless, the IRTA gave the station an official affiliation with the National Research Council. With BCI as a federally recognized entity, Barbour and Zetek were able to secure free entry through customs and access to Canal Zone hospitals for BCI visitors, as well as the purchase and delivery of supplies by the Panama Canal's commissary department. As a result, BCI researchers even had access to ice, which slowed the decay of specimens and made it possible to develop film in the tropical heat. Not merely a convenience, however, this status proved essential for operating a station in the Canal Zone, where so much of daily life was mediated through the U.S. federal government.[51]

At Soledad, Barbour had the luxury of offering Atkins-funded student scholarships to support travel for research related to tropical agriculture.[52] BCI offered no such scholarships and consequently favored more established researchers over students. Nevertheless, Barbour was able to draw on his contacts to offer visiting scientists the use of extra space on military transports. He also obtained a limited number of passes on United Fruit Company steamers each year for authorized BCI visitors.[53] Why would United Fruit subsidize a station where experimental plantings were explicitly banned? The company valued its relationship with Barbour and biology at Harvard. His own seemingly arcane research indeed had practical applications. For example, he helped to establish a "serpentarium" at United Fruit's agricultural experiment station and garden at Lancetilla, Tela, Honduras, to investigate the problem of snakebites on the company's vast plantations.[54] Not only did Barbour himself consult on biological problems, but he and his stations also helped groom future company scientists. Over the years, Harvard graduate students such as George Salt and Philip Darlington put tropical experience gained at Soledad to work in the employ of United Fruit.[55] Indeed, scientists often circulated among BCI, Soledad, and Lancetilla, as well as other experiment stations on United Fruit Company lands.[56]

Zetek's local contacts within the Canal Zone were also significant. His entomological work for the Isthmian Canal Commission and the USDA—where he continued to draw his salary—provided him with an intimate knowledge of the who's who of the Canal Zone administration. Active in the Panama Rotary and Union Clubs, as well as founding the Panama Canal Natural History Society, Zetek engaged in social activities that gave him access to politicians and professionals in the tight-knit and exclusive society of the U.S. Zonians. Although from a modest background, his marriage to Maria Luisa Gutierrez, a Panamanian woman of a prominent family, also gave him access to Panama City's elites.[57] As Fairchild's son Sandy (himself a BCI visitor and, later, assistant director of the Gorgas Memorial Laboratory, the research arm of the Gorgas Memorial Institute) recalled, Zetek "knew everybody . . . and knew who he could bribe or brow-beat and who he had to leave alone."[58] Zetek convinced the Panama Railroad Company to give BCI researchers a rail pass and the $50 steamer rate normally reserved for Canal Zone employees.[59] Not only the station's first laboratory building but also much of its other infrastructure came from the stores of obsolete army and canal equipment.[60] While Zetek begged, borrowed, and stole to keep BCI afloat, he vociferously declined even potentially lucrative projects that he felt would alter BCI's natural landscape. When he got wind of a proposal from Goodyear to put a rubber test plot on the island, he was furious. He insisted that they "keep everything out that was not originally here."[61] Nevertheless, he was happy to work with lumber companies; he landed on a scheme to test material for termite resistance in exchange for free treated lumber to build scientists' housing.[62] BCI thus became accessible for basic research in large part because of Barbour's and Zetek's involvement in the economic exploitation of environments elsewhere.

The travel concessions and amenities that Barbour and Zetek acquired by navigating the Canal Zone's web of red tape and reciprocal relationships were crucial to putting BCI on the map for U.S. scientists. Reduced rates significantly lowered the barriers for scientific visitors during the station's early years and through the Great Depression. The zoologist Robert K. Enders estimated that his expenses for his first visit in 1932, lasting one hundred days, totaled no more than $450.[63] In contrast to Soledad, which especially catered to researchers affiliated with Harvard, BCI attracted visitors from well beyond the East Coast elite. In the 1920s and 1930s, about twenty-five researchers came each year; the number of scientists on the island at any one time hovered around a half dozen.[64] As at other stations, this flow was seasonal. Few visited during the academic year. Much as they would rather be in the field, at

"Woods Hole, Barro Colorado Island, Soledad and everywhere else the story is the same. Teachers have to teach and students have to stu," Barbour joked, "during these winter months."[65] Visitors enthused that BCI made working in the forest "easy and safe" and a "simple matter" even for those who had never been to the tropics before.[66] University of Michigan PhD student Josselyn Van Tyne completed eleven months of research on toucan behavior at BCI. He lauded its "quite unique" conditions: "I know of no other place in the American tropics where the biologist can live and work in comfort and safety in a virgin rain-forest jungle."[67] Stuart Frost, a Pennsylvania State College entomology professor with no tropical experience, chose to use four months of his first sabbatical to visit BCI in 1929. He reported back to his colleagues that on BCI "the scientist finds shelter, a comfortable bed, good food, tables and all the necessary equipment for general investigation. Knowing that all the essentials and comforts of life are abundantly provided for, he can direct his entire time and attention to the sole purpose for which he came."[68] The station and its simple amenities, Barbour explained, "made it possible for the teacher of biology with [a] small salary to have the thrill of Wallace, Bates, and Spruce when they first set foot in the Amazon jungle."[69]

BCI's reputation grew as widely read books like Allee and Allee's *Jungle Island*, Chapman's *My Tropical Air-Castle* and its sequel, *Life in an Air Castle*, joined visiting scientists' lectures, films, and essays about their visits.[70] Chapman even brought a piece of the island back to New York in the form of a habitat diorama at the American Museum of Natural History.[71] Journalists delighted in the novelty of the "Noah's Ark laboratory," where one could find "a world of tropical life . . . stranded to preserve a memory of other days."[72] Facing a barrage of interest from tourist parties and eccentric travelers, Barbour and Zetek found ways to police BCI's boundaries. (The acceptance of one Boy Scout as a temporary assistant had Barbour worrying to Zetek that they would "have whole tribes in on us [if] we don't look out.")[73] To get permission for an extended stay, one had to be a "credentialed scientist"—that is, to have a formal relationship with a scientific institution and a recognized status within the scientific community. From the start, Zetek made exceptions for local dignitaries whose goodwill was needed to keep the place operating smoothly. Zetek regularly played eager host to Canal Zone, U.S. Army, United Fruit, and Panamanian high officials and their families on day-trips to BCI—Panama's president Florencio Harmodio Arosemena signed BCI's guestbook with a flourish in 1930.[74] Wealthy friends of Barbour or Fairchild also added BCI to their tropical sojourns—among them the conservationist Gifford Pinchot and, on more than one occasion, Mrs. Theodore Roosevelt.[75] By 1930, the annual number of such selected tourists had surpassed one hundred; Zetek tried to mitigate their effect on the island's environment and scientific atmosphere by restricting such visits to one day a week.[76]

Not everyone found BCI so welcoming. Latin American researchers were not barred outright, yet only a handful visited BCI before World War II. In the first place, there were consequences to the choice of a station site away from Panama's urban center and within the U.S.-controlled territory of the Canal Zone. Indeed, IRTA's Albert Hitchcock had predicted that a station located in Panama City "would be more likely to receive the sympathetic cooperation of Latin-American scientists and their institutions" than one located within the U.S.-controlled Canal Zone. Others on the committee were less concerned by the need for such a "political advantage."[77] Latin American scientists who did visit BCI rarely stayed to do research on the island. Most circulated through as part of the groups of politicians, officials, and family friends of Zetek. The influential agronomist Carlos Chardón and the delegates of the Puerto Rican Agricultural Commission to Colombia, for example, took a day trip to view the island with members of the Panamanian Department of Agriculture.[78]

These circumstances were in sharp contrast to Harvard's station at Soledad, where beginning in 1927 Barbour officially took charge as the station's "Custodian." (He also became director of Harvard's Museum of Comparative Zoology that year.) Although, as historian Stuart McCook has observed, it "remained a fundamentally North American institution," the Soledad station's boundaries were far more porous than those of BCI.[79] The Harvard station engaged in exchange with Cuban experiment stations, and, under Barbour, in the 1930s it recognized several Cuban botanists as official "Collaborators." Among these were not only distinguished scientists like Brother Léon (Joseph Sylvestre Sauget), Juan Tomás Roig, and Julián Acuña Galé but also up-and-coming young botanists like José Pérez Carabia, the future cofounder (1967) of Cuba's National Botanical Garden. Several also studied in the United States through Barbour's recommendation to the Guggenheim Foundation.[80] Especially during the politically volatile 1930s, the Soledad station was under much stronger pressure to cultivate relationships with Cuba's national communities of planters, scientists, and government officials. It was also bound by Cuban law, however much that law was influenced by U.S. interests. Insulated by the Canal Zone, BCI was under no such obligations to the Republic of Panama. Despite his Spanish fluency and connections within Latin American scientific communities, Barbour made little effort to publicize BCI outside of English-language journals and initiated no programs of collaboration and exchange comparable to those at Soledad. Cuba's scientific community certainly was more prominent and had stronger institutional foundations than those of Panama, which had so recently separated from Colombia and its metropolitan intellectual centers.[81] Barbour also generally assumed that a station with a basic research orientation would be of little interest to Latin Americans, who he felt favored research with direct applications.[82] A variety of factors thus came together to make BCI relatively unknown among researchers from Panama or elsewhere in Latin America. Not only did this deprive Latin American researchers of the benefits of the station, but it also meant that BCI's U.S. visitors had little incentive to cultivate local scientific ties (or even learn Spanish). The island would remain a U.S. scientific enclave well into the second half of the twentieth century.

BCI's accessibility, however, was a matter of perspective even among those from the United States. Women had made up a significant portion of the researchers at Cinchona, Kalacoon, and Kartabo, but barriers were erected against them at BCI. Although Barbour was open to the idea, the station's prominent early supporters, Wheeler and Fairchild, were adamantly opposed to providing lodging for women visitors overnight. In their eyes, the inclusion of women would leave BCI vulnerable to accusations of being merely popular or recreational—or even morally suspect. Their fears stemmed in large part from the vicious rumors that circulated around William Beebe's stations, likely inflamed by the recent, sensational press about Beebe's divorce and private life.[83] "I understand that men and women are pretty promiscuously intermingled at a certain zoological station in British Guiana," Barbour wrote, admitting the need to avoid "the breath of a scandal."[84] Fairchild worried that admitting women would make the station vulnerable to "drifting onto the shoals of popularity." The station's promoters took pains to distinguish BCI from "the other station," avoiding any hint of tropical sexuality and emphasizing it as a space for "real research men," not popularizers like Beebe with his female entourage.[85] "I do not believe the same curiously stimulating atmosphere can be maintained in a body of men and women that can be in a body of men alone," Fairchild opined.[86]

Forced to commute to and from the mainland each day, women researchers found that "early morning bird observations, nocturnal insect and rodent collecting, and round-the-clock flowering plant studies remained out of reach." Historian Pamela Henson points out, "As day trippers, they also missed the common meals, evening discussions and camaraderie in the main building on the island."[87] Many male researchers objected to the policy; their wives were often also invaluable research assistants, and the restrictions on women presented a frustrating obstacle to young family men who wished to pursue an extended visit. Flummoxed after a conversation with Zetek, the up-and-coming primatologist Clarence Ray Carpenter wrote to Barbour, "Frankly, I had been led to believe that the mention of a wife was a source of considerable irritation, regardless of the fact that they do sometimes seem rather necessary and I notice that most men have them."[88] Nor did all women acquiesce quietly. Upon being informed by Zetek that BCI had no facilities for ladies, for example, Winifred Duncan, a zoologist who had previously worked at Beebe's station, "replied that she was no lady and would sleep right on the trail if need be." Zetek could not maintain complete control over access to the island. Denied the multiple-day visit she wanted, Duncan hired her own boat and explored BCI in Zetek's absence. "I know that had we met you would have been disabused of the idea you seemed to have that I was a tourist or a curiosity seeker," she wrote, as she left a handsome tip for the Panamanian caretaker, Donato Carillo, who had helped her around the island.[89]

Working Landscape

To maintain BCI as a scientific space, Barbour and Zetek worked to direct a controlled flow of researchers to the island. Nevertheless, "a fauna and flora on a jungle island, plus a naturalist, are not enough for an efficient working unit," explained Phil Rau, who studied BCI's social wasps and stingless bees for five weeks in 1928: "Another ingredient is required for the making of natural history. This is the provision of shelter, food supply, open trails and safe drinking water to make material accessible and health assured for the investigator. The comforts of civilization in a tropical jungle are not a part of the natural order; some one or some few persons must plan and toil to preserve what the island has for study, and to create a hospitable and comfortable atmosphere, that the visiting investigator may carry on his work without the distractions of the personal problems of well being."[90] For their "toil," Rau thanked not only Barbour and Zetek but also "Donato and his helpers who faithfully strive to make a pleasant home for the visitors."[91] Maintaining the site for science and scientists depended on the labor of a staff of Panamanian workers. Workers ferried visitors back and forth to the mainland, prepared their meals, cleaned their rooms, and dealt with loads of their dirty laundry. They built the station's infrastructure and kept it running—with hard use under humid field conditions there was always an electric generator, a water cistern, or a boat launch to repair. Workers guarded BCI's shores, cut the island's first trails, and over the years kept them free of the constantly encroaching

Laborer getting a haircut on the Barro Colorado Island dock, undated. Photographer unknown. Digital reproduction provided by the Smithsonian Tropical Research Institute, file referer no. 33798.

vegetation. They also collected and preserved specimens, cared for living study animals, monitored scientific apparatuses, and identified organisms for inexperienced visitors. Staying on the island for longer periods of time than most scientists, their labor was integral to the production and reproduction of knowledge at the station.

For this reason, creating and controlling a labor force were a central preoccupation of BCI's managers. "Give us this day our daily Zetek," Barbour would call out each morning to his secretary Helene Robinson upon entering his office at the Museum of Comparative Zoology. Amid the updates, gossip, and pleas for money, Zetek's long letters grumbled about his difficulties with workers, gloated over the achievements of his current favorites, and chronicled surprisingly intimate details about their lives.[92] If the sheer volume of such correspondence between Zetek and Barbour is any indication, BCI's workforce was the most important—and sometimes troublesome—factor in keeping the station functioning. Although never employing much more than a handful of workers at any one time before World War II, BCI had a substantial workforce compared to its closest analogues in British Guiana, Jamaica, or Cuba, where visitors hired seasonal assistants or drew from an existing pool of garden or plantation workers. Of course, everywhere, all but the smallest biological field stations rely on the labor of nonscientists, but the colonial contexts of early twentieth-century tropical stations left a significant imprint on precisely how labor was organized and how employees and researchers related to one another.

To visiting scientists, BCI might seem a world apart, but a closer look reveals a version of Canal Zone society writ small. Zetek's labor management strategies took for granted the territory's broader racially segregated labor hierarchy. In the Canal Zone, the U.S. administration divided the labor force into a privileged "gold roll" and a second-class "silver roll" of workers. These classes were afforded differing pay scales and decidedly unequal living arrangements. In the early years of canal construction, the two payroll designations were ostensibly used to distinguish skilled from unskilled labor, and later they tracked U.S. versus foreign employees. By the time BCI was established, however, the gold- and silver-roll system had hardened into a much more rigid method of racial segregation. Even in the close quarters of the island, Zetek, too, maintained the color line. Workers used separate drinking water and latrines. Longtime assistant Fausto Bocanegra remembered having bread and coffee outside while "gringo" scientists ate scrambled eggs for breakfast in the dining hall.[93]

Zetek felt that he could relate to his workers because of his Catholic faith and command of Spanish (both rarities among the Zonians). He would later boast that worker turnover was low on BCI because "we treat our help as humans."[94] Nevertheless, like managers elsewhere in the Canal Zone, Zetek relied heavily on distinctions of race and ethnicity to classify and subdivide his workforce.[95] BCI drew from a large pool of West Indian former canal workers and their descendants for general construction and maintenance work. For many of these men, work on the island was just one among a long series of odd jobs and seasonal employment.[96] For work in the forest itself, Zetek imagined "native" Panamanian men as better adapted. To clear trails and to assist scientists outdoors, he hired men that he identified as Indians. "Silvestre [Alvires] is at home in our woods," Zetek remarked of one recent hire, also noting his natural talent at "baling hay"—pressing plants to make herbarium specimens—which he did much better than had a young U.S. volunteer.[97] When it came to positions with responsibility over other workers, however, he preferred to hire whites, if possible. Zetek's first foreman was an Australian, Frank Drayton, but after several disagreements he hired Donato Carillo, "a native of Chiriqui."[98] He served as Zetek's right-hand man and caretaker of the island through the 1920s and early 1930s. Encountering difficulty finding a suitable replacement after Carillo's death, however, Zetek proclaimed his wish to replace "half grown Panamans" with Italians.[99] He eventually found Francisco "Chi Chi" Vitola, a Colombian of Italian descent, to take the position. Zetek felt "fortunate in having an Italian white. . . . He is GOOD."[100] Labor on BCI was also highly gendered. In the station's early years, there was indeed one woman who stayed regularly on the island overnight: the cook, who also saw to housekeeping duties in the scientists' dormitories.

Zetek's experience as a Canal Zone entomologist also deeply informed his labor management practices. Fearing that a case of malaria would ruin the station's reputation for safety, Zetek was anxious to keep the island free of insect-borne disease. He closely monitored the movements and health of BCI's workers, requiring them to have regular blood tests to prevent the malaria parasite from being passed to mosquito vectors on BCI. Zetek prohibited workers from traveling freely to the mainland and to towns where disease was prevalent. Employees had to be willing to spend long periods of time on the island, away from urban centers and with limited access to their families. This enforced isolation proved a strain on labor relations. To avoid the problem, Zetek even went as far as to recruit assistants from a nearby orphanage, believing that a lack of family ties would keep them from frequent travel to the mainland.[101] On BCI as in the rest of the Canal Zone, efforts to control Panamanian people took place in the name of sanitation and safety for white newcomers.[102]

Zetek's authority was nevertheless far from absolute. He remained primarily employed by the USDA and usually visited the island only once or twice a week. Workers sometimes told him only what he wanted to hear.[103] Although he attempted to restrict workers' movements, the cook Nemesia Rodriguez frequently "ran away" from BCI. She knew Zetek would find it difficult to replace a good cook who knew the island's routine by heart. She threatened that she would not come back to the island and thus put herself in a better position to negotiate wages and conditions.[104] Although a wage laborer, she essentially employed a strategy of *petit marronage*.[105] Others might simply leave to seek better wages and conditions. The threat from organized labor was less immediate—an intentional product of Canal Zone officials' exploitation of racial and ethnic divisions.[106]

Nemesia Rodriguez, Barro Colorado Island's cook during the 1920s, returns the photographer's gaze. Photographer unknown. Digital reproduction provided by the Smithsonian Tropical Research Institute, file referer no. 33803.

This set BCI apart in significant ways from the stations in British Guiana and Cuba. While Beebe's Kartabo station had had comparatively few employees, striking dockworkers in Georgetown interfered with the shipment of live animals to the Bronx Zoo in the early 1920s, following the formation of the British Guiana Labor Union. Combined with wild price fluctuations and irregular mail caused by the shipping stoppage, Beebe privately admitted that he found the colony's situation "maddening"—whether such conditions contributed to Beebe's exit from Kartabo is unclear.[107] Barbour also faced strikes and labor unrest at Harvard's station at Soledad, which drew its workers from the larger Atkins sugar enterprise. A series of general strikes, followed by brutal suppression from the Cuban army, swept the country during the 1930s. During the short-lived Revolution of 1933, the station was threatened with damage for not complying with strikers' demands, and garden workers were "forced to attend a Communist meeting."[108] With the ensuing abrogation of the Platt Amendment and the enactment of new labor laws, the station was compelled to acquiesce to demands for higher wages.[109] Such concerns rarely arose at BCI. Labor in the Canal Zone, however, sometimes became expensive. Workers, Zetek complained, could become "spoiled" by wages offered by "mining and oil people" elsewhere during boom times.[110] While BCI benefited in many ways by piggybacking on colonial infrastructure in the Canal Zone, the institution was sometimes at a disadvantage when it came into competition with larger corporate or territorial interests.

Many aspects of labor on BCI had analogues on the mainland, but field stations are also idiosyncratic workplaces. Zetek needed laborers who appeared comfortable working with scientists—which seems to have meant showing both skill and deference. Employees had to be willing to take part in strange activities, such as collecting animal blood samples, pressing plants, or feeding good food to captive monkeys.[111] While some workers balked when the ecologist Warder Allee requested help reaching the canopy, a man named Santiago constructed a ninety-foot ladder of railroad spikes up the trunk of a tree; he was a former rubber collector and an experienced tree climber. "Not many white people get to see the jungle from above," wrote Allee, who used the position to measure vertical differences in light, moisture, and temperature.[112] U.S. scientists' labor was in this case built literally on top of Panamanian labor.

Not simply caring for visitors' bodily needs or relieving them of unwanted drudgery, Panamanian workers were central to the production of knowledge on BCI. This was not necessarily through visiting scientists' appropriation of preexisting "local knowledge."[113] Most of BCI's workers were immigrants to Panama, or at least hailed from distant parts of the country. They did not always have a deep knowledge about the particular species inhabiting the island prior to their arrival there. Those employed for many years, however, did develop an intimate knowledge of the plants, animals, and landscape surrounding the station. They did so through long residence and interaction with visitors. Given the job of caring for caged study animals, Bocanegra worked out a proper diet for them on his own. He took pride in the skills he mastered, including how to collect and preserve plants and where to find animals on the island. In turn, he became a "teacher" to new visitors. The station's staff made the island and its life accessible to scientists; as Bocanegra explained of his interactions with one visitor, "I knew the forest and he didn't."[114]

The maintenance of BCI as a scientific space itself depended on the labor of its workforce. The degree of control possible on BCI was the key to the unique epistemological claims that the station increasingly enabled scientists to make. Although armed only with a stick or machete, workers patrolled BCI's perimeter against the trespass of locals poaching peccaries, deer, and agouti or gathering ipecac roots.[115] Their efforts maintained the island in a "natural" state (although such activities long predated the station) but also kept it as a space specifically for scientists. "One of the greatest draw backs of biological investigation in any locality is the human population," explained the Harvard mycologist William "Cap" Weston. Whereas in "civilized regions" scientists doing fieldwork might be met merely with amusement, "in tropical countries," he claimed, "sophisticated indifference is seldom encountered."[116] Researchers had to contend with the intense curiosity or even outright hostility of local people in many tropical field areas, he explained: "Barro Colorado, however, is wholly free from these difficulties and interferences, for there is no native population save the selected, limited, and sympathetic staff. . . . With this personnel, experiments, delicate arrangements of apparatus, and work in process of completion, may be left . . . with the certainty that they will remain scrupulously undisturbed . . . cameras may be set out, instruments may be placed, plants may be tagged or measured, and left utterly unguarded and unconcealed . . . they will remain untouched until one's return days or even weeks later."[117]

The station's disciplined workforce enabled scientists to discipline BCI's landscape. It made the island more laboratory-like, permitting scientific instrumentation to spill out from the lab building and into the forest itself. Crucially, however, it also enabled the type of "prolonged intensive investigation" that would become the station's hallmark.[118]

"Primitive" and "under Control"

Having spent twelve dry seasons on the island, ornithologist and BCI promoter Frank Chapman reflected on the qualities that made it a unique research site. "For the naturalist," he explained, "Barro Colorado's greatest attractions are, *first*, the prevailing conditions are primitive; *second*, they are under control."[119] The island was a nature reserve where entry was permitted only to authorized, credentialed scientists and a disciplined staff. This control over space enabled the increased use of instrumentation in the field. As a controlled scientific space, BCI in some ways became a "jungle laboratory."[120] Indeed, a list of equipment scientists employed in BCI's forest during the 1920s and 1930s reads like a laboratory supply catalogue: recording anemometers, thermometers, evaporimeters, atmometers, illuminometers, barometers, hygrometers, and pH colorimeters. Yet, by equally emphasizing BCI's "primitive" conditions, Chapman in fact drew attention to BCI as a site where organisms could be observed in all their complexity within their natural environment, not reduced to their simplest variables, as in a laboratory setting.

BCI's early years have been characterized as an era of "largely descriptive" work.[121] This belies the innovative practices and technologies for intensive, long-term observation that were developed there during this period. BCI was neither simply a laboratory in the field nor a refuge for "traditional" natural history. Because the station offered conditions at once "primitive" and "controlled," Chapman explained, "one may therefore study the island's life in the environment of which it is an expression, and, returning, resume his researches at the point where they were discontinued."[122] BCI was a key site in the development of a new form of long-term, place-based environmental research that helped reveal the diversity of tropical life in ways never before possible.[123]

Chapman himself was a leading advocate of observing living birds in place over time, rather than relying primarily on specimen collections. Until the turn of the twentieth century, ornithologists had insisted that the scientific study of birds required a gun, not opera glasses.[124] Chapman wanted professional ornithologists to expand beyond their focus on taxonomy and into studies of life history, behavior, and "the relation of a species to its habitat"—a task he and other ornithologists recognized as especially important in the tropics, where the habits of native birds were unfamiliar to northern researchers.[125] This meant the necessity of combining prolonged study with new technologies, like improved binoculars. Upgrading from an 8-power to a state-of-the-art 24-power tripod-mounted pair, Chapman marveled at being able to observe oropendolas weaving their hanging-basket nests from a hundred feet away: "The birds, wholly unaware of my presence, seemed to be within reach of my hand."[126] To document his ongoing observations, Chapman relied on an even simpler technology: he used his binoculars in combination with a custom-built "desk-chair," making prolonged periods of note-taking more comfortable.[127] Chapman saw himself as chronicling an ongoing natural history. He envisioned his own multiyear observations as part of a longer-term, collective effort to understand birds' lives and ecology: "I appointed myself . . . official historian to the Oropéndolas of Laboratory Hill. The story of their lives was not to be written in one nesting season or in two. Three years at least would be required to determine whether an observed act was an incident, a coincident, or a habit. Even then I should have recorded the habits of only a colony as a contribution to a work on the customs of the species; but it would be a beginning."[128] He relished the ability to witness the intimate details of the birds' lives without apparently disrupting their natural behaviors. His work was not experimental in nature, but this form of intensive, prolonged observation of living organisms depended on scientific instrumentation.

Within BCI's permanent and protected confines, binoculars allowed Chapman to survey tropical wildlife "up close" year after year. Other instruments became integrated into the forest itself, transforming it into a landscape of surveillance. Chapman was not only an enthusiast of binoculars; on BCI in the 1920s he also pioneered the first scientific use of camera traps—trip-wire-triggered flash photography.[129] Reclusive animals thus could "capture" themselves in photographs long before monitoring technologies like radio collars became commonplace.[130] Camera traps extended the power of human observation into the darkness of night and recorded more elusive behaviors. They were an early form of remote monitoring, allowing researchers to document from afar—sometimes down to the individual—the presence of ocelots, pumas, tapirs, peccaries, and even stray human trespassers. Most important, camera traps could contribute to a survey of the island's species without killing animals. Chapman called the result a "census of the living, not a record of the dead."[131] The basic technique, enhanced by infrared sensors and digital cameras, is now ubiquitous in wildlife biology, a standard tool for censusing biodiversity and population sizes.[132] These machines in the forest seemed paradoxically to allow more natural interventions. Through this remote technology scientists could observe organisms more closely and with less disturbance than if they had been physically present. Camera trapping enabled scientists to document wildlife on BCI while preserving it for ongoing, future documentation.

Tight restrictions on vertebrate collection—a stipulation of BCI's status as a Canal Zone reserve—also encouraged intensive practices for assessing the presence of species and their population numbers. Robert K. Enders was among BCI's most frequent long-term visitors, spending months at a time on the island surveying mammals during the 1930s. For this purpose, binoculars and cameras were not enough. Knowing future workers' need for a species inventory and reference collection, Barbour and Zetek secured permission for Enders to collect specimens. Enders attempted to limit his use of traps and guns, however, focusing on surveying the island's diversity rather than creating a collection for its own sake: "Mammals were observed as much as

A puma captured by one of Frank Chapman's camera traps, 1920s. The tripwire is visible across the trail. American Museum of Natural History Library, image no. 277350.

possible, only enough specimens being taken to insure identification. Where this was not necessary because of size or visibility, none were collected." Enders explained that previous research in Panama had "been mainly from the systematic viewpoint" and that, although his goal was to make an inventory of BCI's mammal species, he also collected observations of life history, ecological relationships, and population size.[133] Traps were no cure-all, in any case; a tropical forest's many arboreal species evaded trapping, and baited traps became useless in the presence of raiding ants or sudden natural abundances of forest fruit. On BCI, Enders came to see an intimate relationship between his census and an understanding of the island's ecology. "The problems of mammal distribution on the Island," he noted, "are intricately associated with food, the distribution of primary and secondary forest, scrub and clearing, water courses, and season."[134] Observations made at different times of year could thus yield very different impressions of the relative abundance of species.

Instead, Enders depended on sustained, systematic, repeated observations—enabled by the station—to overcome the difficulties of censusing species in a tropical forest. Enders was aware of the limitations of his first census but was "doubtful if such an intensive study of a small area has been attempted." Moreover, he noted BCI's advantages for such work: "The concentration of mammals is almost unbelievable; and the facilities for observation are unequalled."[135] By 1935, he had catalogued fifty-three mammal species on BCI—about half of today's total, but remarkably complete considering that the remainder are largely bat species discovered only since the 1950s with the use of mist nets and ultrasound detectors.[136]

Conditions on BCI increasingly enabled such "census of the living" approaches. Chapman encouraged Clarence Ray Carpenter, then a postdoctoral student of the pioneering ethologist Robert Yerkes, to visit BCI for field studies of howler monkeys in the early 1930s.[137] Primatologists realized that monkeys and apes failed to reproduce the full sweep of natural behaviors in the laboratory, yet few reliable field studies of primates existed. Yerkes sought to rectify this situation by sending his students throughout the world to complete rigorous, long-term studies, but those he sent to study chimpanzees and gorillas encountered difficulties locating and making continuous observations of wild populations.[138] Spending a total of eight months in close observation of howler monkeys and other primates on BCI, Carpenter was able to succeed where Yerkes's other students failed. This was largely due to the circumstances he found at the station.[139] Not only could he rest assured that his population would remain on the island, but trails also enabled him to approach monkeys without disturbing them. At the same time, laboratory facilities and the presence of other researchers made collaboration possible, for example, with the Harvard anatomist George B. Wislocki to relate periods of estrus to dominance behavior.[140] During his time at BCI, Carpenter developed innovative practices to make field observations of behavior more systematic and reproducible. Among these were technological interventions, such as the experimental playback of recorded primate vocalizations. More significant, though, were methods to discipline the researcher's own acts of observation, for example, the habituation of primates to observation from a distance (rather than from behind a fixed blind) and dyadic analysis, whereby the observer documents complex group behaviors by taking notes on pair interactions.[141] Carpenter also developed a technique for accurately estimating wild population sizes—his counts were the only ones Enders completely trusted for his own mammal census.[142] The practices Carpenter developed on BCI would become primatology's standard by the 1960s.[143] Carpenter's conception of a model field site was profoundly shaped by his time on the island—to a degree that primatologists found his ideal of an approachable yet "undisturbed" animal population challenging to fulfill elsewhere.[144]

Like instruments, binoculars, and cameras, the island's network of trails was crucial to documenting life on BCI. As Chapman put it, "A tropical forest without trails is like a house with locked doors—we may force an entrance but we shall not be welcome."[145] Trails allowed silent movement and close observation. Freshly cut trails attracted insects, and new plant species were often discovered along them.[146] The accessibility the trails offered was crucial to Enders's mammal census and Carpenter's observations of primates. Chapman's camera traps also showed that wildlife were as fond of trails as humans were. By 1930 Zetek had already overseen construction of thirty-three of the eventual forty kilometers of trails.[147]

BCI's trails spider-webbed the island, increasingly acting as a grid to systematize observation over time and across space. This framework of trails made the life of the forest "legible" to scientists, although in quite a different manner than a forester's simplified rows; biologists aimed not at sustained yields of timber but at sustained and repeatable observations of the forest's complex ecology.[148] The first trails developed organically, according to the needs of BCI's earliest scientists, the constraints of topography, and the availability of local labor, but soon Zetek worked to curate the network to achieve coverage of all parts of the island: "I considered the future needs of the island, and I believe the 'system' is hard to improve."[149] In multiple ways, the trail system worked to document the history of the island. Zetek and later administrators named each of the trails as a memorial for a BCI scientist (Barbour, Wheeler, Lutz), benefactor (Lathrop, Armour, Harvard), or—with the use of first names subtly marking status—employee (Nemesia, Donato, Fausto).[150] These trails also allowed scientists to compare the present with the past, making it possible to return to precisely the same point in the forest after a period of absence. Zetek labeled each trail with a tin sign stamped with its name and numbered at hundred-meter intervals—beginning with "0" closest to the laboratory and increasing outward. This system not only prevented visitors from losing their way in the forest but also allowed researchers to give the organisms they observed or collected a "coordinate" (e.g., "Zetek Trail 600") to which other researchers could return in the future—reference points embedded in the landscape. One trail marker became the site of a visual pun: a Yale visitor painted "Yale 6" under the "Harvard 0" marker.[151] The joke gestures

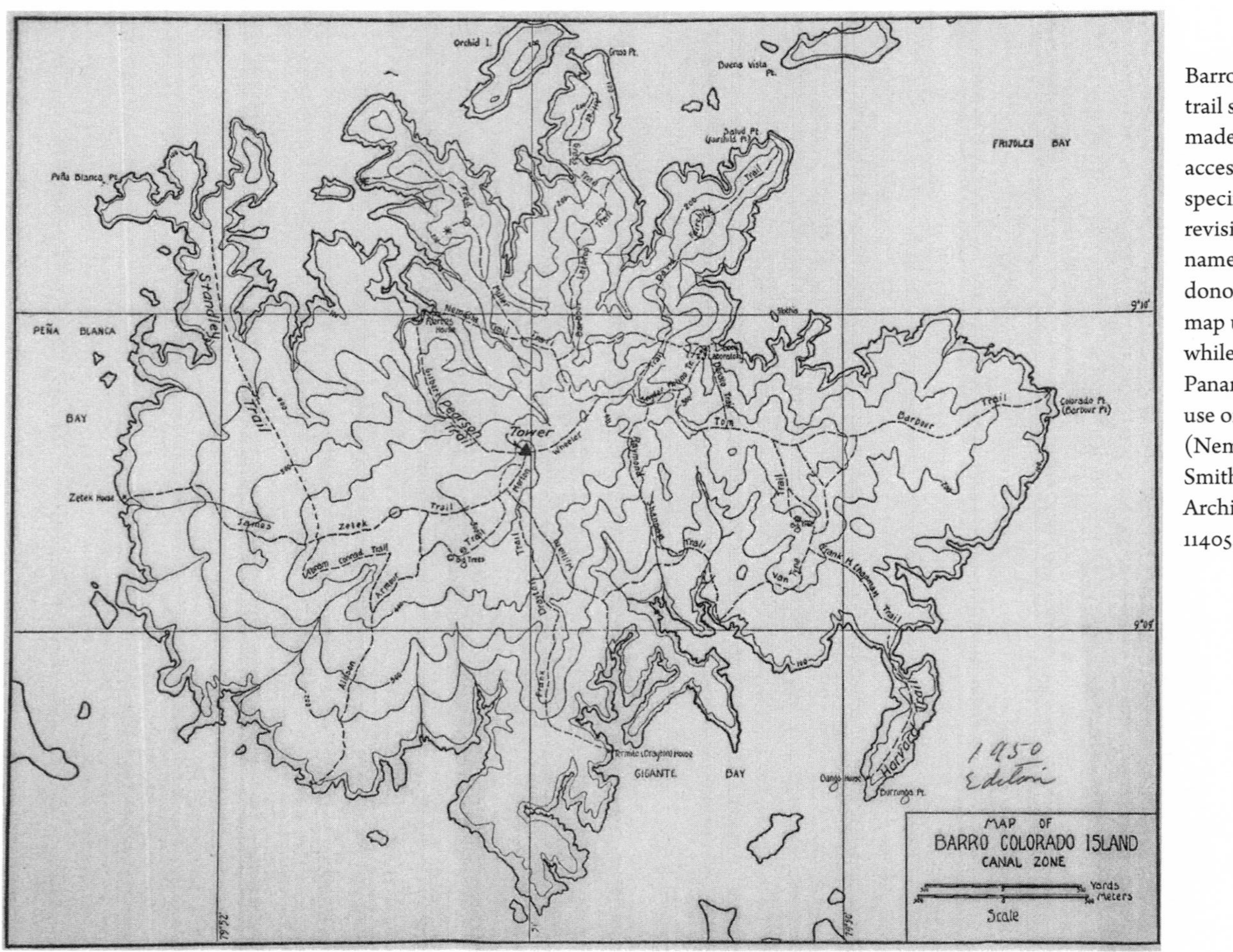

Barro Colorado Island trail system, 1950. Trails made all parts of the island accessible and enabled specific locations to be revisited. Note that trails named for scientists or donors are labeled on this map using their full names while those named for Panamanian employees use only first names (Nemesia, Donato). Smithsonian Institution Archives, no. SIA2013-11405.

at friendly rivalries among the station's visitors, united by a similar social background and academic culture, but the trail system as a whole also brought unity to their collective work of documenting BCI's life.

Within the laboratory building, a cabinet in the station's library held five- by eight-inch index cards documenting the species so far identified on the island. "The SYSTEM used is simple," Zetek explained.[152] In addition to genus, species, and family names, each card recorded when and where—usually the trail coordinate—it was first seen or collected, the names of the scientists who collected and identified the species, whether it was a new species first discovered on BCI, and cross-references to related publications held in the station's library. Perhaps the most important space within the laboratory building, the library and its catalogue allowed visiting researchers to place field observations within the context of what others had seen. As a small-scale "center of calculation," it shrank the distance between field observation and the collation of knowledge. "Eventually Barro Colorado will be a completely card-indexed island, with every plant, bird, animal, and insect on it in the laboratory files," predicted *New York Times* Panama correspondent C. H. Calhoun in 1929.[153] By 1948 Calhoun reported to readers this "unique record . . . cover[ed] 4,924 species of plants and animals," but the official estimate reached 7,000.[154] Between the difficulty of keeping the index up-to-date and the emergence of new information technologies, Calhoun's totalizing prediction never quite came true. Besides, there could be no final list—censusing and monitoring the living ecology of the island was an ongoing, collective project.

A Unit of Jungle

BCI's insularity helped scientists to systematize their observations. As an island, it also raised questions. From the station's beginning, its proponents had claimed that BCI was a representative sample of a typical tropical forest.[155] The sheer number and diversity of species found on the island seemed, to naturalists like Chapman, a testament to the island's natural state: "The life of Barro Colorado is primeval. We can see the trees and plants and it is said that they may number a thousand species. Over 230 species of birds have been observed and either through observation or automatic flashlight photographs we know that the list of mammals is essentially complete." Of known Panamanian species, only the jaguar had not yet been documented on the island by 1929:

> If other forest-inhabiting groups are as well represented we may regard Barro Colorado as an essentially complete faunal unit—a little world in

> itself. Herein lies its inestimable value to the ecologist—he who studies the relation of an animal to its environment, both physical and organic. For it is obvious that, if we would learn the part that any animal plays in the drama of life, we must ask that the setting be undisturbed and all its fellow actors be present. . . . The dead trees of Gatun Lake are not the haunts in which to study the color of a Parrot in relation to its habitat.[156]

But had BCI's cast of characters really been so unaffected by its dramatic change of scene? At first, naturalists had ascribed some of the richness of the island's fauna to a concentrating effect of Lake Gatun's inundation. The Allees' popular account described it "as well stocked with animals fleeing from the water as a tropical Noah's Ark."[157] Zetek also encouraged the "ark" view in his discussions with station visitors, journalists, and benefactors, arguing that "the animals which could migrate practically had to go to Barro Colorado. . . . As a result we have a rich assembly of jungle animals."[158] At first Chapman, too, was persuaded, but he soon began to question the power of the flood to gather animals together.[159] While journalists relished the Darwinian imagery of monkeys "forced to swim for their lives" and stranded jaguars "devour[ing] their weaker fellow-refugees," Chapman noted that Lake Gatun had filled gradually over a period of four years.[160] Not only was the inundation not sudden, but geography also permitted animals to migrate as easily toward the high ground of the mainland as to that of the new island.

Yet if insulation had not altered BCI's composition, might the subsequent protection of the island from hunting and deforestation not be "favorable for its increase"? If BCI's fauna was indeed changing, the implications would resonate far beyond the island's muddy shores. Chapman asked:

> To what extent can this increase continue? Does the island already hold as many vertebrates as it can support? If the existing fauna is balanced will not eventually some species become unduly abundant? Is the island large enough to possess and support reserves of life which may be drawn on to maintain a balanced fauna? These . . . questions . . . are of more than local importance; they are of wide vital significance in the study of life at large. And if the circumscribed conditions which prevail on Barro Colorado enable us to answer them, we shall have made an important contribution to our knowledge of those causes that determine the comparative numbers of species, and lead to their success or failure.[161]

BCI's "circumscribed conditions," Chapman argued, would enable scientists to answer fundamental ecological and evolutionary questions. It could act "as a natural laboratory" for understanding a tropical forest as an ecological community, including the factors that controlled the size and composition of its fauna.[162] Echoing Chapman, Weston explained that the island's "abundant" animal life faced "no interference from outside, their numbers increase subject only to the natural laws of limitation which the different forms in this complex association exert upon each other."[163] The formation of BCI presented a "unique opportunity for the study of Natural Science," Fairchild likewise explained, because "the wild life thus isolated could be studied as a unit of jungle."[164] BCI, in effect, acted as a six-square-mile quadrat. Offering controlled access to nature and nature's laws, the island promised knowledge not just of a portion of Panama but about tropical forests in general.

In 1929, Chapman's questions were not yet answerable. "It will take close, expert observation, extending over a period of years to make, for example, even an approximately accurate estimate of our large mammal population . . . [and] to recognize fluctuations in their numbers," he predicted. "The fact however, that it may be done," he argued, "is evidence of the exceptional advantages that Barro Colorado offers for extensive field-work."[165] In fact, Enders's efforts to census and recensus BCI's mammals were explicitly intended to test the concentrating ark hypothesis. Seeing little change in BCI's fauna by 1935, Enders argued against any artificial concentration on the island, which he doubted could persist over two decades.[166] By 1937, however, he was convinced that "profound changes" were indeed occurring "in the relative abundance of various species."[167] Large mammals had decreased. The number of pumas and ocelots had dropped dangerously low for survival. Tapirs and peccaries were increasingly rare. At the same time, the population of monkeys, armadillos, and rodents like the agouti and paca were on the rise. Enders blamed these changes on the incursion of poachers and a resulting surge of prey populations. Later investigators would use the early surveys of Enders, Chapman, Carpenter, and others as a baseline, asking their own questions about how species diversity and population numbers on the island changed over time. In any case, it was becoming increasingly clear that BCI's forest community was not nearly as stable and unchanging as the image of the primeval rainforest suggested.

An Ecology of Place

If animals had not actually concentrated on BCI, researchers had. Close to four hundred visiting scientists made their way from the station's dock up the two hundred stairs to the top of laboratory hill (or, like Zetek, rode the mechanical lift) between 1923 and 1940.[168] Chapman was not idly boasting, then, when he claimed that "during their fifteen years as a Laboratory, the four thousand acres we know as Barro Colorado have contributed more to our knowledge of tropical wild-life than any other area of similar extent in America—perhaps in the world."[169] By 1940, nearly five hundred papers had been published on BCI research.[170]

While some historians have seen the station as lacking "a dynamic intellectual community" or "a coherent research program" during its early years, the bounded space of the island lent a coherence of its own.[171] BCI scientist Egbert Giles Leigh has called this phenomenon a "unity of place"—each research project provided "both the empirical background and intellectual foundation for further work, and essential context for other, very different projects."[172] Enders's mammal census depended on Chapman's camera traps and Carpenter's study of howler monkey behavior. Observational studies of wild termite colonies complemented long-term outdoor tests of chemically treated wood and microscopic analyses of termite gut protozoa. Standley's *Flora,* supplemented by the observations of ecologist Leslie Kenoyer (a former student of Henry Cowles), provided the botanical background for generations of animal researchers—"no local flora of tropical America is better known," Standley advised.[173] The ecologist Orlando Park chose BCI as the tropical site for his comparative study of nocturnalism in animal communities, not only because he was a graduate student of Warder Allee or because the station enabled controlled measurements over extended periods of time, but also because the community and its "general ecology have been treated by many."[174] Research begat more research. BCI's long history as a research site permitted biologists to interpret their findings within a broader context of the site's ecology and natural history—including life histories, plant-animal interactions, climate cycles, and past land use. Over time, the complex interconnections of animals and plant life on BCI increasingly fostered connections among researchers and their seemingly disparate projects.

Research at BCI was place based.[175] Commanded by no formal research program, its circumscribed and disciplined space nevertheless afforded seminatural intellectual boundaries. Beebe had also attempted to tame a tropical forest's complexity by subdividing space, beginning with his four square feet of Brazilian forest floor and culminating in his studies of a quarter of a square mile of forest around Kartabo, British Guiana.[176] His work and writing inspired many BCI researchers, including Allee in his study of vertical forest stratification, Park in his studies of daily cycles of activity, and, even more directly, Alfred Emerson, who studied termites at BCI after being introduced to the subject by Wheeler and Beebe at Kartabo. Research at Kartabo, how-

Alexander Petrunkevitch (Yale), Paul D. Voth (University of Chicago), Orlando Park (Northwestern), Eliot C. Williams (Park's graduate student), and Fernando Jaen (left to right) at Shannon Trail 1.5, 4 August 1938. The rope in the foreground was attached to a pulley to enable some access to the forest canopy. Photographer unknown. Digital reproduction provided by the Smithsonian Tropical Research Institute, file referer no. 33869.

ever, had been largely tied to the New York Zoological Society and to Beebe himself. It never took on an institutional life of its own after he withdrew in 1926. Emerson, in fact, appealed to Barbour to attempt to save Kartabo by uniting it with BCI into a network of tropical stations, "not in competition with one another, but all cooperating for the good of the science."[177] Barbour preferred to concentrate scarce resources on maintaining BCI.

The concentration of research was not simply a natural product of BCI's artificial insulation. As Chapman wrote, "The damming of the Chagres made Barro Colorado an island but it did not make it an island laboratory."[178] BCI was transformed into a scientific site by the removal of Panamanian settlers and through descriptions of the site as undisturbed and representative of tropical nature. It was maintained for science by the labor of Panamanian workers and through the development of a host of new techniques and technologies for the prolonged observation of tropical life. BCI became accessible and observable, but only in certain ways and only to certain classes of people. The station's proximity to the Panama Canal put it within the reach of a wide range of U.S. visitors, but who visited BCI was not simply a result of its geographic position amid a torrent of travelers of all kinds. Barbour and Zetek moderated the stream of visitors to maintain the island's natural appearance, scientific reputation, and cash flow (goals not always in perfect alignment). Seeking to encourage the flow of some visitors and restrict that of others, they drew not only on their preconceptions about the gender and race of legitimate researchers but also on their connections to individuals within the powerful government agencies and corporations that structured life in the Canal Zone. Scientists on the island were free to pursue basic research, but this freedom was negotiated, bought by Barbour's and Zetek's substantial involvement in the economic development and management of environments elsewhere in the tropics. BCI remained a microcosm of life in the U.S.-controlled Canal Zone, even as it increasingly became a model tropical forest.

CHAPTER FOUR

The Question of Diversity

A few decades ago it was fashionable for ecologists to study communities in the arctic on the grounds that these would be very simple communities and hence easy to understand. Many excellent ecologists still follow this belief, but there are others who feel that it may be easier to understand the extremely complex communities of the tropics. This sounds paradoxical: How can a more complex community be easier to understand?

—Robert H. MacArthur, 1972

On a fall evening in 1939, the Chicago "Ecology Group" met at the home of Warder C. Allee. Almost fifty graduate students and faculty from the University of Chicago, Northwestern University, the Field Museum of Natural History, and other nearby institutions had assembled for the informal seminar, convened every other Monday by Allee and his colleagues Alfred Emerson, brothers Thomas and Orlando Park, and Karl Schmidt.[1] This evening's gathering had been designated "Barro Colorado Night."[2] They heard Victor Dropkin, Emerson's graduate student, speak about the social biology of termites and his experiments on their symbiotic gut protozoa at Barro Colorado Island (BCI).[3] After Dropkin's talk came a presentation of slides and Kodachrome movies captured by zoology professor Ralph Buchsbaum and his wife and collaborator, Mildred Buchsbaum; the two had studied the ecology of invertebrate distribution on the island.[4] Local midwestern environments inspired much of the Chicago school's community and population approach to ecology, yet the tropics held a special place in their work. In *The Principles of Animal Ecology*—the key synthetic work that Allee, Emerson, the Parks, Schmidt, and Buchsbaum were then just beginning to draft—tropical rainforests stood as the epitome of an ecological community at its most complex, reaching the height of "evolutionary diversity."[5] A "free-for-all discussion" followed the presentations.[6] Someone prompted a show of hands, revealing that at least as many of the Ecology Group members present had been to BCI in Panama as had attended the Marine Biological Laboratory at Woods Hole, Massachusetts.[7] "Now when you can compete with the Marine Biological Station, of many years' reputation, you really are 'getting places,'" Dropkin wrote to BCI's manager, James Zetek.[8] By the close of the evening, Dropkin reported, "we were all ready to stand up and cheer for BCI."[9]

Tropical research was beginning to make a mark on the landscape of ecology within the United States. When Daniel T. MacDougal had first campaigned for a tropical station at the turn of the century, research experience in the tropics was a rarity among U.S. botanists and zoologists. No longer were they so parochial. This expansion of tropical experience came from a variety of sources, not the least of which were U.S. government and corporate efforts to develop the natural resources of the tropics.[10] Nevertheless, by the start of World War II, BCI anchored an emerging community of biologists with experience studying basic natural history, behavior, and ecology in a tropical forest. The station played a growing role in biological research and training not only at Chicago and the Ivy League schools but also at the American Museum of Natural History, the Smithsonian, and an array of universities throughout the United States. Visitors recommended the site to their own colleagues and students—the Swarthmore students Robert Enders brought to BCI became notorious as "Panamaniacs" for their enthusiasm.[11] BCI researchers Clarence Ray Carpenter, Nicholas Collias, and T. C. Schneirla influenced a generation of behavioral ecologists with ideas and practices shaped at the station.[12] Even those who never visited could hardly avoid it. Images and examples from Panama circulated through the U.S. biological community as part of talks like Dropkin's and the Buchsbaums', as well as in publications, including the discussions of tropical communities in *The Principles of Animal Ecology*. Not all BCI alumni went on to make tropical research the focus of their careers, but they were nonetheless a widespread and influential group. Their experience at the station shaped both their field practices and their views about how tropical communities fit into the bigger picture of biology.

A lone voice of dissent broke the celebratory atmosphere of Barro Colorado Night, however. Thomas Park "created quite a stir," explained Ralph Buchsbaum, "by suggesting that the research at BCI was of questionable value because it was not 'quantitative.' "[13] Park was an early proponent of statistical approaches to ecology and was best known for using experimental populations of *Tribolium* flour beetles to model ecological phenomena.[14] In Buchsbaum's view, he showed "unconcealed contempt for any kind of descriptive work in zoology." Park had touched a nerve: "From Orlando Park and Allee down to Mildred and myself, we all jumped on him." Emerson, who had himself experimented on termite colonies at BCI, was particularly quick to protest against Thomas Park's charge, "call[ing] attention to many papers of a careful, quantitative nature and . . . citing someone's estimate that BCI was turning out more publishable research per dollar than any other biological station."[15] Others took the opposite tack, reasserting the need for natural historical work in a tropical region where so many organisms had yet to be systematically observed in the wild. Mildred Buchsbaum quipped that BCI was "a place where zoologists might take some time off from laboratory research to learn a few things about animals."[16] Conversation seems to have devolved quickly into a debate over the relative status of biological research pursued in the laboratory or the field. The defenders of BCI and tropical fieldwork ventured home that evening with a feeling of triumph, but had they perhaps missed the crux of Thomas Park's criticism?

Unlike his brother Orlando, who compared whole, natural communities of nocturnal animals on BCI and near Chicago, Thomas Park worked with the deceptively simple two-species experimental system of *Tribolium* on a substrate of flour. While both brothers sought to understand interspecies competition and coexistence, Thomas Park's approach—inspired by the widely influential population studies of Ronald Fisher and Park's mentor Sewall Wright—enabled him to elucidate and test statistical models of population dynamics. What was at stake was not the lab versus the field per se (he encouraged his own students to apply Fisherian methods to natural populations).[17] It was the complexity of the system. How could the study of complex tropical communities—where biologists had encountered difficulty even determining the total number of species present—be made mathematically rigorous?

Over the next two decades, mathematical approaches of the sort Thomas Park called for would come to dominate ecology. Yet, far from entailing a retreat to the laboratory or even to relatively simple, species-poor communities in nature, the rise of quantitative population biology and systems ecology approaches after World War II was accompanied by an explosion of tropical field research. Not only at established research stations like BCI and Soledad but, as this chapter examines, throughout a network of field sites that emerged after World War II, biologists would seek out tropical environments as testing grounds. The complexity of the tropical communities biologists encountered at these sites spurred the development of new quantitative and theoretical perspectives. The complexity of the tropics could be tamed, they found, by reducing diversity itself to a quantitative measure.

This chapter thus begins by tracing the reconfiguration of the institutions and networks of tropical biological research during and immediately after World War II. Then, to understand how and why this context gave rise to new approaches to tropical diversity, it follows the travels and ideas of three key figures: the geneticist and evolutionary biologist Theodosius Dobzhansky, the founding systems ecologist Howard T. Odum, and Robert H. MacArthur, whose work would transform population and community ecology. The influence of Dobzhansky, Odum, and MacArthur reached all corners of biology during the second half of the twentieth century. They are well known as theoreticians, yet the pivotal role of the tropics—as both a field site and a field of thought—in their work has not been fully appreciated by historians of biology. All three, despite their inclinations toward abstraction and simplification, saw the complexity of tropical life not as a barrier to rigorous science but as a challenge—a core theoretical problem to be solved. Some working tropical field biologists, like the attendees of Barro Colorado Night, would react with skepticism to the new efforts to reduce the complexity of tropical life to a few mathematical variables. By 1960, however, species diversity would become a central question in biology: what were the evolutionary and ecological causes of the great numbers of species in the tropics? Tropical diversity would become a phenomenon to be measured, modeled, and explained.

Mobilizing Tropical Biology

As the Chicago Ecology Group debated the virtues of tropical fieldwork, war had already begun to rage in Europe. U.S. biologists' mobilization for war would also literally mobilize them, enabling them to circulate to a wider array of tropical places throughout the circum-Caribbean, as well as deeper into South America and the Pacific. In the wake of World War II, quantitative and systems theoretical approaches would gain impetus in ecology, as in other sciences.[18]

In the short term, however, the war drained researchers from established institutions, disrupting ongoing scholarly projects and publications. At Chicago, for example, army duties in 1943 left Ralph Buchsbaum unable to contribute to *The Principles of Animal Ecology*, which ultimately appeared in 1949 without him. Academia saw an exodus of manpower to the armed forces. As the number of visitors to his stations at BCI and Soledad plummeted, Thomas Barbour continued his work at Harvard's Museum of Comparative Zoology with much of his staff absent "somewhere in the Pacific."[19] No U.S. station's changing fortunes were as dramatic as those of Buitenzorg, Java, where Dutch botanists were interned as Japanese forces occupied the institution for three years following their invasion of the Dutch East Indies in 1942.[20] Nevertheless, with travel in the highly strategic Panama Canal Zone increasingly restricted, BCI temporarily closed its doors to civilian visitors in 1941. Cut off from its supply of daily fees, the station's future was far from certain. Indeed, wartime budget tightening helped put an end to the long-struggling subtropical marine station at Dry Tortugas, Florida.[21] William Beebe also faced the loss of yet another beloved field site, this time in Bermuda, as his seaside station was forced to yield to the construction of a U.S. airfield in 1941. Just as the forest around Kartabo had been cleared for rubber during World War I, the vibrant reefs near his latest station were dredged as the exigencies of war set a higher value on local land (and sea) than basic science could pay.

Shifting economies also fostered the creation of new institutions. Buoyed by high demand for oil and expanding refineries in Venezuela, Laurance Rockefeller (a trustee of the New York Zoological Society) offered Beebe a chance to return to rainforest studies. In 1942, Standard Oil funded a seven-month expedition, using its subsidiary Creole Petroleum's booming base at Caripito, Venezuela, near the Caribbean coast, as a temporary station.[22] By 1945, the company would help him set up a new station to the west at Rancho Grande—the abandoned, half-constructed "castle" of former Venezuelan president Juan Vicente Gómez—in what is today Henri Pittier National Park.[23] In this new venture, Beebe envisioned both a return to intensive forest ecology studies and the future establishment of a new "pipeline" for the zoo, producing a "continuous flow of animals suitable for exhibition, when the end of the war again permits the safe passage of oil tankers."[24] The endeavor was also framed as a matter of inter-American cooperation, winning it additional funds from the Guggenheim Foundation and the Committee for Inter-American Artistic and Intellectual Relations.[25]

Alongside the activities of private foundations, this period likewise saw the U.S. Congress approve a slate of projects to promote agricultural cooperation in an effort to strengthen the economic and political ties that might keep Latin American countries out of the Axis sphere of influence. One initiative created the first agricultural attaché to Brazil and several other countries, loaned U.S. Department of Agriculture (USDA) scientists to "survey . . . tropical resources," and mandated a new "Tropical Forest Experiment Station" in Puerto Rico under the U.S. Forest Service.[26] Created to rationalize timber production and combat the deforestation that had increasingly plagued the U.S. territory during the Great Depression, it would also become an important center of ecological research in the coming decades. As we will see, it was in the nearby El Yunque forest that a young Howard Odum would fall in love with the rainforest during a 1944 army training exercise. He would return in the 1950s to begin a study of this tropical ecosystem on an unprecedented scale.

Demand for commodities like rubber and quinine, as well as the realities of jungle warfare, began to place a new premium on biologists' expertise on tropical organisms and environments. At first most drafted biologists wound up doing work with little relation to their areas of expertise—"whom you knew was as important as what you knew" in determining service assignments, recalled Richard Howard, a tropical botanist who enlisted shortly after completing his PhD at Harvard.[27] Soon, however, biologists' tropical experience was recognized as a valuable national resource. Armed with knowledge of tropical plants gained at Soledad while on an Atkins graduate fellowship, Howard found himself teaching jungle survival skills in Florida and writing manuals for the Arctic, Desert, Tropic Information Center.[28] Paul C. Standley, who had completed BCI's first flora, also produced guides to edible, medicinal, and poisonous tropical plants for the Navy Bureau of Medicine and Surgery.[29] Even the BCI primatologist Clarence Carpenter produced training films, including one instructing shipwrecked soldiers to take a cue from local monkeys in identifying sources of food.[30] War in the tropical Pacific also intensified demand for entomologists' knowledge of insect vectors of disease. Philip Darlington, who had also been an Atkins Fellow and an employee of the United Fruit Company, interrupted his studies of beetle mimicry and island biogeography to put his expertise to work as head of a Malaria Survey Unit in the Pacific.[31] Later becoming a sought-after expert on neotropical birds, Alexander Skutch was likewise just one of many scientists to serve in the search for rubber and natural rubber substitutes during the war. Skutch had extensive tropical experience at the stations at Cinchona and BCI, as well as working for the United Fruit Company throughout Central America, but his work on a joint Peruvian-USDA rubber survey took him for the first time into the forests of the Amazon.[32] The U.S. military thus both tapped into existing scientific networks in the Caribbean and at the same time drove expansion into the Pacific and Amazonia in search of tropical commodities.[33]

With its strong basis in tropical agriculture and horticulture, Harvard's station at Soledad was particularly well situated for redirection toward the war effort. Formally renamed the Atkins Institution in 1932, the station and its endowment had been steadily expanding thanks to Barbour's grantsmanship. The garden's grounds had grown to over 220 acres in 1939. To complement the Harvard House lab, a new dormitory, Casa Catalina (named to celebrate Katherine Atkins's ongoing support), was constructed. The number of visiting scientists in 1940 climbed to forty for the first time in the station's history.[34] When the United States entered the war the following year, however, the number was cut by three quarters. Atkins fellowships were suspended—too few Harvard graduate students remained to take them. The station continued its routine operations, however, including the distribution and exchange of seeds and plants. Responding to the military's dire need for rubber, Soledad cooperated with the USDA and the Bureau of Economic Warfare to provide garden space for the cultivation of disease-free *Hevea brasiliensis* rubber plants for distribution to new plantations, as well as experimenting with potential rubber substitutes, like the rubber vine, *Cryptostegia grandiflora*.[35] The station was even enlisted in a landscaping program for the Marine Corps at Guantanamo Bay; between 1943 and 1948 the garden propagated more than 2,500 ornamental trees to beautify the base and shade U.S. marines from the punishing tropical sun.[36]

Despite BCI's high reputation among academic biologists, its financial and institutional footing was far less stable than Soledad's. Unable to secure an endowment, Barbour continued to personally supplement BCI's budget. Of the two stations under his direction, he considered BCI his more "troublesome offspring."[37] Although only in his fifties, Barbour was also in increasingly poor health during the war years, and he wanted to ensure that the institution could survive without him. Drawing on his Washington connections—including his own brother, New Jersey senator William Warren Barbour, and Alexander Wetmore, then assistant secretary of the Smithsonian and a longtime BCI visitor with extensive Latin American experience—he succeeded in having BCI reorganized federally in 1940. It would now be operated as the Canal Zone Biological Area (CZBA).[38] An act of Congress placed the island under the management of a board of directors composed of the secretaries of war, agriculture, interior, and the Smithsonian Institution, the president of the National Academy of Sciences, and three appointed biologists.[39]

This new federal status came with authorization for an appropriation of up to $10,000 per year, but the promised funds were never supplied. Already somewhat sporadic, the subscription fees of academic institutions also dried up with the U.S. entry into the war. James Zetek, always a shrewd opportunist, managed to keep BCI afloat by recasting the island as a site for field-testing equipment under tropical conditions. With a massive buildup of U.S. troops in the Canal Zone, Panama was important not only as the site of strategic infrastructure but increasingly as a "test tube republic."[40] Standing in for tropical environments around the world, areas in Panama and the Canal Zone served as training grounds for jungle warfare and as test sites for the deployment of chemical weapons and defoliants, including mustard gas, phosgene, and Agent Orange.[41] (Puerto Rican troops stationed in Panama were also a special focus of U.S. gas testing.)[42] Isla San José, one of the Pearl Islands (Las Perlas) off Panama's Pacific shore, saw extensive bombings between 1944 and

1947. In these, CZBA board member Wetmore was complicit—he took part in a baseline survey, declaring that the islands' plants and animals would not be significantly affected if testing covered less than a third of the island's surface.[43] Nothing so dramatic took place on BCI itself, but Zetek was able to gain enough funding to survive the lean years of the war and its immediate aftermath by courting the U.S. military and Eastman Kodak Corporation, which studied the deterioration and corrosion of equipment and photographic film on the island under "tropical conditions."[44]

In 1944, Barbour suffered a blood clot on the lung, complicated by pneumonia, which ended his travels. On 8 January 1946 he died of a cerebral hemorrhage.[45] In his memoirs he had written that his contribution to establishing BCI was "the best job I had ever done in my life."[46] With his passing, and through the crucial assistance of Wetmore, now the Smithsonian's secretary, the CZBA was transferred to the more permanent care of the Smithsonian Institution on 16 July 1946.[47] Upon reopening that year, the station would see an almost immediate return to prewar numbers of visitors, followed by an exponential increase beginning in the 1950s. The continuity so important for long-term, place-based science was maintained, and the broader expansion of tropical biology after World War II in certain ways resumed a trend already well established by the 1930s. Yet mobilizing for war had also pushed U.S. biologists well beyond the more established Caribbean stations into sites in South America and the tropical Pacific, and in larger numbers than ever before. At the same time, it tended to reorient patronage networks toward the U.S. government, a trend that would continue into the postwar era.

Endless Raw Material

Even before the war's end, biologists began to imagine an expanded role for tropical ecology in rebuilding the world during peacetime. Notably, Orlando Park, speaking as the outgoing president of the Ecological Society of America (ESA) in 1944, envisioned a key role for tropical ecology in a "general program for postwar hemispheric cultural solidarity and economic cooperation."[48] In his view, the "development of the American tropics" was all but "inevitable." Park foresaw an era of advanced industrialization in the region (albeit in the "distant future"), requiring on the way developmental stages of disease control, environmental engineering, and the intensification of agriculture. By allying more closely with tropical medicine and agriculture, he argued, ecologists could contribute at all of these "stages," which he predicted, however, would actually occur in complex combination. If development was unlikely to be a purely orderly process, ecologists' expertise could at least make it more rational by helping to "prevent the repetition of past mistakes in conservation or lack of conservation made elsewhere."[49] To do so, he recommended that ecologists cooperate with U.S. national agencies on conservation problems, take increased advantage of BCI after the war, and establish fellowships, research grants, and (echoing Barbour) a journal devoted to "tropical biology in its broadest aspects."[50]

Emphasizing what he saw as the special biological problems of economic growth in the tropics, Park positioned ecology to take advantage of an impending postwar surge of development in the region. He presented the "American tropics" as both an intellectual frontier and a proving ground for professional ecologists in the United States. Despite the nod to hemispheric cooperation, the ecology of the "American tropics" still figured in his scheme as the domain of United States rather than Caribbean and Latin American scientists. In urging ESA's membership to "be prepared to assume our full responsibility in American biology in the study of the ecology of the tropics," he sounded very much like Daniel MacDougal had on the cusp of the Spanish-American War. Park explicitly framed tropical ecology as both a duty and a chance at "international distinction" in science—if U.S. ecologists could benefit the region, there would also be much in it for their profession.[51]

"Before the tropics can be adequately controlled," Park argued in his ESA speech, "there is a great deal of basic work to be done." He pointed to the "diversified environment" of the tropics, which "supports a tremendous number of organisms . . . represented by an unusually large number of species," and suggested—as had many before him—the endless variety of biological problems available for study. The sheer numbers involved meant that taxonomy and species ecology (or autecology, the study of individual species' life history, behavior, and physiology in relation to its environment) would remain fundamental areas of research for the foreseeable future. In addition, however, Park noted, "since the tropical species populations do not exist in a vacuum, nicely isolated from each other, but rather form *natural survival units* which we call communities, we have our biological work pyramided in complexity by the complex interrelationships which are developed in every community."[52] Seeing ecological communities as themselves products of evolutionary processes, Park insisted that species, population, and community ecology were intimately connected. They all must thus be tackled head-on, through the development of "a full scale program in tropical ecology"—a project requiring the institutional and intellectual maturation of their field.[53] Just as economic development could not be expected to proceed in an orderly, linear fashion, Park argued that to make scientific progress biologists working in the tropics would also have to attack ecological questions at all levels simultaneously, from the base to the point of this "pyramid" of complexity.

Park spoke as ESA's president, but his position was not necessarily shared by the majority of ecologists. Notably, the influential British animal ecologist Charles Elton, whose trophic "pyramid of numbers" Orlando Park played off of in his speech, had long maintained that the study of simple communities (like Thomas Park's beetles) would more effectively advance ecologists' search for general community properties. Elton himself worked primarily in the Arctic, where, as his mentor Julian Huxley put it, "the ecological web of life is reduced to its simplest, and complexity of detail does not hide the broad outlines."[54] Elton noted, moreover, that the simple communities of brackish water, temporary pools, or arctic environments could be worked by "one man" without "mental strain" or an "appalling amount of mere collecting . . . and the trouble of getting the collected material identified."[55] Huxley and Elton praised the work of British colonial institutions in the tropics, including the Amani experiment station in Tanganyika Territory and the Imperial College of Tropical Agriculture in Trinidad, but they seem to have seen these primarily as sites where basic ecological principles would be applied, not as wellsprings of new insights.[56] Orlando Park's own brother Thomas, of course, also continued to insist that the analysis of populations was a necessary step before problems at the more complex level of whole communities could be understood.[57]

Orlando Park, however, stood with a significant group of biologists—particularly those interested in the relationship between evolutionary and ecological processes—who believed that tropical environments held unique research advantages. This position was perhaps best outlined by Marston Bates, a protégé of William Morton Wheeler and Thomas Barbour (and a veteran of Barbour's stations), then serving as director of the Rockefeller Foundation Yellow Fever Laboratory at Villavicencio, Colombia. In a piece on "The Advantage of the Tropical Environment for Studies on the Species Problem," Bates acknowledged Elton's case for arctic studies and the potential for tropical work to be "discouragingly difficult." He argued, however, that tropical environments offered "overwhelming advantages," particularly for the study of evolutionary phenomena. "Our ideas of natural selection, for instance, are in large part based on a concept of the tropical rain forest that may have little basis in fact," he argued. The "jungle" of popular imagination was a harsh world. Upon seeing a strangler fig's "struggle for existence" against its host tree, naturalists might assume that the severity of competition reached an extreme in the tropical rainforest. Yet, the reverse might also be true: if tropical environments provided an abundance of light, moisture, and food, perhaps the force of natural selection there was actually quite weak. In reality, Bates argued, "we know almost nothing about population pressures, competition and other factors that may actually control survival and dispersal in the rain forest environment."[58]

That same environment, he maintained, offered "endless raw material" that could be used to address these problems.[59] Studies of geographic variation had taken place mainly in the temperate zone, where form often seemed to correlate with climate. Many tropical species exhibited polymorphism, however, and very similar species frequently could be found coexisting under apparently identical environmental conditions. Such counterexamples to familiar temperate phenomena offered opportunities for ecologists and evolutionary biologists interested in finding solutions to core problems of variation, competition, and environment. Bates, Orlando Park, and a growing number of U.S. biologists argued that, rather than obscuring the fundamental biological picture with its "complexity of detail," the tropics' vast numbers of species and variety of biological phenomena could be a rich scientific resource.

Evolution in the Tropics

It was exactly such raw scientific material that the Ukrainian American biologist Theodosius Dobzhansky sought as he set out for Brazil in 1943. While Dobzhansky initially intended to mine Brazil for new sources of experimental genetic material, what he found there would lead him to articulate tropical species diversity as a central research problem for evolutionary biologists and ecologists. Famous for his 1937 *Genetics and the Origin of Species*, a foundational text of the modern evolutionary synthesis, Dobzhansky sought to apply the abstract mathematical population genetics of Ronald Fisher, J. B. S. Haldane, and Sewall Wright to populations in the wild.[60] Unlike most U.S. geneticists, Dobzhansky saw genetic variability in nature not as an experimental inconvenience but as a source of evolutionary insight; indeed, he began his 1937 book with a chapter on "Organic Diversity" in which he discussed the problem of how differences among individuals, populations, and species arise. In the early 1940s, Dobzhansky was preoccupied in particular with Wright's concept of genetic drift, the idea that, especially in small populations, random fluctuations in gene frequencies might have significant evolutionary consequences. *Drosophila* fruit flies seemed especially susceptible to drift. They experienced massive die-offs in the winter months. When populations exploded with the return of warm weather, any drift effect would theoretically be magnified.[61] Yet what happened, Dobzhansky wondered, in places "where winter never comes"?[62] Would genetic drift be eliminated? He had made collections in Mexico and Guatemala during the 1930s and had even arranged for South American *Drosophila* to be shipped to him via the expanding networks of Pan American Airways. In general, population geneticists had confined their studies to temperate species.[63] Pointing out this geographic bias, Dobzhansky sought support from his longtime patron, the Rockefeller Foundation, in 1942.[64] An opportunity for fieldwork in Brazil arose as wartime interest in cultivating relations with this rubber-producing country spurred the expansion of a variety of the foundation's development projects there.[65]

Dobzhansky was especially pleased by the prospect of travel in the Amazon, a region made iconic for evolutionary biologists by the writings of Alfred Russel Wallace. His so-called Grand Plan in Brazil was to compare the genetic structure of temperate and tropical populations, which first meant widespread collecting to identify promising *Drosophila* species for long-term experimentation.[66] At the same time, taking a page from Wallace, he reflected on tropical nature as a whole and the misapprehensions to which travelers were prone: "Aside from doing *Drosophila* I have been just looking at the tropical nature, observing things at large, and trying to store these observations and impressions for thinking them over at leisure. Tropical forest is an environment so different from ours of the temperate zones that all biological conceptions or preconceptions with which a 'temperate' biologist starts have to be critically examined."[67] His letters show him quite self-consciously following in the footsteps of the traveling naturalists of the past century, but his tropical encounters were also refracted through the lens of the modern evolutionary synthesis. Tropical nature offered a means to test conceptions of evolution and genetics formulated by biologists working narrowly within the temperate zone. Identifying himself as a " 'temperate' biologist," he presented tropical travel as a way to overcome provincial biases, while at the same time using tropical evidence to affirm his own position on the synthesis of natural selection and Mendelian population genetics.

He soon confronted some of his own preconceptions, however. Brazil was in fact far from lacking in seasonality. Brazilian *Drosophila* populations were profoundly affected by the fruiting periodicity of their favored food sources, in turn partly shaped by the alternation of rainy and dry seasons.[68] Nevertheless, working closely with his wife, Natasha Dobzhansky, and University of São Paulo geneticists Crodowaldo Pavan, Antonio Brito da Cunha, and others, he identified several promising strains, including *D. willistoni, D. prosaltans,* and *D. tropicalis,* investigating their food preferences, chromosomal polymorphism, and variation within island and continental populations.[69] Dobzhansky would travel to Brazil four times between 1943 and 1956, staying for many months at a time and publishing dozens of papers with his collaborators on problems of the relation of ecology to population genetics.[70] The overall thrust of this work was the great genetic variability within tropical species. Far from becoming narrowly specialized to face a supposedly monotonous and unchanging environment, the populations they studied displayed great genetic variation and hence the adaptive versatility needed for life in what were in actuality quite heterogeneous and mutable ecological settings.

Dobzhansky went to Brazil for *Drosophila,* but a brief foray seeking much larger quarry—trees—would have a greater effect on the future direction of tropical biology. This work would crystalize the question of why the tropics harbor so many of Earth's species, making it a core biological problem and one amenable to a mathematical and evolutionary approach. Dobzhansky got from flies to forests in large part because of the institutional networks that enabled his travel through Brazil. To access environments in almost every Brazilian state, he took advantage of the country's nationwide aviation infrastructure, in particular the air force's postal service, which provided free access to the hinterland as part of a project of internal colonization.[71] On the ground, Dobzhansky relied on the help of both Brazilian government officials and an international network of scientists. Among the most important of his contacts was Felisberto de Camargo, director of the new Instituto Agronômico do Norte at Belém do Pará, which emphasized rubber research and the agricultural development of the Amazon. Camargo was a graduate of the University of Florida with connections to the Ford and Rockefeller Foundations. With him as an "energetic" host and guide, Dobzhansky's group gained access to a variety of plantations, experiment stations, and backcountry outposts from which to base their collecting excursions—including Fordlândia, Henry Ford's ill-fated Amazonian rubber colony that Camargo boasted of reviving on a sounder basis of agricultural science.[72]

Historians and scientists debate how significantly Dobzhansky's visits shaped the discipline of genetics in Brazil.[73] His interaction with Brazilian foresters, however, certainly pushed him toward the problem of tropical tree diversity, a topic otherwise unusual for the geneticist. With Pavan and two of Camargo's foresters, George A. Black (a Berkeley graduate) and João Murça Pires (who would become a world authority on Amazonian plants), Dobzhansky coauthored two papers, in 1950 and 1953, on the problem of estimating tropical tree diversity.[74] This research would also contribute to his more expansive and influential 1950 essay, "Evolution in the Tropics."

Although the number of tropical tree species had long impressed explorers, the researchers noted the scarcity of "quantitative data describing this diversity."[75] Past workers, from MacDougal to Beebe and the promoters of BCI, had illustrated tropical diversity with broad comparisons of the number of species in temperate and tropical locales, but few provided detailed data amenable to statistical analysis. One exception was the work of the British botanist Paul W. Richards, who during a 1929 expedition in British Guiana had adapted some of the methods of the colony's forestry department to more rigorously describe forest structure. Richards's assistants had demarcated several plots, twenty-five feet in width by two or four hundred feet in length, measuring the trees' diameters and counting species and individuals, one by one. To identify species, he employed an unnamed "Indian assistant" who had extensive experience counting economically important trees on previous forestry valuation surveys.[76] Although the results did not always accord exactly with botanists' delineations of species, Dobzhansky explained, "there is no botanist in the world who can name all the trees in an Amazonian rain forest just on the basis of looking over their trunks and their foliage, as would be taken for granted in our not so exuberant northern woods."[77] In Brazil, working with plots one to three and a half hectares in size on land owned by the Instituto Agronomico just outside of Belém, his group followed Richards's methods, relying on four tree specialists of indigenous ancestry who were employees of the institute.[78]

Interested in the vertical stratification of the forest community, Richards had actually clear-cut sample strips to precisely measure the height of the felled trees, examine the spatial distribution of epiphytes and climbing plants, and collect samples of fruit and flowers to check his tree specialists' identifications against herbarium specimens. He used these data to produce profile diagrams, a technique that was quickly taken up by both foresters and ecologists.[79] It would also form a core method of his landmark 1952 book, *The Tropical Rainforest: An Ecological Study*. Slated to appear in the mid-1940s but delayed by World War II and ensuing years of paper shortages, Richards's long-awaited book combined his own fieldwork in British Guiana, Sarawak, and Nigeria with a masterful synthesis of the perspectives of researchers who had long worked separately in different parts of the tropical world and published in different languages.[80] Many readers took note of the final pages, in which Richards warned that present rates of destruction would leave little rainforest left for ecologists to study by the end of the century. The primary

Key to symbols: (*a*) *Am*, *Amaioua guianensis*; *As*, *Aspidosperma excelsum*; *B*, *Byrsonima* sp.; *C*, *Cassia pteridophylla*; *Ca*, *Catostemma* sp.; *D*, *Duguetia neglecta*; *E*, *Eperua falcata* (soft wallaba); *Eg*, *Eperua grandiflora* (ituri wallaba); *Ec*, *Ecclinusa psilophylla*; *Em*, *Emmotum fagifolium*; *Es*, *Eschweilera* sp.; *L*, *Licania heteromorpha*; *M*, *Matayba inelegans*; *Ma*, *Marlierea schomburgkiana*; *M*ₛ, *Matayba* sp.; *O*, *Ocotea* sp.; *Or*, *Ormosia coutinhoi*; *P*, *Pouteria* sp.; *Sw*, *Swartzia* sp.; *T*, *Tovomita cephalostigma*; *U*, Unidentified. (*b*) *Ab*, *Aniba bracteata*; *Bp*, *Beilschmiedia pendula*; *D*, *Dacryodes excelsa*; *Fo*, *Faramea occidentalis*; *Gm*, *Guarea macrophylla*; *Ii*, *Inga ingoides*; *L*, Lauraceae (unidentified); *Pm*, *Pouteria multiflora*; *Ps*, *Pouteria semecarpifolia*; *Q*, *Quararibea turbinata*; *Sa*, *Simaruba amara*; *Sc*, *Sterculia caribaea*; *Ta*, *Tapura antillana*.

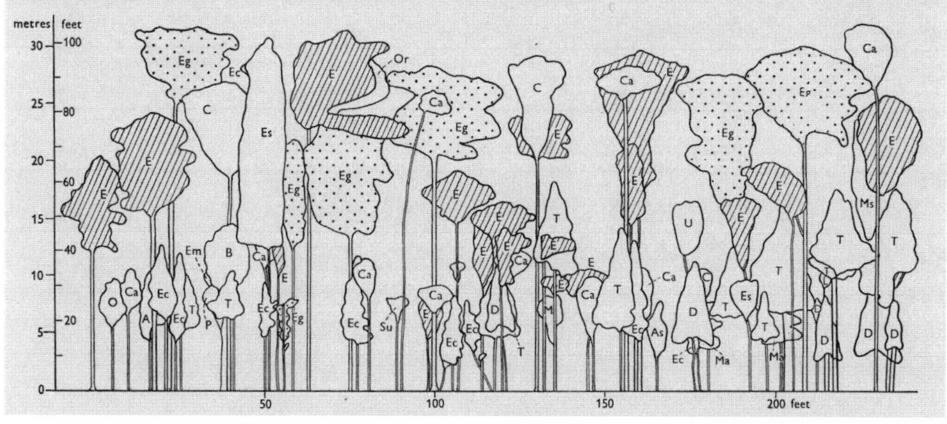

A profile diagram depicting the community structure and species diversity in Wallaba forest, Barbara Creek, Mazaruni River, British Guiana. Richards, 1952 (after T. A. W. Davis, unpublished). Reprinted with the permission of Cambridge University Press.

impact of the book, however, was empirical and methodological. It was lauded as presenting a coherent and comparative picture of global tropical forests for the first time since the work of Eugenius Warming and Andreas Schimper in the 1890s.[81] As simplified representations, but ones grounded in precise measurement of actual forest samples, Richards's profile diagrams played a key role in his analysis of forest structure. Tropical forests might harbor quite different suites of species, but the diagram technique enabled him to compare their structural physiognomy in a way that appeared rigorous and objective. With it he could collate communities from different localities across the global tropics to see underlying structural similarities, such as stratification.

Dobzhansky's group, in contrast, was less concerned with the physical structure of a forest community than with the population structure of its tree species. They too made a key methodological innovation, however, in the application of statistics to estimate the total number of species in a community—what they referred to as its *species diversity*. The coauthors noticed that in their three-and-a-half-hectare plot, a full 45 out of the total 179 tree species were represented by only a single individual tree. This strongly suggested that their sample plot had not captured all of the species present in the community as a whole; a larger plot would certainly have uncovered additional, very rare species. "To obtain even a very rough estimate of the total number of species in the forest community which we have studied," they thus reasoned, "a hypothesis must be made concerning the relationships between the common and the rare species."[82] A few precedents existed for statistical investigations of the relationship among sample size, the total number of species in a given area and time span, and species' commonness or rarity (or what is now termed *relative abundance*). The first effort had been made at the turn of the century by the Danish botanist Christen C. Raunkiær, who explored the relationship between quadrat size and the number of species sampled.[83] In 1943, Fisher, in collaboration with A. Steven Corbet and C. B. Williams (both of whom had experience working with large tropical insect collections) developed their "index of diversity"—the first quantitative measure of species diversity.[84] Modifying Fisher, Corbet, and Williams's index, Frank W. Preston, a British American engineer and amateur ecologist, suggested that common and rare species could be described by a lognormal probability curve.[85]

Combining practices drawn from forestry and population biology, Dobzhansky and his collaborators fit their data to Preston's model. By this method, they were able to predict that an additional seventy rare species might be missing from their sample of Amazonian trees. They could also show, in tabular or graphical form, that superficially similar plots might differ substantially in actual species composition (see table 4.1). These results carried significant practical and theoretical implications. Estimates of species diversity and the presence of economic species were highly dependent on the size and location of sample plots, thus, Murça Pires, Dobzhansky, and Black noted, "the importance of studies in this field for rational exploitation of the forest resources of Amazonia."[86] At the same time, the study demonstrated the extremely low population density of most tropical trees—many were represented by less than one individual per hectare. Such extremely limited effective population sizes might have a "profound influence on evolutionary patterns," Black, Dobzhansky, and Pavan suggested.[87]

Dobzhansky summed up the findings of the Brazilian *Drosophila* and forest work and placed them into a broader biological context for the wider readership of *American Scientist* in his now-classic 1950 article, "Evolution in the Tropics." He was not the first to describe "the greater diversity of living beings" found in the tropics as "the outstanding difference which strikes the observer."[88] In the article, however, he assembled some of the best modern data available to characterize the overall "progressive increase in diversity of species from the Arctic to the equator"—what would by 1960 be labeled the

TABLE 4.1 Species diversity of three forest plots presented in tabular form

Individuals	*1*	*2*	*3*	*4*	*5*	*6*	*7*	*8*	*9*	*10*	*11*	*12*	*16*	*17*	*20*	*26*	*Total*
Species: Plot I (Tefé)	42	14	9	3	..	..	2	1	2	1	2	1	..	1	1	..	79
Species: Plot II (Belém)	18	7	2	5	3	1	..	..	..	1	1	1	1	..	..	..	41
Species: Plot III (Belém)	31	11	10	2	2	..	1	..	..	..	1	..	1	2	..	1	62

Source: After Black, Dobzhansky, and Pavan, "Some Attempts to Estimate Species Diversity."

"latitudinal diversity gradient."[89] Moreover, Dobzhansky reframed this long-recognized gradient specifically as a problem amenable to the population-centered and statistical approaches of the modern synthesis. "Since the animals and plants which exist in the world are products of the evolutionary development of living matter," he wrote, "any differences between tropical and temperate organisms must be the outcome of differences in evolutionary patterns." This led him to ask: "What causes have brought about the greater richness and variety of the tropical faunas and floras, compared to faunas and floras of temperate and, especially, of cold lands? How does life in tropical environments influence the evolutionary potentialities of the inhabitants? Should the tropical zone be regarded as an evolutionary cradle of new types of organization which sends out migrants to colonize the extratropical world? Or do the tropics serve as sanctuary for evolutionary old age where organisms that were widespread in the geological past survive as relics?"[90] In effect, he set the bounds of a new research program, arguing, "These and related problems have never been approached from the standpoint of modern conceptions of the mechanism of evolutionary process."[91] Answers could be found through a combination of the principles of population genetics and natural selection.

Offering a possible solution, Dobzhansky engaged in a discourse of tropicality. "The contradictory epithets of 'El Dorado' and 'Green Hell,' so often used in descriptions of tropical lands" were each rooted in biological reality, he claimed. The physical environment of the tropics might be comparatively mild, he argued, but "competition within a complex community of organisms" was strong. "Tropical environments provide more evolutionary challenges," he explained, but challenges that "stem chiefly from the intricate mutual relationships among the inhabitants" rather than the physical challenges of colder latitudes.[92] His position shared a superficial similarity to that of Wallace and many of the early physiological plant ecologists who emphasized both the tropics' lack of a winter check on plant growth and the region's historical freedom from glaciation. Dobzhansky, however, argued that it was not an absence of environmental challenge but, rather, a difference in the quality of the challenge. He posited that the drastic climate shifts of the Ice Ages produced "more or less random destruction" in colder regions, eliminating swaths of the genetic raw material that evolution had to work with. In the tropics, natural selection could instead "acquire a differential character." Rather than facing periodic catastrophe, populations primarily had to respond to "the manifold reciprocal dependences in the organic community." Under these circumstances, he argued—in an interesting reversal of the usual discourse of tropical primitiveness—selection could become "a creative process which may lead to emergence of new modes of life and of more advanced types of organization."[93]

Not all biologists would accept this interpretation—indeed, Dobzhansky offered tantalizingly few details about the very biotic interactions that his hypothesis brought to the fore. Nevertheless, his articulation of the question of tropical diversity was a significant move, positioning an understanding of life in the tropics as central to a general understanding of evolution. He had pointed to the ways that evolutionary and ecological processes operated together and across scales of biological organization, including the genetics of populations, of species, and of the complex community. The geneticist, ironically, had made a strong case for the study of tropical community ecology. While Dobzhansky suggested a broad theoretical framework for approaching the question of tropical species diversity, Richards's work helped lay a general descriptive and empirical groundwork for future tropical forest studies. Together they put the study of tropical life on a new footing in the early 1950s. The next decade would see ecologists flock to the problem of species diversity, seeking its causes in community interactions.

A Proving Ground

Dobzhansky and Richards both presented methods of manipulating quantitative data that could be used to elucidate the structure of complex tropical communities. Profile diagrams and graphs and tables of species abundance (which Richards's book also employed) provided ways to make complex local ecological communities quantifiable and comparable. The new methods elucidated the similarities and differences among tropical forests around the globe, as well as the general latitudinal increase in species richness from the poles to the equator, suggesting broader theoretical frameworks to understand these patterns. Ecology in general was becoming increasingly theory oriented and mathematical by the 1950s.[94] Playing an outsized role in this trend toward abstraction were the British-born Yale professor G. Evelyn Hutchinson and his host of graduate students.[95] Hutchinson helped revolutionize limnology—the study of the ecology of lakes—by drawing on his training in chemistry, mathematics, and a systems approach. Like Elton (although he criticized Elton for defects of mathematical rigor), Hutchinson emphasized the study of simple communities to understand fundamental ecological principles. Several of his students, however, sought out complex tropical environments as sites to inspire and test ecological theory. Among these were two of his most influential protégés: Howard T. Odum and Robert H. MacArthur.

Odum and MacArthur are usually seen as representing two diverging branches of ecology.[96] In the 1950s and 1960s, MacArthur's work spearheaded a new, quantitative community and population ecology. Meanwhile, Howard Odum, working closely with his older brother and fellow ecologist Eugene P. Odum, played a key role in the development of systems ecology by applying a systems theory perspective to the study of the flow of energy and nutrients through ecosystems. While the contributions of both ecologists to these two major theoretical approaches have been explored, historians have not interrogated the specific and significant role that the tropics played in the development of their thought, and particularly their approaches to complexity. Odum and MacArthur were both first introduced to the tropics—and particularly to forests in the circum-Caribbean—early in their careers through networks of U.S. field science that were being reconfigured in the wake of World War II. Not only did they both return to the tropics throughout their lives to test their developing theories, but ideas and assumptions about tropical nature—in particular its complexity, diversity, and apparent stability—can also be seen woven deeply throughout their work. In part because both men aimed toward abstraction and universalization, the relationship of their ideas to either the region as a whole or the history of particular field sites has been obscured.

One exception, perhaps, is the close association of Howard Odum and Eugene Odum's ecosystem approach with Enewetak Atoll, in the Marshall Islands and part of the Pacific Proving Grounds. The brothers' 1954 study of coral reefs irradiated by U.S. nuclear weapons testing on this tropical Pacific atoll has become a textbook example of an ecosystem study. The ESA Mercer Award–winning project not only helped establish the field of radioecology but also was instrumental in cementing support for ecological research within the

U.S. Atomic Energy Commission (AEC). Historians and ecologists recognize it as a watershed both for the way it incorporated methods from the physical sciences into ecology and for how it tapped into new avenues of federal support for such research. Moreover, it demonstrated the power of the brothers' self-described "top-down" or "holistic" approach to ecology (which they also promoted in their *Fundamentals of Ecology*, a textbook that would overtake Allee et al.'s *Principles of Animal Ecology* after its second edition).[97] Rather than attempting to understand the reef community by studying each part—the coral, algae, and fish species—they used measurements of biomass and overall respiration to estimate the metabolism of the whole community, approaching it as an ecological system interconnected by flows of energy and nutrients. Indeed, neither brother was overly concerned about their lack of background in coral biology or their inability to actually identify most of the reef's species during their brief, six-week study at Enewetak. Their primary concern, and the power of their field methods and theoretical perspective, was in characterizing system-scale ecological structures (the trophic pyramid, or food web) and processes. This approach, in fact, was what enabled them to demonstrate for the first time that algae and coral shared a symbiotic relationship. Despite the apparent preponderance of animal life (coral), plants (algae) formed the majority of the community's biomass—the base of its trophic pyramid.

The Odums' work was innovative, but it also drew on existing generalizations about coral reefs, and tropical environments more broadly, in ways that have been little examined.[98] The Odums worked at Enewetak primarily because of the chance opportunity offered by the AEC, but the complexity and presumed stability of tropical coral systems were also highly attractive to both ecologists, who believed that fundamental ecological principles would be best revealed by studying ecosystems that had reached a "steady-state" equilibrium, or "climax." The diversity of the reef—a "tremendously varied . . . community . . . famous for its immense concentrations of life and its complexity"—in itself suggested to the Odums that it had achieved this state.[99] Emphasizing the "constant environment" of tropical waters, they also assumed that the reef remained "unchanged year after year, and reefs apparently persist, at least intermittently, for millions of years."[100] Such conceptions shared much with prevailing views of tropical rainforests. Richards's standard text, for example, likewise attributed the rainforest's diversity of species to its presumably great geologic age (not strictly following the nuance of Dobzhansky's argument on this point). Indeed, the Odums' 1955 paper explicitly linked rainforests and reefs, citing the diversity inherent in the structure of both communities: "A reef community resembles a complex, tropical, terrestrial community in that it is not dominated by one or two producer species, but there is considerable diversity and variation from place to place."[101]

Significantly, the Odums also noted the apparent paradox of the coral "growing luxuriantly" despite the water's "impoverished" plankton and nutrient content.[102] This description closely mirrored Richards's account of the "impoverishment" of tropical soils and the rapid turnover of nutrients within a "closed cycle"—what he called a state of "complex equilibrium."[103] Whereas Richards drew no broader generalizations, the Odums suggested that complexity, mutualism, and rapid metabolic turnover were essential factors in the long-term stability of ecosystems. Howard Odum, more specifically, also believed that their findings supported an idea (drawn from engineering and thermodynamics, and the work of Alfred Lotka) that increasingly obsessed him: that steady-state ecosystems have adjusted to achieve the "maximum output of energy for their own survival" given the "input" available. In other words, he believed that ecosystems evolved to make maximally efficient use of the sun's energy—what he called the "maximum power principle."[104] The Enewetak study, then, also epitomized key ideas about energy and ecosystem self-regulation that would thread through their future work, but particularly that of Howard Odum.

The Odums sought ecological principles that transcended any single site or type of community, but the tropics thus held special significance for Howard Odum as the global region of maximum insolation. The connection was also personal, however. During his World War II service as a U.S. Army Air Force meteorologist, the young 2nd Lt. Howard Odum had come into contact with a wide variety of tropical environments throughout the Caribbean and Latin America, completing training and weather studies in Cuba, Dutch Curaçao, British Dominica, Colombia, Panama, Puerto Rico, Peru, and Ecuador, including the Galapagos Islands.[105] In fact, while stationed in the Panama Canal Zone (among the world's greatest highways for birds) as an instructor at the U.S. Army Air Force Tropical Weather School at Howard Field, Odum gathered observations for his first scientific paper. In it, he argued against a common hypothesis connecting day-length change and bird migration, noting that near-constant daylight hours in the tropics could not stimulate their return north.[106] This early effort demonstrates not only his burgeoning interest in explaining natural historical facts through attention to complex physical systems but also his early concern with the relationship between tropical facts and universal theories. According to a later collaborator, Odum caught the "tropical bug" from his very first tropical experience in

1944—an intensive, three-month training at the Institute of Tropical Meteorology at Rio Piedras, Puerto Rico.[107] This course not only introduced him to systems thinking and an energy organizational approach but also brought him for the first time into a tropical rainforest at El Yunque, part of the experiment station recently established by the U.S. Forest Service.

The rainforest captured Odum's imagination, and his search for fundamental ecological principles would eventually lead him back to Puerto Rico and the El Yunque forest. After the success of the Enewetak study, Odum began to envision ways to put a similar "top-down" approach into practice to understand the functions of a rainforest ecosystem. By 1956, he was director of the University of Texas Institute of Marine Science at Port Aransas.[108] Because his funding from the state could not be used for research outside of Texas, Odum approached the Rockefeller Foundation with his idea for comparing tropical rainforest and tropical plankton communities using equivalent methods.[109] He explained: "At first glance the massive towers of the tropical rain forest are in great contrast to the practically invisible sprinkling of phytoplankton in the clear blue tropical ocean waters. Yet if the theory of optimum efficiency for maximum power adjustment has any validity, both communities are adapted to similar total photosynthetic power output." Both communities were presumably adapted to similar solar energy input, yet they exhibited strikingly different biomass structures. The ecological functions of these contrasting communities could be studied, however, by measuring or estimating what he called "holistic community characteristics"—including, as at Enewetak, the amount of chlorophyll present, the pyramid of biomass, and photosynthetic productivity.[110] Such measurements would enable ecologists to "develop a comparative science of world ecosystems," cutting through the weeds of site specificity to reveal underlying functional patterns.[111]

As part of this project, Odum also proposed measuring a new variable: the species-variety index, a measure identical to the species diversity index originated by Fisher, Corbet, and Williams and used by Richards and by Dobzhansky's team in Brazil. Odum found this measure "most convenient," enabling the expression of "the diversity of the community in a single number that is independent of the area or number of individuals counted" and, notably, without the need for precise taxonomic identification. By relating this index to the other community characteristics, Odum proposed that his study could illuminate understanding of the latitudinal diversity gradient, recently highlighted by Dobzhansky. "The very old problem of species diversity has its heart in the tropical rain forest and the rate of production," Odum argued. "Two diametrically opposite views have been held," he wrote. "One is that the greater the production, the greater the diversity of life which may be supported. By this view the diversity of life is a function of the total quantity of life. The reverse opinion [is that] the greater the production, the lower the diversity because of the competition."[112] He framed the problem as an old one, but it appeared in a wholly new guise here, formulated as a choice between a positive or negative correlation between diversity and productivity. Along with his other metrics, Odum argued, diversity indexes "permit comparison of communities with minimum effort to test these ideas."[113]

Even the most complex tropical communities could be tamed by applying a whole-ecosystem approach. Standard methods were already available to make each of these "holistic" measurements for plankton communities. To measure a rainforest, Odum proposed choosing a "typical rain-forest site" and felling a sample plot. This was akin to the method Richards used to make his profile diagrams. Because of his interest in trophic structure rather than physiognomy, however, the range of measurements to be made was much greater—not only recording tree species, height, and diameter but also, for example, collecting and measuring the "dry weight of producers, herbivores, and carnivores" to estimate biomass and the chemical extraction of chlorophyll in the field (one method involved a meat grinder and acetone).[114] He also planned to estimate plant photosynthesis and respiration by enclosing sample leaf clusters in plastic bags to measure the gas produced, and proposed one day scaling up the method to enclose a portion of forest within a giant plastic cylinder. Through intensive measurement and control of space, Odum believed he could reduce the complexity of the rainforest to a few key, manageable variables. Moreover, he presumed that the laborious work of cutting vegetation and operating large experimental structures would be "feasible because of inexpensive local labor" available in a tropical country.[115]

Nothing at such a scale had ever been attempted. The Rockefeller Foundation was at first hesitant to support Odum's proposal. The costs and other practicalities of his plan were vague, and many of the terms he used (*productivity*, *efficiency*, etc.) were not yet in wide circulation in ecology. Rockefeller funding officers also expressed concern that, at thirty-two, Odum was too young and inexperienced to run such an ambitious project. In fact, it turned out that Odum was not even sure precisely where the study would take place. Despite the variability across global tropical forests, made evident, for example, by Richards's extensive work, Odum specified only the need for "a typical rain-forest site."[116] A Texas colleague had suggested a locale near Vera Cruz, Mexico, where Rockefeller also had an agricultural group. Sites near the Rockefeller virus labs in Trinidad or Belém, Brazil, also emerged as possibilities. In fact, Odum initially proposed using the renowned BCI station—rather unlikely given the strict prohibitions there against tree cutting and other large-scale environmental modifications. One Rockefeller officer balked, wondering "how they would get permission to cut down rainforest, etc."[117] Nevertheless, in part because of his brother's growing reputation at the AEC, the foundation decided to take a chance on Howard Odum.

Puerto Rico soon emerged as a solution to the problem of a field site. After access to sites there was assured, Odum would cite a variety of natural reasons to favor the island: as a relatively small island, it had a simpler flora and fauna, which would make it easier to study, and its more uniform seasonal rainfall brought it closer to "a desired steady state." At the same time, the island remained a U.S. territory, although having declared commonwealth status in 1952. Working there was "simpler diplomatically, linguistically, and logistically."[118] Perhaps most important to the choice were Puerto Rico's scientific institutions, which were expanding alongside the slew of postwar projects of intensive U.S. industrial investment known as Operation Bootstrap. Among these were the experiment stations of the U.S. Department of Agriculture and Forest Service, as well as other new facilities affiliated with the University of Puerto Rico, including the Institute of Marine Biology (established 1954) and the Puerto Rico Nuclear Center (established 1957).[119] Scientists with training in the United States played a prominent role at these institutions, and it was through such connections that Odum's former professor at the University of North Carolina, Robert E. Coker, put him into contact with the Harvard-trained biologist Juan A. Rivero, director of the Institute of Marine Biology. With the help of Rivero and other Puerto Rican and U.S. scientists affiliated with the University of Puerto Rico and the U.S. Forest Service, Odum carried out his comparative studies of a variety of marine and terrestrial environments beginning in 1957. This work took him to a range of tropical communities across the island, including coral reefs, turtle grass flats, the Bahia Fosforescente (Phosphorescent Bay), and a red mangrove forest. It also led him back to the El Yunque rainforest.[120]

El Yunque had a long history of land use by foresters and other scientists before Odum's arrival, a fact attested by the presence of Mount Britton, the nearby peak named for the New York botanist Nathaniel Lord Britton who had initiated the scientific survey of Puerto Rico in 1913.[121] Located in the Luquillo Mountains on the northeastern part of the island, the forest had been a Spanish Crown Reserve before the 1898 U.S. invasion. President Theodore Roosevelt designated it a national forest in 1907; it was, and remains, the only tropical rainforest in the U.S. National Forest System.[122] Following the failure of a variety of Civilian Conservation Corps reforestation projects during the Great Depression, the Forest Service initiated the Tropical Forest Experiment Station in 1939, renamed the Institute of Tropical Forestry in 1961. The institution aimed to find forest management strategies more appropriate to tropical conditions in Puerto Rico and in the lands of U.S. allies throughout the hemisphere. Leslie Holdridge, a professional forester known among tropical ecologists today for his system of ecological classification, played a key role in expanding the institution's efforts in scientific research.[123] Largely through his efforts, in 1940 the station initiated *The Caribbean Forester*, an important multilingual technical journal that greatly increased communication among foresters (and to some extent ecologists) across national and imperial boundaries throughout the circum-Caribbean.[124] Holdridge at the same time helped establish a series of experimental plots to investigate promising tree species and reforestation methods. As we will see, he would also later play a key role in establishing La Selva in Costa Rica as a major ecological field site. By the time of Odum's 1958 visit, however, the forester Frank H. Wadsworth had taken over this work as the station's director and supervisor of the national forest.

El Yunque's designation as an experimental forest, as opposed to a nature reserve like BCI, enabled Odum to begin to put his ambitious, quantitative field methods into action. At the same time, working at such a site put Odum, like Dobzhansky and Richards, into close contact with foresters. From Wadsworth, Odum learned about the natural history and recent human history of the area and was exposed to methods such as Holdridge's classification system (notable for recognizing far more diversity in tropical than in temperate life zones). Wadsworth's concerns about resource use and conservation also meshed with Odum's own increasing interest in ecosystem management. Odum began to consider the problems that foresters encountered in their attempts to manage the forest to produce usable timber. Wadsworth, notably, had encountered difficulty increasing production because of the slow growth rate of El Yunque's trees. From his systems perspective, and armed with estimates of the forest's biomass and metabolism, Odum surmised that slow growth was not due to a "lack of light, lack of nutrients, or inadequate photosynthesis."[125] Rather, it was a function of the ecosystem's climax state.

Odum spoke a language much like that of forest managers, but where they saw waste and inefficiency in tropical nature, Odum saw an ecosystem reaching its highest potential. Most of the rainforest's productivity was channeled into leaf and root activity for respiration; as a result, he argued, "very little production is left for any net growth."[126] The principles of systems ecology suggested that total production could not be increased above the limit placed by solar energy. The addition of fertilizers, for example, could not solve the problem. Odum's findings at El Yunque increasingly seemed to confirm his ideas about energy and ecosystem evolution. "If, in the rainforest's long evolutionary history, natural selection has already worked toward maximum yield," he wrote, "the ordinary values found in this ancient ecosystem may imply inherent limits of a thermodynamic nature."[127] Odum, like Dobzhansky, saw the tropical rainforest as representing an evolutionary extreme. For Odum, however, it also signaled a natural limit for environmental management. Overall productivity could not be increased, he found, "since the [rainforest's] efficiency may already be as high as has ever been measured in other systems."[128] A role remained for managers, he suggested, however: "Man can not improve on the photosynthetic efficiency of natural communities although he may divert the maximum power output from a product of little interest to him such as maintenance of diverse creatures to a harvestable product."[129] For Odum, the efficiency of tropical nature could not be improved, but ecosystem science could show humans a way to harness it better—perhaps by rerouting its energy away from diversity maintenance and into a more useful product.

Odum's emphasis on ecological control, productivity, and efficiency mirrored Operation Bootstrap's vision of management for a productive, efficient Puerto Rican economy capable of yielding high returns on U.S. investment. Described as a "technocratic optimist," Odum was a firm believer in the idea that ecological theory could and should be applied for rational resource use.[130] His work promised the ability to bring intellectual order to the "unruly complexity" of tropical ecosystems during the same era that politicians and investors worked to rationalize and reform Puerto Rico's agrarian economy into a modern industrial one. Indeed, Odum's connections to Operation Bootstrap were more than rhetorical. During this period of influx of U.S. dollars and experts, Odum himself would leave the University of Texas to become chief scientist at the Puerto Rico Nuclear Center in 1963. At that time he would begin his even more ambitious "Rain Forest Project," generously funded by the AEC. The project's centerpiece was the irradiation of a small portion of the experimental forest, at a site called El Verde where the U.S. Forest Service had a small field station. A 10,000-curie cesium gamma source, placed by helicopter, emitted enough radiation over three months in 1965 to destroy foliage within a thirty-meter radius.[131] If scholars have seen postwar Puerto Rico as a social and economic "laboratory," this corner of El Yunque became an ecological laboratory as well.[132]

Howard T. Odum's crew prepares the cesium source to irradiate the forest at El Verde, December 1964. Odum and Pigeon, *Tropical Rain Forest*, C-28.

The AEC injected the funds Odum needed—ultimately surpassing a million dollars—to realize his vision at the scale of "big ecology." Drawing on his military background, he by all accounts ran the project like a military campaign, "assisted by a platoon of scientists" from a variety of disciplines and from both the United States and Puerto Rico.[133] With this manpower and money, he was able to finally implement his giant cylinder experiment, enclosing an entire sixty-foot-wide patch of forest in plastic sheeting to measure whole-canopy respiration, photosynthesis, and primary productivity. Despite

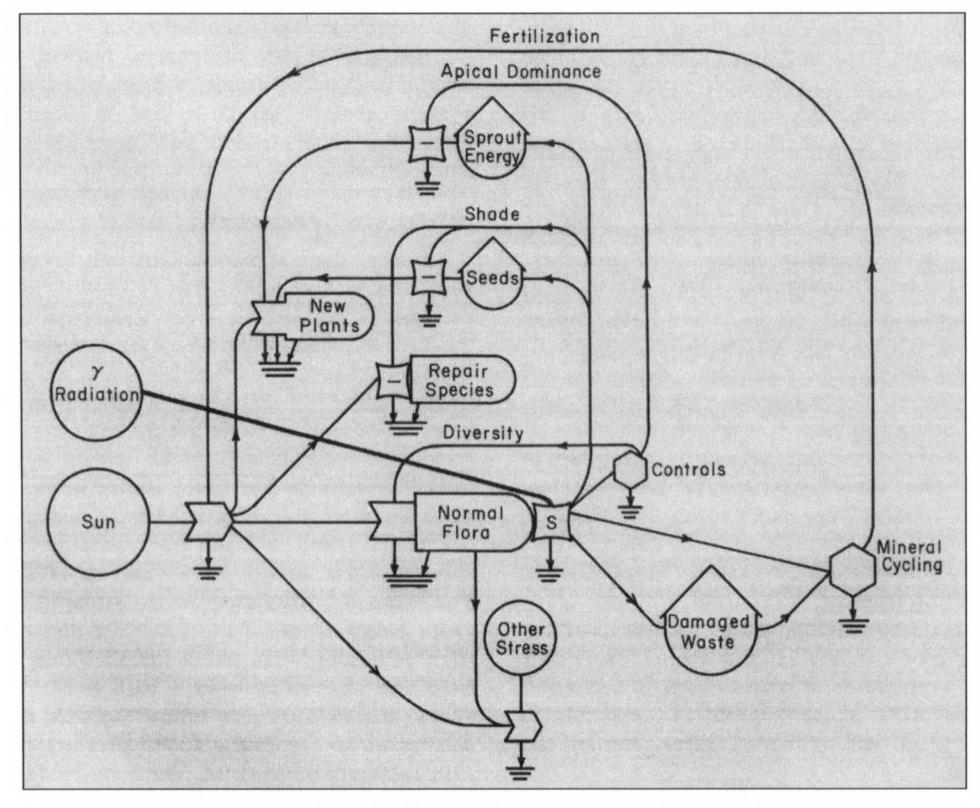

Energy diagram of El Verde "showing repair mechanisms at the ecological level" with "diversity" as part of the system. Odum and Pigeon, *Tropical Rain Forest*, I-257.

the radioecology angle, the Rain Forest Project was fundamentally a continuation of the earlier Rockefeller-funded research program. The study of the rainforest's recovery after irradiation, however, added a new layer, enabling Odum to model the ecosystem's resilience under stress—a property he related to its species diversity. His assumption of a relationship between diversity and stability was then shared implicitly or explicitly by many biologists, and Odum's systems theory seemed to give it the force of an axiom. Visually, in the energy circuit diagrams he produced to describe the functions and flows of a rainforest ecosystem "machine," diversity appeared as a key component, to be dialed up or down.

Systems ecology thus permitted not just intellectual control over an otherwise bewilderingly complex community but managerial control as well. In a way that had been foreshadowed by Orlando Park's 1944 ESA speech, the Rain Forest Project was deeply connected to U.S.-led development schemes and the projection of U.S. power. (This was no coincidence, perhaps, given Park's position promoting ecology in the AEC by this time.)[134] Not only were equivalent systems principles at work in economists' rationale for diversifying Puerto Rico away from a precarious dependence on sugar monoculture, but the project was also directly linked to a broader vision of regional economic and military control. For the AEC, the project's most immediate rationale was understanding the potential environmental effects of the proposed Project Plowshare "Pan-Atomic Canal," a nuclear-excavated sea-level isthmian canal intended to replace the aging Panama Canal.[135] The possibilities for recovery after nuclear warfare or reactor accidents in "underdeveloped tropical areas" also ran through the project.[136] At the same time, the Puerto Rico Nuclear Center itself was tied to a project of Puerto Rican economic and technological modernization. Odum, as chief scientist for the institution, saw ecological engineering as central to this mission. "If the century of advanced research in land management of temperate regions is mainly inapplicable," he asked rhetorically, "what knowledge and methodology might give us early plans for best land usage?"[137] The Rain Forest Project's systems approach to understanding El Verde, he promised, could shed light not only on how to manage Puerto Rico's other remaining forests but also the "worldwide problem of tropical forest regime management."[138] As in Panama, U.S. scientists and the military saw Puerto Rico as a stand-in for environments around the tropical world. In fact, the Institute of Tropical Forestry collaborated with the Department of Defense to test Agent Orange and various "tactical herbicides" for use in the Vietnam War within the El Yunque forest, including sites near El Verde.[139] These became integrated with the Rain Forest Project as Odum compared the recovery of irradiated and herbicide-treated plots. Puerto Rico's colonial status made it an acceptable site for destructive experimentation in the service of greater U.S. interests, military and scientific alike.

Odum downplayed this darker side, emphasizing the peaceful applications of ecological knowledge. Indeed, he came to believe the rainforest could provide a model for rational long-term social progress and environmental development—a kind of cybernetic vision of tropical paradise. "The rain forest achieves complexity, high metabolism, and stability over geological time periods without surges and waste," he wrote. "Can we find in this example the clues for designing our own equally effective systems of man and nature?"[140] Odum's core interest became what by the 1980s would be called *sustainable development*. Nevertheless, rather than the maintenance of species diversity for its own sake, diversity interested Odum primarily as a component of a stable system, able to recover from stress and persist into the future. "How much diversity is desirable for stability and control?" he asked.[141] His systems perspective had led him to make a forceful connection between species diversity and the problems of long-term resource management. In making this conceptual linkage, Odum foreshadowed the movement for sustainable development and the formulation of *biodiversity* by conservation biologists later in the century.

No Refuge in Complexity

If Howard Odum became increasingly interested in the role of diversity in ecosystems by the 1960s, this was also in large part due to the work of Robert MacArthur, who played a critical role in moving species diversity to the center of new theoretical debates in ecology. MacArthur, like Odum, was a student of Hutchinson whose substantial contributions to ecology were drawn from outside the traditional bounds of the discipline. Unlike Odum, however, MacArthur was inclined toward neither environmental management nor large-scale field experimentation. His insights were mathematical, taking abstract concepts from such fields as information theory and using them to build testable models in population and community ecology.[142] It was MacArthur's work that would ultimately answer Thomas Park's call to place the study of community dynamics on a quantitative basis. His inspiration and empirical data, however, came not from the simple biological systems of the lab or high latitudes but primarily from the complex, diverse communities of the tropics. During his tragically short life, MacArthur would reshape ecology, applying a hypothetico-deductive approach to such problems as niche partitioning, the latitudinal diversity gradient, and most famous, with Edward O. Wilson, island biogeography. The real places of study at the foundation of his theorizations often remained obscured by his infamously abstract, concise style of writing and citation, yet both the data and core questions that ran through his work were rooted in biologists' tropical encounters. MacArthur saw science as, fundamentally, a search for simple patterns underlying the complexity of the natural world. This interest in ecological complexity made the tropics highly attractive. MacArthur "visited the tropics as often as he could, and delighted in the endless facts of natural history," recalled his collaborator Wilson, but ultimately these fell under the blade of "his cartesian scalpel."[143] He had little patience for the mere accumulation of facts, writing provocatively, "Not all naturalists want to do science; many take refuge in nature's complexity as a justification to oppose any search for patterns."[144] For MacArthur, the complexity of tropical nature was no refuge; it was a challenge.

MacArthur wrote his first scientific paper while serving with the U.S. Army at Fort Huachuca in southern Arizona between 1954 and 1956. He had been drafted shortly after completing his master's degree in mathematics and joining Hutchinson in zoology at Yale.[145] Appearing in *Ecology* when MacArthur was just twenty-four years old, "Fluctuations of Animal Populations and a Measure of Community Stability" presented a mathematical formulation of a diversity-stability relationship for the first time. In their *Fundamentals of Ecology*, the Odums had posited that the larger the number of energy pathways available within a food web, the more stable the community.[146] In his spare, three-page paper, MacArthur borrowed an index from information theory, as developed by the communications engineer and cryptographer Claude Shannon, to give a precise measure of community stability in this sense. This index originally served to quantify the information content of a message based on its probability structure (entropy); here it measured a community's complexity in terms of the numbers of species and possible trophic links among them. Whereas Fisher, Corbet, and Williams's index of species diversity simply fit a curve to empirical data of the number of individuals and species in a given sample size, MacArthur's use of Shannon's index grew out of a theoretical model of community interactions—and indeed, he presented it as a way of testing the principles of this model.[147] Such an approach was highly unusual in ecology at that time. While MacArthur would mine mathematics texts for ecological applications throughout his career, it seems to be more than a coincidence that he looked specifically to information theory during this period. He was stationed at Fort Huachuca just as the army opened its Electronic Proving Ground, which would become the center of the U.S. government's development of electronic weapons and intelligence systems. As in other areas of biology, the continued and increasing closeness of academia and the military after World War II fostered sometimes unexpected connections with the physical sciences and mathematics, especially the fields of cybernetics, systems, and information theories.

At first glance, however, little about this now-classic paper might appear to be "tropical." It was not based directly on fieldwork, in the tropics or elsewhere. (Indeed, MacArthur was in Arizona because its arid and decidedly nontropical environment made it ideal for electronics testing.) At its heart, however, MacArthur's 1955 paper was concerned with precisely the questions Dobzhansky had raised in "Evolution in the Tropics": what was the nature of latitudinal patterns of diversity, and how did they relate to the complexity of community interactions? MacArthur's paper began in abstract ecological space but ended with two "plausible" deductions about the nature of communities living in Earth's most extreme environmental regions:

> A. Where there is a small number of species (e.g. in arctic regions) the stability condition is hard or impossible to achieve; species have to eat a wide diet and a large number of trophic levels (compared to number of species) is expected. . . . Populations will vary considerably.
>
> B. Where there is a large number of species (e.g. in tropical regions) the required stability can be achieved along with a fairly restricted diet; species can specialize along particular lines and a relatively small number of trophic levels (compared to number of species) is possible.[148]

In other words, based on their relative species diversity, populations in arctic communities should be predicted to undergo wild fluctuations, while populations in tropical communities should remain comparatively steady while showing high degrees of specialization. MacArthur suggested the power of his mathematical model by deducing properties of arctic and tropical communities that already appeared intuitive to naturalists.

The tropics figured in MacArthur's debut publication as a theoretical space—one extreme on a spectrum of species diversity and ecological complexity. Although entitled "Population Ecology of Some Warblers of Northeastern Coniferous Forests," his dissertation research would bring him to the tropics in person. After returning to Yale, MacArthur took up another problem related to species diversity, focusing on a field study of five very closely related species of warblers. Given their similarity in size, shape, and food preferences, he asked, how was it that they were able to coexist within the relatively homogeneous forests of New England? Why didn't one species outcompete the others? MacArthur had been a keen birdwatcher since childhood, and he was quickly able to observe how the birds divided up their habitat, and avoided competition, by foraging in different parts of conifer trees, nesting at different times, and engaging in territoriality. The community's species diversity thus was a product of the behavior and life history of its constituent populations. The resulting 1958 *Ecology* paper, rich both in natural historical detail and in theoretical insight, would receive ESA's Mercer Award and remains a textbook example of niche partitioning. Often overlooked, however, is the fact that MacArthur did not examine the birds only in New England; he also followed them to their wintering grounds in Costa Rica. (Appropriately enough, his research was supported in part by the Frank M. Chapman Memorial Fund of the American Museum of Natural History; Chapman, too, had traveled south in pursuit of migrating birds.) While the warbler paper is best known for examining factors contributing to the maintenance of local diversity, MacArthur recognized a need to compare the birds' behavior in the tropical portion of their range.[149] Most immediately, his purpose was to determine whether the warblers' behaviors depended on their environmental surroundings, but the bigger question of the greater diversity of tropical species lay not far behind. Indeed, latitudinal diversity was a major topic of conversation among MacArthur and his friends during graduate school—"the subject commanded many hours of our attention," one later recalled, "even when we were supposed to be preparing for our comprehensive exams."[150]

MacArthur spent about three weeks during the winter break of 1956–57 in Costa Rica, a location determined in part by the birds' migration patterns, but also by existing U.S. scientific networks. He consulted with the well-traveled ornithologists James Bond and S. Dillon Ripley. (That Bond was at once the foremost authority on Caribbean birds and the namesake of the fictional secret agent 007, while Ripley had been a real-life spy for the U.S. Office of Strategic Services during World War II and would later become secretary of the Smithsonian Institution, suggests the multiple intersections of science and international relations during this era.)[151] Like that of other U.S. biologists, MacArthur's fieldwork was mediated by a constellation of sites and scientists with close ties to resource extraction in the region. Costa Rica had a long relationship with U.S. business and science; indeed, it was a launching pad for both the United Fruit Company and the Rockefeller Foundation's International Health Commission.[152] The locations of Costa Rica's most significant field sites were at the same time a product of its own government's efforts to incentivize the internal colonization and development of its rural hinterlands, and of its friendly policies toward white U.S. and European immigration.[153] MacArthur's primary contact in Costa Rica—acting as his "naturalist and guide"—was Leslie Holdridge, the forester who had helped to develop the Tropical Forest Experiment Station in Puerto Rico, where Odum would soon begin his rainforest work.[154] After several years in Haiti, Guatemala, and Colombia, Holdridge had settled down in Costa Rica in 1949 to take a position with the recently established Inter-American Institute of Agricultural Sciences at Turrialba, a research and teaching institution formed at the behest of the United States and operated by the Organization of American States.[155] He married a Costa Rican woman and, through his connections with a Costa Rican land investment group, purchased land of his own in 1953. He called it "Finca La Selva," or "Jungle Farm."[156] Situated in the country's northern Caribbean lowlands, the 613 hectares served as a private weekend retreat and an experimental plantation. There, he developed methods of mixed planting with the aim of producing tree crops without destructive clearing. He also enjoyed playing host to foreign naturalists.[157] Some were "so enthusiastic," he later recalled, "that we gradually went into the business of having scientific visitors," MacArthur among them.[158]

During his trip, MacArthur also met with another expatriate U.S. biologist, the eccentric and ascetic botanist-turned-ornithologist Alexander Skutch. Skutch, too, had taken advantage of cheap land prices to acquire farmland in Costa Rica in 1941. Like William Beebe during World War I, Skutch sought "jungle peace." After years of working for the United Fruit Company and USDA, he retreated to the seventy-eight hectares he named "Los Cusingos" (after the fiery-billed aracari, a local toucan) in southern Costa Rica. There, he devoted himself to the study of living birds and a simple life in harmony with nature, based on his reading of the Indian philosophy of ahimsa, or nonviolence. He married the naturalist Pamela Lankester (herself the daughter of an English immigrant naturalist), adopted a Costa Rican son, and supported his family through subsistence agriculture, plant collecting, and his prolific natural history writing.[159] With religious fervor, Skutch answered the call of ornithologists like Beebe and Chapman, whom he had met at BCI, to study living birds rather than collect them. His description of the behavior and life histories of tropical birds were, and largely remain, unmatched in sheer quantity, duration, and detail.[160] He was in many respects out of step with mainstream ornithology, repudiating the practice of bird banding as brutal, for example. Nevertheless, he was respected for his disciplined observation of previously unknown phenomena of behavior and ecology, most famously that of cooperative breeding or "helpers at the nest."[161] Until his death in 2004 at age ninety-nine, Skutch would live at Los Cusingos in relative isolation, periodically receiving biologists like MacArthur on "pilgrimage" to meet him and encounter the bird life of a tropical forest.[162]

MacArthur's first tropical sojourn was brief. The emergence of sites owned privately by U.S. scientists living abroad, however, permitted access to land on either end of Costa Rica, as well as to expert resident knowledge. MacArthur's own observations were somewhat inconclusive; he could locate only one of the five warbler species, and the large size of rainforest trees made recording feeding zones more difficult than among the northern conifers. Nevertheless, he could rely on Skutch's judgment about warbler behavior, which confirmed his hypothesis that each species' disparate foraging behaviors were intrinsic, not varying greatly between winter and summer, the tropics or temperate zone.[163] Not only the empirical details of tropical natural history, however, but also the controversy over Skutch's ideas about the relative roles of cooperation and competition in maintaining bird diversity in the tropics seems to have sparked MacArthur's imagination. Skutch had engaged in a fierce debate with the British ornithologist David Lack (famous for studying competitive exclusion in Galapagos finches) over Lack's insistence that birds maximized clutch size to produce as many young as they could adequately nourish. Skutch contended that tropical birds generally produced small broods, and that they did so not for lack of resources, which were abundant in the rainforest, but to avoid attracting predators—an idea based both on meticulous long-term observation and his convictions about natural harmony. Skutch also identified a latitudinal decrease in clutch size toward the tropics, coincident with the increasing diversity gradient.[164] MacArthur would go on to a postdoctoral position with Lack, whose emphasis on competition corresponded more with his own, but he remained fascinated by the patterns Skutch had identified.[165] Questions about the role of competition and natural selection in the evolution of behavioral ecology in general were becoming increasingly sharpened as tropical life history data accumulated and comparisons with the north temperate zone could be made.

Shortly after his return from Costa Rica, MacArthur attacked the problem of how competing species divide resources in his 1957 paper, "On the Relative Abundance of Bird Species." The paper famously put forth a statistical model comparing the environment to a "broken stick" of nonoverlapping niches. It drew on the "considerable amount of raw data" published on the commonness and rareness of species in communities, and the model seemed to fare well when tested against some censuses of tropical forest birds.[166] In effect, it tied Hutchinson's niche concept mathematically to a prediction of expected species diversity. In this paper and his other early publications MacArthur did not simply take a quantitative approach to population and community dynamics. More specifically, he placed species diversity at the center of fundamental theoretical questions about the structure of ecological communities, their stability or dynamics, behavior, and the role of competition over resources.

The research possibilities of the problem of diversity, as illuminated by MacArthur's approach, however, became a much larger part of the consciousness of the U.S. biological community after Hutchinson's 1958 address to the American Society of Naturalists, "Homage to Santa Rosalia; or, Why Are There So Many Kinds of Animals?" Contemplating the coexistence of two species of aquatic insects in a small pond near the sanctuary of Santa Rosalia in Sicily, Hutchinson had asked himself "why there should be two and not 20 or 200 species" living there. Extending the question to the whole planet, he wondered why there were an estimated one million total species living on Earth rather than, say, an order of magnitude less or more. Hutchinson challenged biologists to seek a unified theory of species diversity. He outlined major factors at work in generating and placing limitations on diversity, including the energetics and stability of complex food webs, the size and heterogeneity of the environment, and the dependence of animal diversity on that of plants. Not surprisingly, he suggested a central role for competition and niche diversification, and in doing so he drew liberally on MacArthur's models. "Homage to Santa Rosalia" in essence articulated a research program for diversity studies, and while the paper may have begun in Sicily, its logic led to the tropics.[167] Any universal biological theory of diversity clearly would depend on understanding the distribution of life in the tropics, where communities seemed to approach the upper limits of species diversity.

The 1960s would see an explosion of publications on the role of community interactions in the maintenance of local species diversity and on broader geographic patterns, including diversity on islands and the latitudinal diversity gradient, problems for which tropical data were crucial. Ecologists' skyrocketing interest in diversity was thus made possible in large part by the theoretical tools MacArthur contributed but was at the same time underpinned by a growing network of tropical sites and scientists. MacArthur's own research and collaborations traced out this network. He would eventually send his own students to Costa Rica, where he had first observed rainforest birds with Holdridge and Skutch. He would use censuses from El Verde, Puerto Rico, in collaboration with the ornithologist Harry Recher, a contributor to Odum's AEC Rain Forest Project.[168] He would also coauthor several important papers on habitat patchiness and theoretical limits on diversity with Richard Levins, then employed at the University of Puerto Rico.[169]

The older stations at Soledad and BCI gradually regained their footing following World War II and Barbour's death and, as described in chapter 5, reasserted themselves as keystone institutions for tropical biological research and education. Soledad's most famous alumnus, Edward O. Wilson, would collaborate with MacArthur on their influential and controversial theory of island biogeography.[170] BCI's magnetic pull was also undiminished, being both well established institutionally and rich in accumulated data about tropical forests and islands. MacArthur became a seasonal fixture at BCI during the 1960s. He worked there with an array of collaborators, including fellow former Hutchinson student Peter Klopfer, Joseph Connell (his coauthor on *The*

Biology of Populations), his physicist and birder brother John MacArthur, and his own graduate students Robert Ricklefs and Martin Cody.[171] The station's other regulars long remembered the scene of MacArthur and his entourage deep in debate and jockeying for position around the dinner table at BCI.[172] By 1966, Martin Moynihan, BCI's new director, could report, "The broad subject which attracted the largest single group of visitors with similar interests was, not unexpectedly, species diversity."[173]

The power of MacArthur's approach and that of the new quantitative community ecology was in the simplification of complex ecological data to a few key variables, none more significant than species diversity. Scientists, especially those working at BCI since the 1920s, had learned much about tropical community dynamics by studying the natural histories of constituent species and populations, but this was painstaking work. Moreover, in its deep connection to place, how far was it generalizable to other ecological contexts? To get at fundamental patterns, MacArthur stripped away all detail irrelevant to the problem at hand. To model how an island's species diversity could be predicted by its size and distance from the mainland, for example, he and Wilson ignored distinctions of habitat, taxonomic specificity, and history and assumed an equilibrium state.[174] This differed radically from a traditional natural historical approach, as many scholars have noted; Thomas Barbour would have recognized the broad patterns MacArthur and Wilson described across Caribbean islands but could never have accepted as explanatory an account that treated all species as equivalent. Despite MacArthur's central concern with spatial patterns of diversity, his publications often elided specific research sites. *The Theory of Island Biogeography* makes only scant references to BCI and Las Perlas, even though MacArthur himself censused birds on these islands.[175] The iconic graphical representation of the theory deals with abstract space, obscuring place and region, even though MacArthur would later note that patterns in island species diversity were accentuated (and thus easier to see) in data from tropical islands where overall diversity is greater and a lower proportion of birds are migratory.[176] The complexity of local environments, although excluded from his mathematical model of island diversity, in fact figured prominently in MacArthur's many other papers on local and latitudinal diversity patterns (especially those of birds). Yet in MacArthur's hands, environmental complexity itself became abstracted; he carefully chose proxies like foliage height and habitat patchiness, which could be measured across a wide variety of localities and fed into mathematical models.

Although leaving the landscape itself intact, MacArthur's theoretical simplifications were in this sense even more radical than those of Odum, whose

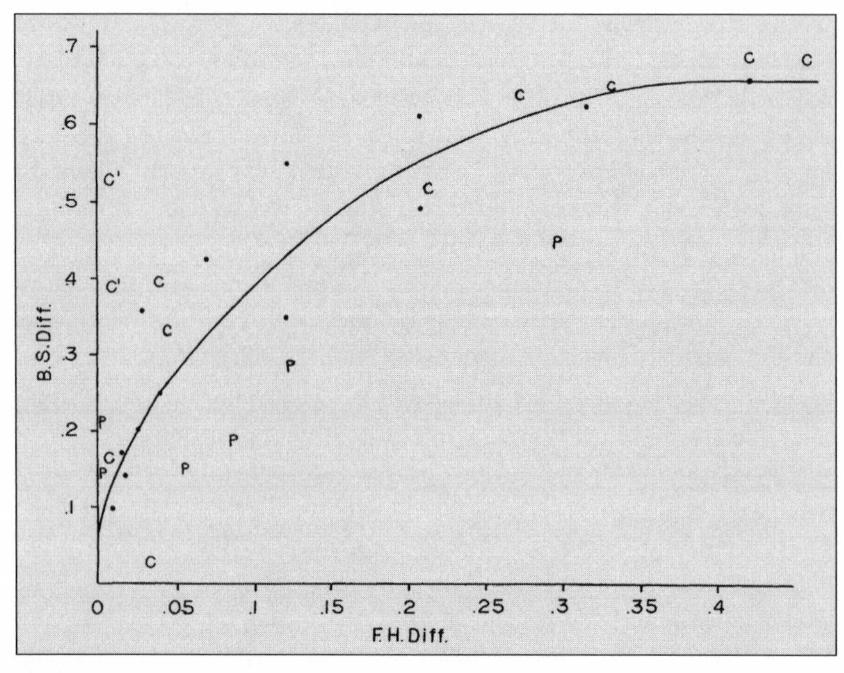

Graph relating bird species diversity (BS) to habitat diversity (in terms of foliage height, FH) based on censuses in the Panama Canal Zone (C), Puerto Rico (P), and the temperate United States (solid points and the curve). MacArthur, Recher, and Cody, "Relation between Habitat Selection and Species Diversity," 323.

ecosystem models became increasingly Byzantine in his efforts to produce a totalizing view. Their strength was not in providing a complete picture of a community in all its diversity—in the older sense of nature's great profusion and variety. Rather, MacArthur sought to elucidate fundamental ecological relationships by relating an index of species diversity to the spatial dimensions of environment and niche. By the time of his tragic death from renal cancer in 1972 at the age of forty-two, the question of the tropical increase in species diversity had failed to yield to any one simple equation. In his final book, *Geographical Ecology: Patterns in the Distribution of Species,* he would admit with some "embarrassment" that many causes were "likely to be true, at least in some places!"[177] MacArthur's contribution was thus not an answer to the question, raised by Dobzhansky, Hutchinson, and others, of why the tropics were so rich in species. Instead, it was a demonstration of the power of a narrower, mathematically defined view of "diversity" to articulate a fruitful research program in ecology. With a quantitative concept of species diversity as a tool, ecologists could confront even the most complex of Earth's communities.

Conspicuous Patterns

MacArthur sought simple ecological laws, but he did not find them by studying simple systems. He counted himself among the ecologists who preferred to study the "extremely complex communities of the tropics." Paradoxically, he explained, it "may be easier to understand" ecological principles in the complex tropics, because "the complex community has strong interactions among species so that the lives of the separate species are less independent than in a simple community. Where there is greater interdependence, patterns may be more conspicuous."[178] MacArthur and an increasing number of ecologists saw the complexity of tropical communities as a scientific resource rather than a stumbling block. Shifting their focus from individual species and their interactions, MacArthur, Odum, Dobzhansky, and Richards used the abstraction *species diversity* to capture tropical life in a number that could be measured, modeled, and manipulated. In this way, the tropics became more than just a source of untapped biological study material. Rather, the very complexity of tropical life itself became a wellspring of a rich array of theoretical questions across biology—about evolutionary mechanisms, ecosystem functions, and patterns of distribution. The concept of species diversity enabled biologists to study the structures and properties of tropical communities without first having detailed knowledge of each of their multitude of component species.

Such an approach differed wildly from the place-based natural historical research ongoing at BCI. There, a deep understanding of a local ecological community was forming through slow accumulation, building on Frank Chapman's intimate portraits of bird life histories, Clarence Carpenter's analyses of monkey behavior, Robert Enders's inventories of the island's fluctuating inhabitants, and hundreds of other studies. In contrast, no birds perched or monkeys swung from the branches of Richards's diagrams. Dobzhansky, similarly, found the actual identity of the species present in Brazilian forest plots less interesting than the patterns of rarity and abundance that they displayed. New graphical and quantitative field techniques brought species diversity to bear to make comparisons between the tropics and temperate zones, as well as among environments across the tropical world. Local studies could now be incorporated into more geographically extensive and biologically "fundamental" theoretical frameworks.

Not everyone welcomed the new approaches. Their focus on diversity, ironically, seemed to sacrifice a rich portrait of diverse species in all their complex interactions. One evolutionary biologist criticized Odum's Rain Forest Project as "'slash-and-burn ecology,'" not only because of the large-scale experimental interventions involved but also because his ecosystem models cut out so much of the biological story.[179] Similar criticisms have been leveled against MacArthur and his followers. In some tellings, they have become the "villains," spearheading the "colonization" of ecology by reductionist theoreticians and technocrats.[180] Indeed, in Donald Worster's classic history of ecological ideas, MacArthur and Odum are featured as representatives of the so-called imperial tradition, a strand of thought seeking intellectual mastery and dominance over nature through a reduction of ecology to mechanistic and economic terms.[181]

These figures have less often been seen as colonizers in the more literal sense. Yet, as we have seen, the development of concepts and techniques for studying species diversity was closely tied to the control of tropical lands. During World War II and into the Cold War, U.S. interests in tropical environments intensified. With new sources of federal and private foundation support, U.S. biologists gained access to an array of new field sites within and beyond the circum-Caribbean. Of course, this expansion was driven not only by U.S. projects but also by a wide variety of internal land colonization and development efforts, as seen in the distinct contexts of Puerto Rico, Costa Rica, and Brazil. On the ground, too, fieldwork depended on building local relationships, including in some cases more collaborative alliances with non-U.S. scientific communities. Nevertheless, in the 1950s the networks of U.S. tropical research still traced the conspicuous patterns of U.S. power. The most popular sites for U.S. researchers remained those within the circum-Caribbean and controlled by the U.S. government, as at BCI and El Verde, or by private U.S. corporations and individuals, as at Soledad and in Costa Rica. Such close ties shaped how quantitative and field approaches from forestry and engineering entered the study of ecology in the tropics. While the theoretical emphasis of early diversity studies tended to obscure both locality and geopolitics, these come sharply into focus when we consider Dobzhansky's visit to Fordlândia or the helicopters dropping radioisotopes and herbicides for Odum at El Verde. Even MacArthur's work, disinterested in the practical applications of ecology and never garnering substantial government or corporate funds, nevertheless relied on a broad network of U.S.-controlled sites for both its formative data and ongoing testing.

In their dependence on tropical land, the new studies of species diversity remained fundamentally continuous with long-standing traditions in tropical biology. The rise of diversity studies in fact encouraged a surge in tropical biological research of all kinds. Natural historical data long accumulated at BCI and other sites found fresh meaning when approached with the question of how species diversity arises and is maintained in ecological communities. As experts on the Earth's most complex and diverse communities, tropical biologists suddenly found themselves positioned at the center of a lively theoretical field, their work now relevant to core debates in areas as seemingly disparate as community ecology, systems ecology, and evolutionary biology. In effect, the Park brothers' quarrel on the eve of World War II found a resolution during the postwar era. Complex tropical communities could indeed be studied in a quantitative way, and research at stations like BCI remained the vital underpinning of such work.

It was an exciting time in tropical ecology and in ecology generally. The question of species diversity had launched tropical research into the scientific mainstream. Yet at the same time, uncertainty lingered in the background. In 1959 revolution swept Cuba. Protests erupted in Panama. Anticolonial and nationalist movements engulfed the tropical world. The survival of the stations that fueled the study of ecological diversity suddenly appeared uncertain. U.S. researchers would no longer be able to take access to tropical lands for granted.

CHAPTER FIVE

A Global Resource

For the future of "one world," more knowledge of the tropics is of great importance—more knowledge of the plants, of animals, and of man. The United States has an obligation to cooperate in gaining that knowledge, especially of the American tropics.
—Conference on Tropical Botany, 1960

Biological diversity must be treated more seriously as a global resource, to be indexed, used, and above all, preserved.
—Edward O. Wilson, 1988

In January 1963, Harvard professor Edward O. Wilson gathered with representatives of several other universities in Coral Gables, Florida.[1] The assembled group would move to create the Organization for Tropical Studies (OTS), a consortium focused on providing U.S. students and faculty with field courses in tropical environments. The conference was part of a series of meetings begun in 1960, sponsored by the National Research Council and National Science Foundation, which had already in 1962 resulted in the formation of a new society for tropical biologists, the Association for Tropical Biology (ATB). This flurry of professionalizing activity in tropical biology clearly reflected the growing intellectual ferment over the problems of the ecology and evolution of species diversity, which centered on the tropics. At the same time, tropical biologists were galvanized to unify professionally in the 1960s—particularly in support of institutions for field education—in direct reaction to threats facing long-standing tropical stations.

Wilson was present at Coral Gables less because of his work on tropical island fauna (his collaborative work with Robert H. MacArthur had not yet been published) than because of Harvard's relationship with Cuba.[2] Indeed, although now a Harvard professor, Wilson had first traveled to the tropics in 1953 as a graduate student in the university's summer tropical botany course, Biology 215, in Cuba. The class had used Harvard's venerable Atkins Institution at Soledad, near Cienfuegos, as its base station. Wilson had been thrilled to finally reach the tropics. He had grown up on images of jungle wilderness in *National Geographic* and the books of William Beebe.[3] The experience, however, was in some ways "not satisfying."[4] He was shocked by the "ecological destruction of the island."[5] The class had "had to travel . . . miles and miles through the sugar cane fields" in search of native Cuban species.[6] "To find the last refuge," he explains in his memoir, "we had to go beyond the reach of bulldozers and chain saws, mostly up onto the slopes of steeper mountains and down the banks of river gorges." Traveling across the island, he recalls, "I began to undergo a fundamental change in my view of the tropics."[7] In Cuba he began to perceive conservation as not just an issue within industrialized nations like his own but as a global problem; he explains, "The world, I discovered, was not in better shape elsewhere."[8] By the 1980s, Wilson would parlay his personal experience of disappointment about the degradation of tropical nature into environmental activism, becoming perhaps the most public face of a new movement not just to study but to preserve biodiversity.

The 1963 meeting, however, had been sparked not by an environmental crisis but by one with political roots. Even as Wilson started his first steps on an "intellectual journey" toward conservationism in Cuba during July 1953, Fidel Castro and a small band of fighters had been preparing an attack on the heavily defended Moncada Barracks—the strike that launched the Cuban Revolution.[9] After the revolutionary government came to power in 1959, the Soledad estate, like other U.S.-owned properties, was ultimately nationalized. The loss of the station was a serious jolt to Harvard and to the community of scientists in the United States who increasingly identified themselves as tropical biologists. U.S. biologists found themselves cut off from one of their oldest and most established institutions for tropical research, and one with a respected and expanding role in graduate education. Like U.S. policy makers, many biologists feared a domino effect—fears that seemed confirmed as Panamanian calls to dissolve the Panama Canal Zone grew louder in 1963 and 1964. As a wave of nationalism and anticolonialism swept the region, even the future of insulated Barro Colorado Island (BCI) appeared uncertain.

During the twentieth century U.S. biologists had gained a foothold in the tropics through their alliances with U.S. interests. Now the ground shook beneath their feet. The geopolitical reconfiguration of the circum-Caribbean necessitated transformations in both the institutions and the rhetoric that U.S. biologists relied on to support their research in tropical environments. Understanding these transformations is important not only to understand the emergence of tropical biology as a discipline but also because tropical biologists' responses to the political and professional crises of the 1960s prefigured

Graduate student Edward O. Wilson with his pet giant Cuban anole, Methuselah, at Harvard's Atkins Institution at Soledad, July 1953. Photograph courtesy of Edward O. Wilson.

and conditioned the rise of the concept of biodiversity in profound ways. U.S. biologists had long justified basic research in the tropics as the foundation for the exploitation of tropical environments; now they pointed to its application in conservation and sustainable use. Once dependent on connections with U.S. interests, now they emphasized globalism and international collaboration. Whereas the species diversity of the tropics had previously been an intellectual puzzle, it was now framed as a solution to the world's intertwined economic and environmental problems. The language and networks that tropical biologists, including Wilson and the other architects of OTS, developed to support basic research and education in their field would transform tropical diversity into "a global resource."[10]

Sixty Years of Soledad

Understanding these developments first requires a closer look at the immediate effects of the Cuban Revolution on U.S. biological institutions. While a shock to tropical biologists, the loss of Harvard's station at Soledad was not instantaneous. Rather, it involved a period of uncertainty and a variety of attempts to adapt to local political circumstances. At the close of 1960, the station's director, Ian Duncan Clement, and his wife, Vivian, were among the last U.S. citizens to leave Cuba.[11] Even as Clement transferred portions of his research to Cambridge and Jamaica, he held out hope that the separation would be only temporary. The station's herbarium, library, and lab were "put in stand-by condition." The Clements and the station's remaining Cuban employees packed up equipment and sent duplicate planting records and inventory cards to Cambridge. They sold off stocks of hybrid corn seed to the new Cuban government. With the help of lab technician Esperanza Vega and two visiting U.S. research collaborators, Clement continued to record observations and hand pollinate experimental cotton varieties through the end of December. The station's horticulturalist, Felipe Gonzalez, and his assistants pruned the arboretum's trees, preparing them "to withstand a period of neglect."[12]

During the interval of active fighting through the rise of the new revolutionary government in 1959, Clement maintained a "public posture and indeed . . . hope . . . of 'business as usual.'"[13] With no political appointees in charge of the station's fate, Soledad had been more insulated from the periodic disruptions that had wracked Cuban agricultural experiment stations during the Batista regime.[14] At the same time, even as the range of tropical sites easily available to U.S. biologists had expanded after World War II, Soledad continued to play a key role, especially for botanists. Its well-established botanical garden and arboretum distinguished it from forest stations like BCI or El Verde; it touted its role as a taxonomic and genetic library of plant diversity for academic researchers and horticulturalists alike. Moreover, with its ongoing experimental work and, beginning in 1950, an annual Harvard course in tropical botany, Biology 215, the station served as a bridge between the academic and economic sides of tropical research. It was the only tropical station operated by a university, and Biology 215 was celebrated as the "first course in tropical botany ever to be offered by an American university."[15] (The Johns Hopkins itinerant laboratory in Jamaica and University of Pittsburgh's summers at Kartabo, British Guiana, had decades earlier provided instruction in the tropics, but with a zoological emphasis.) Indeed, Harvard botanists envisioned the station as poised to assume "an even greater importance in tropical botany," filling a void created by "the deterioration of the other great plant collections of the tropics in the wake of postwar political and economic changes." As Sri Lanka and Indonesia struggled toward independence in the late 1940s, U.S. botanists assumed that the famous British and Dutch colonial gardens at Peradeniya and Buitenzorg were "destined for obscurity."[16] Soledad's position in Cuba had appeared secure in comparison.

When Clement, a Harvard-trained economic botanist, took the directorship of the Soledad station (now renamed the Atkins Garden and Research Laboratory) in 1948, his goal had been to prove "that an academic institution can do practical work." The station was to be "in no ivory tower."[17] While it remained a base for natural history research and collecting by U.S. scientists visiting Cuba, Clement put a renewed emphasis on the station's role in agricultural development on the island and throughout the region. During the 1920s and 1930s, the station had been dependent largely on Thomas Barbour's personal friendships with wealthy and influential individuals, such as the Atkins family and U.S. Department of Agriculture (USDA) plant explorer David Fairchild. Clement expanded and reinforced the station's ties with a range of organizations and corporations throughout the circum-Caribbean. The station exchanged plant material and propagation techniques with institutions in Florida, including the University of Florida, University of Miami, USDA experiment stations, and the Fairchild Tropical Garden (named after Barbour's close friend), as well as with the Rockefeller Agricultural Program in Mexico, the Organization of American States' Inter-American Institute of Agricultural Sciences in Costa Rica, and the United Fruit Company.

Significantly, the station also served as a pipeline of experts to these organizations and companies, with its increasingly formal role in graduate education through student fellowships and Biology 215. "By training and placing men in key positions in other parts of the tropics," Clement explained, the station "can multiply its own modest efforts manyfold."[18] In preparation, students in this class not only were familiarized with tropical vegetation through field trips to representative Cuban natural areas but also focused on the "tropical economic plant products . . . which have become essential to the American economy during this critical period."[19] The Atkins Garden's collection itself served this purpose well, but with the support of United Fruit the class also visited the Escuela Agrícola Panamericana, near Tegucigalpa, Honduras, and the company's Lancetilla Botanical Garden at Tela, Honduras, during the course's first two iterations.[20] Soledad had also long engaged in agricultural exchange with other experiment stations within Cuba.[21] During the 1950s, however, as the postwar sugar boom came to an end, the station increasingly sought to support efforts to diversify Cuban agricultural production and improve rural conditions. Working in collaboration with the Cuban government, the Rockefeller Foundation, and various U.S. corporations, projects focusing on reforestation, nutrition, and developing plants for arid and poor soil conditions were explicitly couched as serving the Cuban national interest.[22] They were also aligned with those of U.S. policy makers who were becoming concerned that Cuba's economic volatility would lead to political instability.

Clement increasingly called attention to this public role as political conditions in Cuba deteriorated following the failed elections of 1954. The station's endowment, he emphasized, provided "for more than the study of tropical botany"; its mission mandated the advancement of "the general welfare of Cuba."[23] If Soledad was to be no ivory tower organization, neither could it afford to be seen as a Yankee enclave. The station had accrued a reputation as a tropical playground for U.S. naturalists—an atmosphere captured by Harvard cartoonist D. A. Mackay's 1941 depiction of the station, replete with palm trees, tennis courts, and stereotyped Cuban peasants.[24] It had never been as insulated as BCI; agronomists from a variety of Latin American countries visited, and the station recognized several official Cuban scientific "Collaborators." Despite the significant power over the Cuban economy exercised by the United States, the station was subject to Cuban laws, including a mandate to hire a certain percentage of Cuban employees. In many ways, the station's operation depended on Cuban goodwill; it could not afford to be seen as a private resort for North American scientists. In 1950, then, Clement decided to open the station's garden to public visitation, despite "all the headaches of left-over picnic lunches and vandalism."[25] A Spanish edition of the *Guide to the Atkins Garden* was prepared by minister of agriculture and station collaborator Julián Acuña Galé and paid for by the Claflin family, heirs of the Atkins estate. The station played host to a growing number of local tourists and groups of Cuban secondary school students.[26] Especially as criticism of the Batista regime grew and the island's politics became more unpredictable, Clement kept aloof from party factionalism, highlighting the station's role in serving the Cuban people as a neutral scientific organization.

As the Cuban Revolution unfolded, Clement remained optimistic. The station had survived difficult times before—economic depressions, world wars, the Revolution of 1933, and more than one hurricane. Clement continued to improve the facilities. Research proceeded, although the number of visitors had dropped sharply by 1957. Biology 215 was held almost as usual through June and July 1958. Field trips focused on western Cuba, however,

Cartoon depicting Harvard's Atkins Institution at Soledad. U.S. scientists appear as recognizable caricatures—Thomas Barbour pickles a lizard, Marston Bates catches butterflies, Oakes Ames holds an orchid as a vulture carries him off. Cubans, in contrast, appear with their faces obscured by wide-brimmed hats. The pioneering Puerto Rican mycologist Margarita Silva-Hutner, the sole other Latin American and only woman portrayed, appears dressed more in evening attire than in field gear and is trailed by male researchers wielding suggestive (if botanically correct) mushrooms. D. A. Mackay, 1941. © President and Fellows of Harvard College. Arnold Arboretum Archives.

away from the insurgency, and were accompanied by an escort arranged by the minister of agriculture. During the final months of fighting, however, the station found itself on the front lines. As Castro's forces made repeated advances from the east, the garden fell behind rebel lines each night; after a time, it remained so during the day. The rebels commandeered some of the station's equipment, but this, Clement reported, "was returned promptly in good condition."[27] The events of these months echoed those of the 1890s, when Cuban insurgents and Spanish forces had swept across the Soledad plantation and Edwin F. Atkins had diplomatically appeased both sides with provisions.[28] In what must have been a surreal atmosphere, the Clements kept to their work. "The shooting war passed through the Garden area fairly rapidly," Director Clement reported stoically, only "a few days' work was lost at the end of 1958 when local travel was made impossible by ground and air action in our area."[29]

When it was over, the garden and laboratory remained in "splendid condition," suffering "almost no damage during the revolution." Harvard authorities attributed this "in part to the high regard which Cubans, irrespective of their political views, have for it."[30] Despite the station's close relationships with U.S. corporations, including the notorious United Fruit Company, Cuban officials accepted this claim of scientific neutrality—or at least they recognized the station's value for their own programs of reform. Clement served as "an unofficial consultant" to the new government, responding to requests for planting stock and technical advice on reforestation, forest education, and agricultural research programs. He even began collaborating in plans to establish an ecological field station in the diverse wetlands of the Ciénaga de Zapata to the west of Soledad.[31] Foreign researchers continued to use the station, and the number of recreational visitors actually increased substantially during this period as the government worked to boost tourism.[32]

Clement could do little in the face of the rapid deterioration of U.S.-Cuba relations during the summer and into the fall of 1960, however.[33] Quietly, he began discussions with the nearby Santa Clara University about taking over the station if Harvard was forced to leave—a possibility that left hope for maintaining international scientific ties.[34] The larger Soledad estate was ultimately expropriated, and the station automatically came under the authority of Cuba's Ministry of Industry.[35] The government made no move to interfere with the station's activities, but control over its operating funds became increasingly uncertain.[36] Clement finally made the decision to abandon the station, making provisions for three Cuban staff members who had been vocal in their opposition to the new regime. Gardener and domestic servant Manuel

López repatriated to Spain, and Esperanza Vega joined Clement in Massachusetts.[37] Felipe Gonzalez transferred to the Fairchild Tropical Garden in Florida, becoming head of the garden's nursery.[38] On 31 August 1961, Harvard suspended its financial support of the station, but Clement held out "hope that it will not be too long before it is once again part of Harvard."[39]

Tropical Problems

In the meantime, Harvard was left searching for a place to hold its tropical botany course. It was not alone. Other U.S. universities had also become interested in establishing their own tropical field courses by 1960. Many hoped to take advantage of the interest of the National Science Foundation (NSF) in supporting education, an interest fueled by the recent *Sputnik* launch. The most ambitious of these was the University of Michigan's proposal to establish CenTrop, a center for tropical research and teaching in southern Mexico, where a variety of Michigan faculty were already actively pursuing field studies. With Soledad out of commission, this "instructional aspect" would distinguish "it from all existing tropical facilities such as Barro Colorado," which maintained facilities exclusively for research, not teaching. NSF responded with interest, but Michigan faced unexpected obstacles.[40]

Whereas Harvard's station was pushed toward increasing cooperation in Cuba by changing political circumstances, the Michigan group, headed by entomologist Theodore H. Hubbell, attempted from its formation in 1957 to build on a foundation of international cooperation.[41] Hubbell and turtle specialist Norman Hartweg had taken a series of preliminary trips through southern Mexico, not only to scout for an ideal locale but also to meet with the distinguished biologists and conservationists Faustino Miranda of the Universidad Nacional Autónoma de México (UNAM), Enrique Beltrán of the Instituto Mexicano de Recurses Naturales, Efraím Hernández Xolocotzi of the Escuela Nacional de Agricultura, and representatives of several other Mexican scientific organizations and government agencies. U.S. researchers could not afford to sidestep Mexico's well-developed scientific community.[42] As early as 1958, they felt they had an "enthusiastic welcome" and assurances from UNAM of help "to secure the necessary lands" for the establishment of a station and surrounding natural park lands in an area near Tuxtla Gutierrez in Chiapas.[43] Discussion dragged on, however. Privately, rumors had it that the sensitivity of U.S.-Mexican relations in the wake of the Cuban Revolution had made some of the Mexican supporters wary. A high-ranking NSF official reportedly heard that they had become "afraid of the reactions of the Castroists in their own government toward cooperation with the U.S."[44] No firm agreements were forthcoming. Cold War politics had thrown both Michigan's new proposal and the well-established Soledad station into limbo.[45]

U.S. demand for access to tropical sites for teaching and research only increased, however. In 1960, the NSF identified tropical biology as a "critical area" for support.[46] It had received not only Michigan's CenTrop proposal but also one from William Robbins, trustee of the Fairchild Tropical Garden and emeritus director of the New York Botanical Garden. Robbins sought support for a conference on tropical botany. His primary goal was to raise the profile of the Fairchild garden as a research institution. He saw an opportunity to promote it as a site where tropical plants could be studied (as the garden's bulletin subsequently described it) "on free soil," safe from the "wave of Castro-like revolution" threatening botanical institutions abroad.[47] Aware of wider interest in the tropics throughout the U.S. scientific community, however, NSF officers David Keck (a friend and associate of Theodosius Dobzhansky) and John Wilson supported the conference on the condition that it be broadened to address the overall status of research and teaching in tropical botany, emphasizing its importance to national interests.[48]

They organized what NSF announced as the "Conference on Tropical Botanical Problems of Concern to the United States," held 5–7 May 1960 at the Fairchild Tropical Garden.[49] It brought together twenty-five scientists representing a variety of universities, botanical gardens, and agencies with longstanding interests in tropical plant life, notably the Smithsonian, New York Botanical Garden, Missouri Botanical Garden, and USDA. The group included two representatives each from Michigan, then still holding out hope for CenTrop, and the University of California, which also wanted to develop tropical courses. Perspectives on Harvard's Soledad station were well represented not only by Clement but also by two former instructors of Biology 215, Richard Howard and Walter Hodge. Besides Clement, then quietly preparing himself for an exile from Cuba of unknown duration, the only other scientists representing institutions actually based in the tropics were Frank Wadsworth of the U.S. Forest Service in Puerto Rico; Leslie Holdridge of the Inter-American Institute of Agricultural Sciences at Turrialba, Costa Rica; and, suggesting the continued relative strength of Anglo-American scientific relations, J. W. Purseglove of the Imperial College of Tropical Agriculture, Trinidad. No Latin American representatives were present.

The meeting was a major step toward cementing the institutionalization of tropical biology within the United States. It became the first of a series of NSF-funded conferences, and seeing potential for coordination within the broader U.S. scientific community, the agency requested that subsequent meetings include zoologists in order to encompass the full scope of biology in the tropics. These meetings would lead to the formation of ATB and OTS, cementing tropical biology as a distinct professional domain within biology. Of course, as early as the 1920s, Thomas Barbour had advocated for "tropical biology" as a central and unifying field, but it had remained relatively peripheral on the overall U.S. biological landscape, spread among the fiefdoms of various research institutions and their handful of stations. The 1960s saw for the first time the emergence of professional societies, formal educational programs, and specialized journals for tropical biology.

Certainly, the growing clamor of interest in questions about the global distribution of species diversity played a major role. The year 1960 alone saw several key publications on the evolution, ecology, and new ways of measuring species diversity, all of which made it increasingly clear that this fundamental biological phenomenon could not be understood without deeper attention to tropical data.[50] The problem of species diversity, broad and multifaceted though it may be, at the same time began to emerge as a unifying research focus for the nascent community of tropical biologists. Marston Bates's popular *The Forest and the Sea* also brought some of these ideas to a wider public audience during this period. Yet, intellectual excitement was not quite enough to catapult tropical biology to this new status. Newly abundant federal funding through NSF was crucial both to launching new institutions and to bringing biologists from a variety of different, often competing, institutions together (not to mention NSF's push to bridge the persistent gap between botanists and zoologists). It was the sudden uncertainty concerning the fate of the Soledad station and U.S. scientists' future access to tropical field sites, however, that provided the catalyst. Biologists captured NSF's attention by arguing for their research as critical to U.S. national interests—interests now under threat in Cuba, the Caribbean, and perhaps the entire tropical world.

The 1960 Fairchild conference, then, was significant not only for its institutionalizing role but also for the discourse of "tropical problems" that its participants articulated. By tropical problems the conference participants first and foremost meant issues for fundamental research—problems, or questions, of broad scientific interest that could be answered only through research in the tropics.[51] The language of "problems" concurred with NSF's delineation of "critical areas" and their concern with identifying research programs or domains of inquiry that were underserved but essential to national interests. At the same time, the discussion of tropical "problems" (the word is used twenty-one times in the conference's brief sixteen-page report) undeniably linked these intellectual concerns to the broader economic, social, and political dilemmas of the region and its relations with the United States. The tropics, with "its vast area, the richness of its flora and fauna, its potential productiveness," the report argued, "[and] the extent of its unsolved problems make it of direct and serious interest to the United States. Economically, it is a present and potential source of raw materials; scientifically, it presents problems absent or inadequately represented in the temperate or arctic zones; politically, it is the portion of the earth where ferment and change are at present most extreme."[52] The conference participants drew on an established discourse of tropicality, referencing the rich untapped natural potential of the region, as well as its dangers and unruliness—human and natural—to be overcome. The "definite differences" of tropical plant life from that of the temperate zones raised attractive and compelling research questions for plant ecology, physiology, and biochemistry: What "laws control the occurrence and distribution of plants in a tropical rain forest"? Why do some species display periodicity and seasonality and others do not? Why are chemical substances like alkaloids, rubber, and a variety of toxins so much more prevalent in tropical than in temperate species?[53]

The Fairchild conference attendees insisted not only that these were key theoretical questions but also that a better fundamental understanding of tropical plant life would necessarily pay dividends in "applied fields such as agriculture, horticulture, or forestry."[54] "An attack on the basic problems," they argued, "properly precedes the solution of applied or short-term problems."[55] Given the vast, unknown potential of tropical plant life, basic research could not be shortchanged in pursuit of quick economic fixes. In insisting on the value of basic investigations of tropical life both for its own sake and as a foundation for more sound applications and resource use, the group's recommendations also made use of arguments long in circulation. The claim that basic biological research could unlock the vast potential of the tropics echoed arguments made decades earlier by individuals like Thomas Barbour and William Morton Wheeler. Significantly, however, whereas earlier commenters would have referenced the luxuriance, variety, or "maximum" forces of life present in the tropics, the Fairchild conference report drew on current research trends to specifically focus on the uncatalogued "species richness" or "species diversity" of the tropics. This was a subtle but important shift, forming the core of later arguments for the conservation of "biodiversity."

This new argument for support of basic tropical research hinged not upon some mysterious or exotic quality of tropical environments but on the seemingly hard fact of sheer species numbers. If the tropics had more species, then it must also have a vast number of unexploited species with economic potential.

This potential must first be identified and unlocked through basic research. The report noted that the "probable number of new species of seed plants still to be discovered and described approaches 150,000; some authorities consider this an underestimate. But one may be sure that the great part of these unknown plants, whatever their number, will be discovered in tropical regions."[56] Of these species awaiting discovery, who could say which might be the source of new medicines or industrial materials in the future? For the botanists present, most of whom were systematists, this signaled in particular a great need for basic taxonomic inventories, and the report strongly advocated support for local floras and monographic works. At the same time, it justified the group's recommendation for the establishment and support of "permanent research stations" adjacent to "relatively large" natural areas, since "it is necessary to protect as many species and habitats as possible."[57] This conservation angle likely derived from the presence of the foresters Wadsworth and Holdridge and the ecologist Stanley A. Cain, founder of the University of Michigan's Department of Conservation and future assistant secretary of the interior for fish, wildlife, and parks. "For the sake of basic scientific knowledge," they argued, and "from the most practical viewpoint, it is essential to avoid the depletion of our present pool of germ plasm, which can serve as an important source of information and can provide possibilities for new resources and new combinations of plant materials."[58]

Although not consistently employing "diversity"—they also interchangeably referred to numbers, richness, and the idea of a gene pool—the Fairchild conference attendees thus had built the core structure of the argument that would later be adopted by the promoters of the biodiversity paradigm. That is, they argued that species diversity, and the environmental and genetic diversity it entailed, was an unexploited resource that only basic research could unlock—research that depended on researchers' access to preserved natural areas. In contrast to the global ambitions of conservation biologists at the end of the century, here U.S. regional interests remained the focus and central justification. A need for the conservation of tropical environments was recognized as "urgent" but framed as necessary primarily in support of basic research and economic development.[59] Only biologists trained in the tropics could solve these tropical problems.

An "Obligation to Cooperate"

Where would these tropical biologists come from? The move to tie the conservation of species diversity to economic development and basic research was innovative, but the central assumption remained that U.S. biologists would be the ones primarily completing the research. The report drew on a Cold War rhetoric of globalism and international cooperation—a peaceful and democratic "future of 'one world'" depended on having "more knowledge of the tropics" to solve tropical problems.[60] This was, however, a globalism with the United States at the helm, and like U.S. botanists at the turn of the twentieth century, the authors of the Fairchild report still envisioned the Western Hemisphere's tropics as their own special domain. U.S. scientists had an "obligation to cooperate" to develop scientific knowledge of the tropics, "especially of the American tropics."[61] The report recognized that the scale and complexity of problems in the tropics "demand[ed] maximum cooperation between men of all nations concerned" and commended the "many bright areas of competence and good facilities" for biology in Brazil, Mexico, and a few other Latin American countries.[62] With not a single representative from these countries among the thirty-six meeting participants, however, the vision for tropical countries' role in "cooperating" seems to have been limited to providing access to land, plants, and facilities. This sense was reinforced by the report's assertion that tropical biology was underdeveloped not only because of its inherent complexity but also because of a less "scientifically oriented standard of culture" present "in tropical and sub-tropical portions of the human population."[63] The Fairchild report foresaw that "the chief burden of investigation and the chief responsibility must eventually lie with the scientists of the countries in which the plants occur." Suggesting that tropical countries were not yet ready to take up this burden, the report argued, "it must be admitted that facilities and trained personnel are not as successfully developed in tropical American countries as they are in the United States." It remained in the meantime, then, a U.S. duty to address urgent tropical problems. The report's authors envisioned fulfilling this duty by broadening the experience of U.S. biologists into the tropics rather than broadening the community of biologists to include researchers from the tropics.

During the conference, Harvard's and Michigan's representatives seem to have downplayed their difficulties in establishing and maintaining sites for tropical training—indeed, the Fairchild report still listed Soledad as a viable station. Following the Bay of Pigs debacle in 1961, however, the chances of returning soon to Soledad appeared increasingly slim. Some in the U.S. biological community certainly responded by shying away from further involvement in the building of biological stations or institutes in tropical countries. Raymond Fosberg, who had represented the U.S. Geological Survey at the Fairchild meeting, suggested that "building up an institute, then losing it to a

Castro, would be foolish" (he also argued against focusing research in any single locality on the grounds that research questions should guide site selection).[64] "The history of such institutions is mostly against the idea, especially the recent history," he told Keck following the conference. Instead he argued for "a strong home institute located in the U.S." that could "operate by means of 'tentacles', temporary establishments, arrangements with tropical institutes, expeditions, etc . . . [with] some center to which the workers can return when they have completed their assignments or when there has been a political upheaval."[65] William H. Hatheway, who had worked on maize as a graduate student at Soledad (the same year as Wilson's visit) before joining the Rockefeller Foundation in Mexico, told Fosberg of rumors that the ornithologist C. G. Sibley had similarly argued against investment in stations in Latin America, contending that "all that was needed was a chain of motels, perhaps along the Pan-American highway." Such an attitude was not looked on favorably by Hatheway or his colleague Efraím Hernández Xolocotzi, the influential ethnobotanist who had also welcomed Michigan's CenTrop plan. They considered the Rockefeller agricultural programs "useful as a sort of model" for an institute for basic tropical research, which should collaborate with existing Latin American graduate schools and economic botany institutions. Under Sibley's "motel" approach, they argued, "gringo botanists would be able to make their collections in comfort," but any pretense of "develop[ing] local talent" would be abandoned.[66]

Most biologists agreed that permanent stations of some sort were in fact necessary. Many types of work, especially ecological research, were not best served by expeditions or temporary field sites. BCI's reputation still stood as a testament to this fact, but it was clear that additional sites were required to reflect a variety of tropical environments and to better accommodate a formal teaching program. At NSF, Keck believed the best way to develop the institutional basis of U.S. tropical biology was to first identify the field stations and other centers for research and teaching already in existence in the tropics. This meant broadening U.S. biologists' scope beyond the U.S.-run institutions that it had long relied upon. Hodge joined Keck in NSF's Division of Biological and Medical Sciences as a special consultant on tropical biology. Between November 1961 and April 1962 they made three tours, together assessing nearly sixty institutions—research stations, university facilities, museums, herbaria, and gardens—in Latin America and the Caribbean "where basic biological research and/or research training may be conducted and where the participation of foreign scientists is invited."[67] Their final list of potential research sites described facilities available, the accessibility of nearby natural areas, and the types of research or teaching programs already ongoing at each institution. It of course included sites maintained by U.S. organizations, such as BCI, El Verde, and several United Fruit Company and Rockefeller Foundation sites. The majority of potential institutions listed, however, were maintained directly by Latin American and Caribbean governments, universities, and other organizations. (Among these, notably, was an institution at Tuxtla Gutierrez, which Mexicans had gone forward to develop without the University of Michigan.)[68] The central goal of the Hodge and Keck report was to obtain access to tropical environments for use by U.S. researchers. Soledad and all other institutions in Cuba were omitted without comment, well out of reach to U.S. visitors by the middle of 1962. In positioning U.S. researchers as "foreign scientists" seeking local invitation, however, its tone and attitude differed appreciably from that of past tropical station builders. In this case the cultivation of international relationships, rather than the acquisition of land, seemed to be at the heart of the enterprise.

Hodge and Keck shared their report in July 1962 at a follow-up to the Fairchild conference, held at the Imperial College of Tropical Agriculture, St. Augustine, Trinidad, which had recently merged into the University of the West Indies. Meeting just a month before Trinidad and Tobago would gain independence from the United Kingdom, the tenor of this meeting was quite different. The group took a field trip to Simla, the estate bought by an elderly Beebe in 1949 after Venezuela's 1948 coup had made the Rancho Grande station impossible to maintain.[69] Beebe died just a month before the conference. If Trinidad itself appeared as a model of "orderly decolonization," a visit to the station where the legendary naturalist had finally found his "jungle peace" also underlined the possibilities for continuity in international science through the region's political transitions.[70]

Significantly, this time sixteen of the thirty-six meeting participants represented institutions based in the Caribbean or Latin America.[71] Of these, half were based in the West Indies (Trinidad and Jamaica) but born in Britain, the United States, or, in one case, Egypt.[72] The Latin American institutions, however, were with only one exception represented by nationals of their countries of residence—Venezuela, Colombia, Peru, Mexico, and Brazil.[73] Although a minority, these participants had a large role in shaping the outcome of the meeting. As Albert C. Smith, director of the Smithsonian National Museum of Natural History, explained to the secretary of the Smithsonian Institution, "It soon became evident that the delegates, particularly those from Latin America, wished to form a strictly international organization. They would not have been content with the type of U.S.-based Institute of Tropical Biology" advocated by some following the Fairchild meeting.[74] On the whole, the Latin American participants—among them Efraím Hernández-X and João Murça Pires (of Brazil's Instituto Agronômico do Norte; see chapter 4)—already had strong connections with the U.S. biological community. They were most interested in broadening and strengthening lines of communication through the dissemination of publications, international exchanges, and fellowships. As a result, the group quickly agreed to come together to found an international society: the Association for Tropical Biology (ATB).[75]

Exceptional Diversity?

In tropical biology education, however, developments toward a new international project were already under way before Hodge and Keck even boarded their first plane. Clement badly needed a replacement site for Biology 215.[76] At the same time, University of Southern California herpetologist Jay Savage proposed a field course intended to introduce U.S. university faculty to the tropics.[77] With the initiation of President John F. Kennedy's Alliance for Progress, "exchange" became a key cultural bulwark against the hemispheric spread of communism and found increasing federal support.[78] In the summer of 1961, both Harvard and the University of Southern California held field courses in Costa Rica—with the latter's Fundamentals of Tropical Biology course funded by NSF.[79] Building on the course's success, NSF agreed to sponsor a conference the following April 1962 titled "Problems in Education and Research in Tropical Biology" in San José, on the campus of the Universidad de Costa Rica. At that meeting representatives of the Universidad de Costa Rica, University of Southern California, Harvard, and several other U.S. universities hatched plans for a joint venture in tropical biological education.

Given the venue and participants, it is not surprising that those present also took a forceful stance supporting U.S.–Latin American cooperation. Clement in particular argued from his experience in Cuba that "any proposed center or institute for education and research in tropical biology must be a cooperative project." There was broad agreement that "the only satisfactory way in which current needs in education and research can be met is through a coordinated cooperative program that includes North American and Latin American institutions and recognizes Latin American higher educational institutions as full and equal partners in any organized program." Savage reported, moreover, that "the sense of the session was that Costa Rica was the only logical place to begin the extraordinary effort required to meet the challenges of tropical biology."[80] Nine months and much correspondence later, the Universidad de Costa Rica, Harvard, and the Universities of Michigan, Southern California, Florida, Miami, and Washington met in Coral Gables at the end of January 1963.[81] There they formally incorporated the Organization for Tropical Studies (OTS). With the transfer of Savage's NSF grant for Fundamentals of Tropical Biology, OTS begin teaching tropical biology courses in Costa Rica in the summer of 1964.[82]

Why did U.S. biologists turn so suddenly toward Costa Rica? From the earliest discussions in San José and Coral Gables to the later promotional material of OTS, the organization's supporters argued that several factors set Costa Rica apart. First, it was "a small but extremely diverse country."[83] If diversity was a key attribute of tropical life, any new educational institution must have access to that diversity, and Costa Rica's manifested in an overall high number of species, as well as in "a large variety of ecological situations in a comparatively small area, from sea level forest to high altitude páramo."[84] The country's small size and variability, moreover, made its "extremely diverse environments easily accessible for class study."[85] OTS's founders never envisioned establishing a single station but, rather, an institute based near the Universidad de Costa Rica with connections to a cluster of field sites (with hopes of eventually acquiring them as formal stations) that would enable comparison of a variety of tropical communities. Costa Rica's topography and position on the isthmus made a variety of conditions accessible locally. Like earlier station builders, the founders of OTS were also very much concerned with ease of access to scientists visiting from abroad, but this concern manifested through a new inter-American rhetoric. Costa Rica was "centrally located within the Americas; it is readily accessible by ship and airplane from the south, by these means and by automobile from the north."[86] If earlier U.S. station builders envisioned the "American tropics" as on their doorstep or in their backyard, this new discourse framed Costa Rica's location as central, linking cooperating American continents. OTS's logo likewise expressed the idea of hemispheric unity visually, with a bilingual motto.

The founders of OTS explicitly emphasized the country's accessibility and environmental diversity in arguing that it was the "ideal location for a basic educational center in tropical biology." The geographic arguments for "ideal" station localities, however, have always been highly flexible. Since the turn of the century, every proposed field station had touted the availability of facilities, its accessibility, natural setting, and—in the tropics especially—the diversity of species and environments available. After all, the description of Costa Rica as small, diverse, "and lying between the two American continents" would seem to apply equally to Panama, already home to BCI.[87] Indeed, before deciding

Bilingual logo of the Organization for Tropical Studies, 1968.
Smithsonian Institution Archives, box 6, Acc91-178.

to abandon CenTrop and join in the formation of OTS in 1963, Hubbell's Michigan group had argued that both BCI and Costa Rica were too distant and expensive to access compared to Mexico.[88] Hodge and Keck had certainly reported favorably on Costa Rican institutions but provided no ranking or suggestion that Costa Rica was far and away the best such location.

OTS's founders and early promoters emphasized that Costa Rica was "chosen primarily for its suitability for work in biology."[89] Nevertheless, it was not only its natural advantages but also its exceptional "cultural, educational and political attributes" that they had considered. "Costa Rica is a model of democracy in Latin America," wrote Hartweg, summarizing the discussion that took place during the 1963 Coral Gables meeting; "its people are peace-loving, cultured and have an unusually enlightened attitude in regard to conservation of natural resources."[90] The country was "stable in politics and economics"—presumably unlike the rest of the region. In emphasizing "the unique character of Costa Rica," OTS's founders had tapped into a discourse of Costa Rican exceptionalism long promoted by the country's elites and U.S. commenters alike.[91]

The idea that Costa Rica was the "Switzerland of Central America" or perhaps "a slice of Iowa misplaced" on the isthmus had its roots in the writings of nineteenth- and early twentieth-century nationalists.[92] (The idea that Costa Ricans were especially conservation minded, a "green republic," was the only relatively new item on this slate of national characteristics—and one that the development of OTS itself would help to shape.)[93] Costa Ricans' national mythos traced their egalitarian, republican spirit to their roots as a society of yeoman farmers of European ancestry—an origin story that contrasted Costa Rica with the hierarchical, slavery-based and racially mixed societies of its neighbors. Historians of Costa Rica have more recently overturned this "white legend," revealing economic inequalities in common with neighboring countries and far more demographic and internal regional diversity than acknowledged by past scholarship.[94] Nevertheless, Costa Rica's exceptionalist narrative was played to great effect during the presidencies of José "Don Pepe" Figueres during the decade following the country's 1948 disputed elections and civil war. Figueres's abolition of Costa Rica's army, close personal ties to U.S. elites, and above all, anticommunist reputation bolstered Costa Rica's image as a beacon of democracy and stability during the Cold War. Exceptionalism became a self-fulfilling prophesy, as Figueres's adept positioning of Costa Rica as a key U.S. ally enabled the country to enact social and economic reforms without facing the interventions suffered by neighboring countries.[95] During this period, negotiations with United Fruit resulted in new funds for infrastructure improvements. Spending on education likewise increased.[96] In 1955, the Universidad de Costa Rica was reformed and reorganized on a U.S. research university model.[97] According to Hodge and Keck's report, it was "the most up-and-coming University in Central America."[98]

OTS's emergence in Costa Rica, then, was less a product of geography or any inherent qualities of the Costa Rican national psyche than of local developments within the country's scientific community. An essential role was played by the existing international connections, ambitions, and interest in expanding the Universidad de Costa Rica's biological program held by the university's professors and administrators—particularly Rafael Lucas Rodriguez Caballero (director of the School of Biology), John deAbate (vice-dean of Science and Letters), and Jose Joaquin Trejos (dean of science and letters and future president of Costa Rica). Jay Savage, who first proposed the Fundamentals course, did so after coming into contact with deAbate and Rodriguez during an NSF-sponsored project that he had been a part of since 1959.[99]

Science in Costa Rica had long thrived on transnational connections.[100] Both deAbate and Rodriguez already had strong relationships with the U.S. scientific community: deAbate was a graduate of Tulane; Rodriguez completed his PhD at the University of California, Berkeley, before returning to his native Costa Rica. Rodriguez had also consulted with the University of Florida's Archie Carr, a pioneering herpetologist, conservationist, and mentee of Thomas Barbour and Theodore Hubbell who had recently set up a field camp at Tortuguero to study sea turtle migration and nesting. They worked together closely during 1956–57 to establish a School of Biology as part of the reform of the Universidad de Costa Rica.[101] Indeed, even before the biology department's establishment, Rodriguez had been involved in founding the university's *Revista de Biología Tropical* in 1953. Begun sixteen years before ATB would establish *Biotropica,* it was the first journal for "tropical biology"—a subject its founders justified by its potential to "merit consideration in major scientific centers. . . . Our Country being rich in study material for research on the biology of tropical regions, this is the field in which our contribution to Universal Science may have some significance."[102] The journal published on an eclectic mix of topics, primarily medical; it functioned to publicize the Universidad de Costa Rica as a research institution, to ensure that the university's research was accessible locally, and to enrich its library's collection through international exchange. Few U.S. scientists (aside from Carr) published in it until the 1980s.[103] Nevertheless, its founders, like the Universidad de Costa Rica professors involved with OTS, framed Costa Rica's tropicality and biological richness as a natural patrimony that could leverage increased involvement in international science.[104]

At the same time, as discussed in chapter 4, an informal network of landholding U.S. and European scientists, including Alexander Skutch, Charles "Don Carlos" Lankester, and Leslie Holdridge, already anchored the itineraries of many visiting U.S. field scientists in Costa Rica. Of these, Holdridge was an especially powerful advocate for U.S. research in Costa Rica. It was Holdridge, present at all of the meetings leading to the formation of OTS, who convinced Clement to shift Biology 215 to Costa Rica in 1961 and who secured key support for OTS from James Bethel, program director for special projects in science education at NSF and forester at the University of Washington.[105] At Turrialba, Holdridge and his colleagues had extensive experience consulting with the Costa Rican government and international corporations. On his own, Holdridge had been hosting scientific visitors, such as Robert MacArthur, at his La Selva estate since the mid-1950s. After collaborating on the instruction of the Harvard course, Holdridge, along with Joseph Tosi and Robert Hunter, established their own nonprofit consulting firm, the Tropical Science Center, to facilitate similar projects.[106] He clearly saw the establishment of OTS in Costa Rica as an opportunity to expand this enterprise and set about inserting the Tropical Science Center as a key mediator between the U.S. members of the organization and Costa Rican institutions.

Although framed as a project of bilateral cooperation between the U.S. and Costa Rica, OTS involved a patchwork of individual and institutional interests in its establishment. The formation of the consortium helped advance Universidad de Costa Rica professors' interests in developing their university's biology program and advancing its international reputation. OTS participation, however, remained greatly skewed toward U.S. actors. Indeed, even the majority of those representing Costa Rican institutions at its foundational meetings were U.S. transplants: Holdridge and his associates at Turrialba. The U.S. participants did not necessarily share all of the same goals, however. Notably, the New York Botanical Garden and Associated Colleges of the Midwest ultimately declined to join OTS despite involvement in the early conferences. Perhaps the involvement of the Universidad de Costa Rica is a less surprising achievement in cooperation than the fact that OTS brought six (and soon more) competing U.S. universities into a lasting partnership. The Universidad de Costa Rica's involvement was nevertheless the keystone to OTS's success. The U.S. wing of the organization depended on the university not only for access to land and facilities but also for the legitimacy of Latin American partnership. "While the initial impetus for the OTS comes largely from North American universities, the University of Costa Rica is a full partner in the venture," the organization's formative documents promised; "it is the intent of the group to involve Latin-American individuals and institutions to the fullest extent possible and to provide the maximum of service to Latin-American and other tropical areas."[107] Assurance of equity was increasingly necessary in the latter half of the twentieth century, not only as third world countries articulated critiques of "scientific imperialism," but also as funding agencies and foundations insisted on the clear indication of benefits to host countries.[108] Many of OTS's promises to the Costa Rican scientific community would long await fulfillment, however. With the formation of OTS it nevertheless became increasingly clear that tropical biologists would have to consider both the diversity of species and the diversity of participation in their field.

Out from "under the American Flag"?

Even as OTS formalized a new era in U.S.–Costa Rican scientific relations in 1963, the future of BCI in Panama faced uncertainty. Panama's president

Roberto F. Chiari had been placing mounting pressure on the United States to renegotiate the canal treaties and the status of the increasingly unpopular Canal Zone.[109] During the lead-up to an important Central American summit, Martin Moynihan, director of BCI—now officially part of the Smithsonian's Canal Zone Biological Area (CZBA)—broached the problem with Smithsonian secretary Leonard Carmichael. "It seems very possible that the Canal Zone will eventually be returned in whole or in part, to the Republic of Panama. Even if this should not occur," Moynihan explained, "the wishes of the Panamanian government certainly will become increasingly influential in determining the policy of the Canal Zone government."[110] How would a scientific institution that for forty years had been dependent on its privileged status within this U.S.-controlled tropical territory survive these changing circumstances?

BCI also now competed with a growing number of stations, field sites, and other tropical institutions for the attention of U.S. biologists, who were increasingly aware of the need to compare a variety of tropical environments and ecological communities. The Hodge and Keck report nevertheless affirmed BCI's status as "without exception . . . still the best field station in the wet lowland neotropics."[111] While the institution had suffered some neglect in the decade after the 1946 Smithsonian takeover, it remained the gold standard for tropical stations. Tropical forests still held special fascination, particularly from the point of view of the growing interest in the ecology of species diversity. In terms of its "location . . . [and] facilities . . . where one can step from his dorm or lab immediately into undisturbed forest on established trails," BCI remained unmatched.[112]

It would also be many years until any new station could approach BCI's accumulated scientific record. This point had been made clear to the Smithsonian administration by the station's cadre of "alumni"—influential supporters who pushed for reinvestment in the institution during the early 1950s. T. C. Schneirla, widely influential for his studies of the evolution and ecology of army ant behavior, highlighted in particular the importance of "the fact that it is an island, a clearly bounded natural research laboratory where systematic, programmatic investigation of tropical life can be made and repeated at intervals in order to learn what changes occur over considerable periods of time."[113] Marston Bates had likewise explained the importance of "long-term projects" to Carmichael: "The maintenance of a laboratory for the study of tropical rain forest conditions is very important for the development of biology, since the rain forest in many respects represents 'optimal' conditions for terrestrial life. Almost all of our knowledge of this environment has been collected on an 'expedition' basis, and we need more sustained observation, and a program of continuing studies."[114] Warder C. Allee likewise voiced his concern that the Smithsonian commit to preserving the station for ongoing study. Alfred Emerson, pointing to the development of new stations in Africa, affirmed, "We obviously need a strong station in the American tropics."[115] "Hundreds of scientific papers have been published based on studies made on the island, yet, as is well known, the surface has barely been scraped in any field of biological research in the tropics," explained Eugene Eisenmann, ornithologist at the American Museum of Natural History. As a Panamanian American, he embodied the complex and intimate relations between the two countries; the station was, he declared, "the only place under the American flag" where biologists could give such intensive, prolonged study to "an essentially virgin tropical rain forest."[116]

But how long would BCI remain "under the American flag"? Panamanian dissatisfaction and periodic protests had been escalating since the buildup of U.S. troops during World War II. Egypt's nationalization of the Suez Canal in 1956 stoked Panamanian nationalism, and during this period retiring station manager James Zetek pondered BCI's past and future. "Too few know," he noted, "the Republic of Panama owns every inch of land of the Canal Zone."[117] Even on isolated BCI, tensions between the Panamanian workforce and U.S. visitors were high. Smithsonian officials heard complaints that "laborers considered the scientists . . . nuisances," interfering with the "little Panamanian community on the Island," along with reciprocal staff objections citing visitors' imperious behavior.[118] Meanwhile, pressure to reform the Canal Zone came both from the Republic of Panama and from within the United States. As the U.S. civil rights movement gained momentum, segregation in the Canal Zone increasingly became an international embarrassment. U.S. cold warriors recognized the liability it posed as they sought to deflect criticism that the Canal Zone was a U.S. colony and thus an example of hypocrisy that fed into communist rhetoric. The Gold and Silver Roll system was quietly dismantled even as the color line persisted in new forms. Carl Koford, a recent graduate of UC Berkeley chosen as Zetek's successor to run BCI/CZBA, arrived in 1956 as Canal Zone officials devised the "Latin American School" system for West Indian students to avoid school integration.[119] Seeing the same system of discrimination in effect on BCI during his first overnight stay, Koford was infuriated. He ordered the integration of the station's bathrooms, water cooler, and dining area. Panamanian workers would now get three eggs at breakfast, just like U.S. visitors.[120] Koford lasted only a year as BCI's director, amid repeated conflicts with the Smithsonian administration over such matters as his intention to buy local Panamanian-raised food rather than "sanitary" canned food through the U.S. commissary.[121] His plan to live with his family in "huts" on the island rather than in Ancón, the Canal Zone's administrative center, had resulted in outcry that he would never be able to fit in with Zonian society as Zetek had.[122] Not simply a matter of personality conflict, these disputes over the proper management of BCI took place within a context during which the future role of the United States in Panama was itself being reconfigured.

This context shaped not only food, housing, and labor at the station but also the reform of its scientific program. Arriving in 1957, Koford's replacement, Martin Moynihan, oversaw the development of the Smithsonian's tropical outpost from the single station on BCI into a full-fledged research institute, including new stations and sites on the mainland and a formal scientific staff. Moynihan had studied at Princeton, Oxford, and Harvard, counting the ethologist Niko Tinbergen and evolutionary synthesist Ernst Mayr as mentors. Whereas Zetek had complained in his final years of the "mental twists" required by modern ecology, Moynihan was eager to make BCI a center for cutting-edge research.[123] As an animal behaviorist, he not only fit well with the long-standing traditions of research at BCI but was also well placed at the burgeoning intersection of evolution and ecology. Under Moynihan's leadership BCI was rededicated as a purely scientific space. Tourist visits had become a significant source of income during the lean final years of Zetek's reign. Seeing them as a threat to both the island's natural state and the station's scientific status, Moynihan brought them to an end. Poaching, which had exploded during Panama's postwar depression, was brought under control as he hired more local guards.[124] He also took advantage of newly available NSF funding for his own animal behavior research and to modernize BCI's physical plant—most significantly including a $100,000 grant in 1965 for the installation of an electric cable, which finally provided the station with a reliable power supply.[125] From the appointment of the first graduate assistant in 1958 to the recruitment of fifteen staff scientists by the time of his resignation in 1974, Moynihan's development of a scientific staff was also key to the institution's reassertion of a central role in basic biological research in the tropics. This staff took on research projects not only on BCI but also at new marine stations on either side of the isthmus and on expeditions throughout the global tropics. The institution's expansive new scope would be signaled by the rechristening of CZBA as the Smithsonian Tropical Research Institute (STRI) in 1966.

There were sound scientific reasons for the expansion of the institution's domain. At least since Paul W. Richards's *The Tropical Rain Forest*, ecologists had been increasingly aware of and interested in the differences among rainforests and other communities in the tropics. A single site, let alone one that was an artificial island, could no longer be rationalized as a representative sample of tropical nature.[126] Whereas Zetek had jealously tried to keep researchers (and their daily fees) on BCI, Moynihan strongly encouraged his staff scientists to travel to better understand the relation of BCI's forest to tropical forests elsewhere. He sent the young mathematical ecologist Egbert Leigh around the world—to Ivory Coast, Madagascar, India, Malaya, and New Guinea—to compare the structures of forests in the global tropics.[127] If theoreticians were going to make generalizations about the tropics, it was important to know how representative data at BCI were. Moynihan also knew that the CZBA itself had to diversify, and his efforts to obtain more land in the Canal Zone reflected his knowledge. The acquisition of new stations and field sites served as "areas of different ecology" for comparative research.[128] This was especially important for the growing number of studies on geographical variations in species diversity; notably, MacArthur and his coauthors compared bird censuses on BCI with areas of differing foliage height in the Canal Zone, as well as with the islands of Las Perlas.[129] It also enabled the expansion into tropical marine studies, as exemplified by Ira Rubinoff's research on the effect of the isthmian divide on the ecology and evolution of fish diversity. New sites enabled more substantial field experimentation—increasingly the norm in ecology—than was permitted on BCI itself. Moynihan also thought that acquiring additional land would "cut out a little competition"—first Michigan's proposed CenTrop and then OTS emphasized their "ecological variety" in contrast to what was available on BCI.[130] The additional space served the institution well, enabling the exponential growth in scientific visitors—from 39 in 1960 to 556 in 1970—which in turn helped to justify the Smithsonian's attention to its farthest-flung outpost.[131]

Yet, at the same time, Moynihan's territorial expansion of CZBA was politically strategic. It had to be: he had arrived at a particularly tense time in the history of the Canal Zone. Clashes between police and protesters during the celebration of Panamanian Independence Day in 1959 spread from the vicinity of the Tivoli Hotel in Ancón across Panama's urban centers. With tensions exacerbated by the recent revolution in Cuba, the United States deployed troops. For the first time a barbed wire fence was erected along segments of the Canal Zone–Panama border. In his acquisition of new field sites, Moynihan was careful to stay firmly within the bounds of the Canal Zone.[132] The first addition was a plot of grassland and second-growth forest within the Navy Pipeline Reservation, east of the canal near Gamboa. Marine stations at the U.S. Army and Navy installations at Naos Island and Galeta Island would open by 1965. Certainly, these sites were far from pristine; locations within the republic might have been scientifically more attractive. Even while remaining within its borders, however, Moynihan in fact was positioning the institution to withstand the eventual reduction or dissolution of the Canal Zone. Moynihan admitted that the Smithsonian's status in the Canal Zone was "slightly dubious" given the most recent U.S.-Panama treaty assuring that Canal Zone activities would be limited to the actual operation of the canal. Nevertheless, he felt "sure that, if we play our cards correctly, we will have no difficulty in continuing operations in this area, no matter what happens to the Canal Zone itself. But we must start making plans for the future now."[133] Military sites, specifically, not only were protected from civilian entry and had unused facilities available because of the postwar drawdown of U.S. forces, but their long-term prospects were also not seriously in question. Even if the Canal Zone should be abandoned, Moynihan had assurances that the U.S. military would maintain these installations.[134] (The lengths to which the United States went to retain the Guantanamo Bay Naval Base in Cuba were a case in point.)

As U.S.-Panamanian relations became increasingly rocky, however, Moynihan also suggested that the Smithsonian look into the Gorgas Memorial Laboratory, which had U.S. government support but had long operated successfully within the Republic of Panama. It could serve as a legal model to ensure the persistence of CZBA, whatever happened to the Canal Zone.[135] Although taking advantage of the U.S. military presence, he also advocated getting "on the best possible terms with the Panamanian government, and attempt[ing] to secure favorable publicity." Few Panamanians were even aware of CZBA's existence. He worried that those who were "probably think that it is run by the Canal Zone government," which was "unfortunate, as the Canal Zone Government is particularly unpopular in Panama—much more so than any other U.S. government agency." He suggested that an offer of fellowships to Panamanian students in biology might be the best route for publicity, especially because university students were "among the most radical elements in Panamanian political life." He doubted whether there would be students actually interested in a fellowship opportunity or "capable of taking real advantage of it."[136] "Frankly," he explained, "the real objective of this plan was to obtain favorable publicity for the Smithsonian Institution in the Republic of Panama."[137] Funds were not immediately forthcoming, but talks were soon set up with the Organization of American States to secure fellowships for Latin American scholars in the future—a first in BCI's forty-year history.[138]

In 1964 tensions came to a head. Violence erupted following the 9 January "Flag Pole incident," a dispute over Panamanians' right to fly the flag of their republic within the Canal Zone. Begun as a clash between Panamanian high school students and Canal Zone police and residents, fighting spread into Panama City and throughout the country. At least twenty-five people died, and hundreds were injured.[139] The following week, Moynihan calmly communicated to Secretary Carmichael, "It seems quite possible that Panama and the United States will start negotiating or discussing the status of the Canal Zone again some time in the near future," asking him to "keep in touch with the State Department, to find out if Barro Colorado is mentioned."[140] Like Clement at Soledad, Moynihan downplayed the risk to the station posed by changing political circumstances. In spite of assurances that BCI was well removed from the conflict, however, "an appreciable number of scientists . . . cancel[ed] their proposed visits" in 1964.[141]

The uncertain future of the Canal Zone had already begun to shape CZBA's local territorial expansion and plans for fellowships for Panamanian or Latin American students. In subtle but significant ways, the reanimation of canal negotiations following the 1964 uprising also contributed to the sharpening of the institution's scientific mission and its reorganization as STRI. First, the change of name itself distanced the institution from the controversial Canal Zone. It signaled instead its affiliation with the Smithsonian Institution, an independent organization held in trust by the U.S. government, but not a government agency, and one enjoying a broadly positive public image. Second, the name change underlined a refocusing of the institution's core mission and scope, and a deterritorialization of sorts. Its work was, in theory, no longer to be delimited by the physical boundaries of BCI, the Canal Zone, or even Panama, nor was its destiny then to be bound to the political future of any particular plot of land. As an institute for "tropical research," its geographic territory was now the global tropics. Its intellectual domain became the full scope of "tropical problems."

Moynihan had in fact at first resisted articulating a specific research program for CZBA beyond "research on tropical biology as a whole."[142] He swam hard against the current of big, team-based projects in biology. Even as Howard T. Odum's Rain Forest Project ramped up in Puerto Rico and the International Biological Program (IBP) began in 1964, Moynihan's philosophy remained simply to attract "first-rate" researchers and allow them to pursue their own interests.[143] The eclectic group assembled at BCI gained underlying unity from the foundation of past research and from Moynihan's own interests and elite scientific connections—he hired heavily from the ranks of students of Mayr and G. Evelyn Hutchinson.[144] Not all observers recognized or were impressed by the coherence of research at the station. Some saw CZBA as an "ad hoc" organization focusing mainly on animal behavior, without "an organized coordinated basic research program."[145] A new Smithsonian secretary, S. Dillon Ripley, arrived in 1964 well prepared to raise the profile of Moynihan's operation, however. He was a veteran of tropical fieldwork himself, and in the wake of the public interest generated by Rachel Carson's 1962 *Silent Spring*, he was especially keen to showcase and intensify the Smithsonian's contributions to ecology and conservation.[146] Ripley initiated a Smithsonian Task Force on Tropical Biology.[147] It aimed not only to foster better internal integration between CZBA and other Smithsonian branches but also to put the Smithsonian in a stronger position relative to competing organizations in the tropics—notably OTS and Odum's Rain Forest Project. In particular, the relevance of the institution's mission to the aims of a variety of new potential patrons would have to be clear. There were not only the NSF, U.S. Atomic Energy Commission, and IBP to consider but also the reemergence of proposals to construct a sea-level isthmian canal—which the Johnson administration wielded as a bargaining chip as negotiations recommenced in 1965.[148] Such a massive environmental engineering project presented biologists with opportunities for "big science"–style funding.[149] Ripley and Moynihan both wanted to situate CZBA as more than an outpost in Panama. It would be an international leader in tropical science.

"Species diversity" held the key. Previously, Moynihan had listed the "ecology of species diversity" as just one of a variety of research areas being pursued by staff and station visitors.[150] When CZBA became STRI in 1966, species diversity was named as its core research program. Amid the growing frenzy of interest in the evolution and ecology of species diversity, what could be a better emphasis for the Smithsonian's "tropical research institute"? Articulating the problem of species diversity—specifically "the relationship(s) between species diversity and evolutionary 'success' "—as STRI's central research program had the advantage of changing little about the research already taking place there.[151] A frequent visitor, MacArthur was named an honorary STRI research associate, and as his and Wilson's fame rose following *The Theory of Island Biogeography*, Ripley would press Moynihan to hire a theoretical ecologist (they landed Leigh, a student of Hutchinson).[152] At the same time "species diversity" could be a broad umbrella, encompassing everything from behavior studies of animals' "diverse social systems" to taxonomic inventories.[153] Nearly any evolutionary or ecological study—dealing with competition, mutualism, predator-prey relations, niche partitioning—could be reasonably framed as relevant to the central biological questions of the production and maintenance of species diversity.

Yet, even more than uniting a vast array of seemingly disparate research projects, framing STRI as an institution devoted to the study of tropical diversity reinforced its stance as a basic science institution—an institution devoted, like the Smithsonian itself, to the "increase and diffusion of knowledge." That knowledge could be expected to be unbiased, to remain apolitical in the midst of the turmoil of U.S.-Panama relations. This did not mean that it would produce useless knowledge. STRI would not be merely a tool of the Canal Zone or U.S. government, but it could serve humanity, and locally, it could serve the people of Panama. In November 1966, STRI debuted—even as Presidents Johnson and Robles continued negotiations—with a Conference on Tropical Biology. Framed around the idea of the "development of a world program in tropical biology," it emphasized the possibilities for cooperation between the new institute, other tropical organizations including OTS and ATB, and a variety of U.S., Panamanian, and nongovernmental agencies.[154] Noting "POLITICAL PROBLEMS," however, the conference organizers took caution to make "no overt mention of the sea-level canal" and warned attendees to take "extreme care . . . to avoid becoming involved with the political aspects of the proposed canal." They explained, "We expect to make it clear at all times that the chief aim of the Conference and of later activities is to advance basic biological research. We strongly hope that where possible, Panamanians will be involved in our program and that its results will benefit the country and people of Panama while forwarding scientific knowledge." At the same time, STRI expected to provide its neutral scientific expertise where requested.

Taking a strong stance of scientific neutrality enabled STRI to benefit from the proposed sea-level canal without tying itself too closely to U.S. interests. Basic tropical science could serve all parties. The organizers insisted: "The probable construction of the canal merely adds urgency to the need for expansion of our long established research areas in that part of the world. We are glad to cooperate with mission-oriented groups already working there. . . . But it is not our intention to tailor our research plans solely to their needs. We are committed to extending basic scientific knowledge for others to apply as appropriate. We believe this approach is the best way for us to assist in any official U.S. Government plans and yet remain true to our principles."[155] Like Clement in Cuba, Moynihan and the Smithsonian administration called on the scientific neutrality of their institution in a time of diplomatic crisis. In the case of STRI, however, it was not agricultural research but basic research on tropical species diversity.

The "Most Urgent World Problem Today"

Over the coming decade, tropical biologists would make the same move on a much larger stage. The reorganization of STRI had been given immediate impetus by the 1965–67 attempted renegotiation of the canal treaties, the associated sea-level canal gambit, and the lure of funding from IBP (itself a Cold War project). The argument that basic research on tropical species diversity could have practical applications in environmental management was convincingly made in the context of the sea-level canal debate. Rubinoff's seemingly arcane work on mechanisms of isolation in Caribbean and Pacific species certainly had found sudden practical relevance, given the sea-level canal's ability to mix marine communities, changing their composition and potentially wreaking ecological havoc.[156] Although wading into politically risky territory, STRI's stance of scientific neutrality paid off. Indeed, the implicit critique of a U.S. project helped to publicly distance STRI from U.S. policy and make credible its claim of loyalty, above all, to science.

This argument worked not only to support a push for biological surveys ahead of the construction of any new isthmian canal but also in the context of the rapid development of tropical countries around the world. During the late 1960s and 1970s, policy makers and the public became increasingly open to arguments for the need to assess the environmental impact of development projects. Basic tropical research was necessary to know what was being lost in the rush to development and, perhaps, to suggest more sustainable paths. In this period, STRI and OTS would often find themselves in competition for scarce funds; nevertheless, the institutions converged in their rhetoric about the necessity of science for the conservation of tropical diversity as a key global resource. In an article for *BioScience* publicizing the resolutions of the Panama Conference on Tropical Biology, Smithsonian scientists Helmut Buechner and Raymond Fosberg argued, "The most urgent world problem today is the establishment of harmonious relationships between human societies and the environmental resources upon which they depend." As U.S. citizens became more environmentally aware, they could not remain parochial in their concerns. Tropical environments were among the world's richest; "therefore, preservation of the productivity of tropical environments is a matter of world concern."[157] Species diversity was not only a resource for tropical nations; the United States and a growing world population would depend on the use and conservation of tropical resources.[158] Explaining the role of OTS, Theodore Hubbell likewise declared an "urgent" need for financial support commensurate to the scale of tropical problems, "first, for the advancement of science, and second, to provide a basis for the intelligent use and conservation of the resources of the tropics. Research on tropical ecology is at least as important for humanity as are studies of the oceans, and overwhelmingly more so than the exploration of space." Harboring the majority of the planet's unknown species, the tropics were the real "last great frontier" for scientific knowledge and for human survival on a planet of scarce resources.[159]

The implication—which, with the emergence of conservation biology, would become increasingly explicit over the coming decade—was that the theories produced by tropical biologists would have practical value as the scientific basis for conservation policy. Synthesizing the conversation bubbling among ecologists, Raymond Dasmann published *A Different Kind of Country* in 1968. The book placed "biological diversity" at the center of a broad conservation philosophy, emphasizing management to maintain diversity in the face of the simplifying and homogenizing tendencies of the modern industrial world. Dasmann's own fieldwork centered on California, but he had extensive tropical experience in Australia, New Guinea, and Dominica, and he had taken part in the 1963 "Fundamentals of Tropical Biology" course in Costa Rica.[160] Tropical rainforests, as "the most biologically diverse environment on earth," thus played a special role in his argument.[161] Drawing on the work of the Odum brothers, MacArthur, Wilson, and others, he described current ecological understanding of the causes of this diversity—the optimal physical conditions found near the equator, the multiplication of niches, and exceptions to the latitudinal gradient governed by factors such as island size. "The significance of this diversity," he argued, however, "may be considered from many points of view." It could be seen as "a vast reservoir of genetic material" and a "biological storehouse." At the same time, it provided a "value to the human spirit" that should be recognized. "Looked at from another viewpoint," he added, "the diversity of the tropics offers clues to proper management of lands and people elsewhere. Biological diversity is associated with stability and adaptability."[162]

Through the late 1960s and 1970s these clues were explored at a slew of special conferences, from the Brookhaven Symposium on Diversity and Stability in Ecological Systems and the ATB Symposium on Insular Evolution in 1969 to the ATB Symposium on Biological Diversification in the Tropics in 1979.[163] The purported stability of diverse systems became a rationale for their preservation and for the consideration of species diversity in communities beyond the tropics. MacArthur and Wilson's theory of island biogeography also rapidly found application as an argument for the establishment of large nature reserves—islands of wilderness in a sea of development.[164] Each of these promising avenues of theory would become the focus of heated controversy over the coming decades.[165] Even the proper measures of species diversity became mired in debate.[166] Rather than fade away, however, the idea that a sound science of species diversity could guide global conservation only became stronger, developing into the core of the new "mission-driven discipline" of conservation biology.[167]

During this period STRI and OTS would become the most influential institutions for tropical field science, training a generation of future tropical biologists. Among these were key figures in the new conservation biology. While repeatedly referencing the intellectual and economic problems of the tropics, however, the leaders of both STRI and OTS carefully maintained a distinction between their own orientation toward basic science and the applications to tropical problems that each promised would ensue. "OTS is not a conservation organization," Hubbell explained. It would address "tropical problems" not through any advising capacity but, rather, "by means of a sound program of basic research and teaching."[168] Smithsonian Secretary Ripley took another tack in support of basic science to understand "the enormous diversity of organisms in the tropics." A congressman asking "why a U.S. institution should go so far afield when its own country is not yet adequately known scientifically" could certainly be answered in "practical terms," but such an answer is "not the real reason. The reason is essentially that given by the mountain climber for his efforts to scale Everest or some other difficult peak—'It is there.' In the tropics a great unconquered mountain of fascinating biological information is there." The tropics' diverse species were a global economic resource, but for biologists they were fundamentally a scientific resource. "The tropics are where the rich harvests are," Ripley proclaimed, "and science is not conscious of national boundaries."[169]

The Rich Harvests

Of course, U.S. scientists indeed had to be quite conscious of national boundaries in order to build and maintain their institutions in Panama and Costa Rica and to work elsewhere in the tropics. Taking a stance of scientific disinterestedness, STRI, OTS, and the growing community of tropical biologists attempted to walk a fine line, pursuing funding from both U.S. and international organizations and at the same time promising benefits to host countries and all of humanity.[170] The science of species diversity appeared to be pure. It was the product at once of deep, abstract theorizing and of daring field scientists, communing with the rainforest at their far-off stations. Yet from these esoteric and exotic origins had emerged compelling applications that could save the world—not just maintaining its aesthetic variety but promising a host of new drugs, foods, and materials. If, as historian Paul Lucier argues, the rhetoric of "pure" and "applied" science emerged in the nineteenth century, manifesting "an essential tension in the relations between the search for knowledge and the pursuit of profit in a capitalist society," for tropical biologists operating during this period of transition to a postcolonial world, such tension would be ratcheted up nearly to the breaking point.[171]

U.S. biologists' presence in the tropics had grown over the century largely through connections to U.S. corporations, government agencies, and the military. Until the emergence of OTS—a direct response to the breakdown of U.S.-Cuban relations—tropical biologists' most important stations were located in colonies or neocolonies that garnered special U.S. military and economic protection. The liabilities of this dependence were made abundantly clear during the period between the Cuban Revolution and the renegotiation of the Panama Canal Treaties. The crisis posed by the loss or potential loss of key stations spurred tropical biologists not only to organize themselves professionally but also to alter their discourse. Tropical biologists increasingly spoke of the need for international collaboration with host countries. They began to emphasize the role of basic research and training in tropical science, not as a matter of U.S. national interest alone, but as a world concern. Tropical biologists' expertise, they argued, was necessary to the solution of problems that could no longer be contained in the tropics. Species diversity was not only at the exciting frontier of science; it was also a global resource—an almost magical resource that, through its preservation, would produce untold riches.

EPILOGUE

Postcolonial Ecology

If biologists want a tropics in which to biologize, they are going to have to buy it with care, energy, effort, strategy, tactics, time, and cash.

—Daniel Janzen, 1986

Wilson, Raven and the other members of the cadre decided that American foreign policy, especially as directed to developing the Third World, would be their patron. Their message would be that a country's quickest route to development would be to exploit, while preserving, its "biological wealth." "Once people saw and understood biological wealth," Wilson says, "things started rolling."

—*New Scientist*, 1986

As early as the 1960s, tropical biologists had begun to argue for conservation and development based on the principles of species diversity. It was not until the 1980s, however, that this argument gained a broad public audience and a real foothold in U.S. and international policy. The idea of a global crisis of species diversity seemed to explode forth from the 1986 National Forum on BioDiversity. Over a thousand participants, including journalists from at least forty major news outlets, and an even larger audience at over a hundred broadcast sites in the United States and embassies abroad watched as a slate of eminent speakers painted a dire picture: species lost before scientists could name them, lush rainforests harboring potential cancer cures clear-cut to make way for cattle pastures.[1]

It was a calculated and self-consciously political media blitz.[2] But it had not sprung from nowhere. As *New Scientist* reporter Christopher Joyce explained, this "new branch of environmentalism, called biodiversity . . . owes its existence to this subculture of tropical biologists, who now find themselves drawn out of leafy obscurity into the glare of environmental politics."[3] To call Edward O. Wilson or Paul Ehrlich obscure was a stretch. (Both were well known for recent public controversies—Wilson for claims about the biology of human behavior in *Sociobiology* and Ehrlich for *The Population Bomb*.)[4] Although overstating their political naïveté, Joyce was correct in tracing the origins of *biodiversity* to the "subculture of tropical biologists." Wilson, Ehrlich, Peter Raven, Thomas Lovejoy, and the other organizers of the conference had not chosen the tropical rainforest as the literal poster child for biodiversity simply because it provided a compelling alternative to the panda, whales, or other charismatic endangered species. They in fact had decades of close, personal experience with these environments through their fieldwork. Biodiversity, and its accelerating loss, was not an idea cooked up in Washington. It was rooted in biologists' tropical stations.

Diversity Found and Lost

Traveling naturalists had long experienced the variety and exuberance of tropical life, but by establishing stations for prolonged field research, tropical biologists were able to pursue practices that laid the foundation for the modern conception of species diversity and its offshoot biodiversity. Beginning at the turn of the twentieth century, stations allowed such researchers as Forrest Shreve at Cinchona, Jamaica, to observe and experiment with living tropical organisms both in a laboratory setting and outdoors within their ecological home. This work began to reveal diversities of adaptation from the level of the organism up through the ecological community. Field stations enabled experimental techniques, but at least as important were an array of new, complementary place-based practices, including behavior and life history studies, repeat population censuses, and fine-scale taxonomic inventories. Harvard's station at Soledad, Cuba, provided botanists with a living library of tropical plant species and at the same time gave Thomas Barbour a base from which he could piece together the history of Caribbean zoogeography. William Beebe favored immersion in tropical life over collecting specimens, yet his square mile of Guianese jungle still promised an inexhaustible supply of new species. In Panama, too, James Zetek's card catalogue of species at Barro Colorado Island (BCI) swelled as visitors poured into the station to study its wild tropical life. Even so, in the mid-1960s experienced botanists still struggled to identify all of BCI's diverse plants. Beginners could not approach BCI's plant species with confidence until the 1970s, when systematist Thomas Croat, with the help of the ecologist Robin Foster, revised and completed the work begun by Paul Standley and Leslie Kenoyer half a century earlier.[5] This gradual shift in the scale of taxonomic study was particularly important for understanding tropical forests, where the numbers of species, and especially of endemic species, can be extraordinarily high. At field stations in the temperate United States and Europe, biologists could take static species lists for granted. At BCI, among "the world's best-known tracts of tropical forest," new species still remain to be discovered.[6]

Biologists' expectations of diversity in tropical nature thus were often confirmed and accentuated by their experiences at stations. More than simply helping biologists to enumerate tropical species, however, stations also forced researchers to come to terms with the complexity they encountered. Expecting an ancient and changeless landscape, they began to uncover surprising dynamism within and heterogeneity among tropical communities. Shreve and Duncan Johnson questioned the nature of the rainforest climax community in Jamaica's Blue Mountains, for example, and Robert Enders's BCI mammal censuses revealed curious changes on the artificial island over time. Expansion into new field sites enabled more careful comparisons across ecological communities in the tropics by midcentury. While some researchers sought ever deeper natural historical knowledge, others developed methods of strategic simplification. In field plots and on paper, rainforests became increasingly manageable through the convenient reductions of profile cross sections, ecosystem diagrams, and especially indices of species diversity.

In combination, these new place-based practices ultimately helped biologists not only to appreciate diversity but also to characterize its loss. During the 1970s, BCI research took on new meaning in light of Robert H. MacArthur and Edward O. Wilson's theory of island biogeography. As early as 1929, Frank Chapman had predicted that BCI's "circumscribed conditions" could help reveal "those causes that determine the comparative numbers of species."[7] Encountering unexpected species richness, early researchers had assumed either that animals had been concentrated there or that its community was representative of mainland forests. As comparative data accumulated, it became clear that the island was actually somewhat less rich in species than the mainland. It had even lost species during the station's lifetime: gone were not only the largest predators but also about forty species of birds. While some of this loss could be attributed to poaching or habitat change as cleared areas reforested, some of the disappearances could not. Based on a ten-year census and analysis of past observations by Chapman, Bertha Sturgis, Eugene Eisenmann, and others, Edwin Willis in 1974 attributed the extirpation of about a dozen of BCI's bird species to the "island effect"—the decay of species diversity induced by the reduction of habitat size, as predicted by MacArthur and Wilson.[8] This finding not only appeared as a brilliant test of island biogeography theory but also suggested the theory's potential for application to the design of nature reserves.[9] Reserves, after all, could be imagined as islands of wilderness surrounded by a sea of developed land. Indeed, for BCI this was no metaphor; it was both an island and a reserve.

As promoted by tropical biologists John Terborgh, Jared Diamond, and Wilson himself, measurements of BCI's extinction rates became canonical evidence in conservation biologists' argument for the need to create very large nature reserves.[10] Indeed, they would feature in Wilson's speech at the BioDiversity Forum.[11] Terborgh also put these ideas into practice, overseeing a new station at Cocha Cashu in Peru's newest and largest national park, Manú, in 1973. In 1979, likewise, Lovejoy established the long-term Minimum Critical Size of Ecosystems Project, a joint U.S.-Brazilian effort to test the application of island biogeography theory to species diversity loss in plots of rainforest fragmented by clear-cutting near Manaus.[12] Rancorous disputes would ensue over the wisdom of applying the theory to conservation around the world, but such findings remained ample proof of the value of permanent research stations and their ability to produce important long-term data, whatever its interpretation or application.[13] The new sites joined BCI and those of the Organization for Tropical Studies (OTS) both as places devoted to basic research and as model rainforests that promised insights for conservation.[14]

Land for Science

Even as they expanded beyond the circum-Caribbean and became more international, stations thus remained the keystone institution for the U.S. community of tropical biologists. By the 1970s, tropical stations were also increasingly places of formal education where future generations of tropical biologists were trained. Indeed, that there was a community of "tropical biologists" in the United States at all by this period was a direct result of the establishment of these institutions earlier in the century. Securing land and patronage with the promise of producing tropical expertise, tropical biology had grown with the expansion of U.S. empire in the circum-Caribbean. This hegemony had been shaken by the Cuban Revolution, however. Following Cuba's appropriation of Soledad, BCI, now part of the Smithsonian Tropical Research Institute (STRI), was left as the only remaining station under U.S. control from that earlier era. It continued to be the most important of these stations, but the institutional and political landscape surrounding it long remained in flux.

The emerging environmental politics of biodiversity must be understood in this context. Long tied to U.S. interests, tropical biologists had to renegotiate their place within the changing structures of regional power. By the 1970s, few at STRI were under illusions about the long-term prospects of the Canal Zone.[15] The 1965–67 Panama Canal Treaties had not been approved, but as a new round of negotiations commenced in 1974, Martin Moynihan passed the STRI directorship on to Ira Rubinoff, who would successfully follow through in formalizing STRI's legal relationship with the Republic of Panama. As in Cuba, the Panamanian government recognized value in maintaining the station as a source of expertise and international prestige, even if STRI's ties to Panama's intellectual community remained relatively weak. Following the legal model of the Gorgas Memorial Laboratory, STRI signed a contract nominally affiliating it with the Panamanian Ministry of Health. This enabled STRI to operate outside of the Canal Zone and ultimately served as a basis for a corollary agreement to the successful 1977 Torrijos-Carter Treaties, which ensured STRI's persistence beyond the dissolution of the Canal Zone on 1 October 1979. The 1989 U.S. invasion only underscored STRI's interests in maintaining distance from U.S. policies.[16] Additional agreements solidified STRI's status in Panama through the U.S. withdrawal of all canal operations in 1999.[17] Some in the broader community of U.S. tropical biologists balked at STRI's gradual realignment as "bending over backwards to go Panamanian."[18] At STRI itself, however, there was little controversy—the return of the Canal Zone to Panamanian authority was simply a fact of life, and the necessity of cooperation with Panama was evident.[19] As biologists used BCI to form a picture of extinction rates in rainforests around the world, Canal Zone residents faced the disappearance of their own tropical "island" of U.S. territory. Some wore T-shirts in protest, labeling themselves "ZONIAN—ENDANGERED SPECIES."[20]

OTS, founded explicitly on the basis of an international partnership among equal academic institutions, would seem to elude the imperial history of older stations, yet there, too, over the two decades before "biodiversity," U.S. biologists had had to confront the politics of managing fieldwork in a foreign country. On the ground, maintaining relationships proved difficult. By 1968, the Universidad de Costa Rica had become "somewhat disenchanted with the OTS."[21] OTS funds for facilities on campus failed to materialize.[22] The pull of existing U.S. networks was strong. During the directorship of William H. Hatheway, OTS became heavily reliant on Leslie Holdridge and his Tropical Science Center for logistics, accounting, and even course design. A new OTS administration wrested back control, but relationships with the Costa Rican scientific community had weakened. OTS courses were not marketed to Costa Rican students, and few U.S. students had contact with Universidad de Costa Rica biology professors.[23] Some dismissed it as a "gringo enclave"; its focus remained on providing U.S. students with tropical experience, and not until 1974 would OTS sponsor a Spanish-language course (offered annually beginning in 1985).[24] Despite ebbs and flows in its relationship with

University of Panama rector Abdiel Adames (left) and Smithsonian Tropical Research Institute director Ira Rubinoff (right) exchange copies of a renewal agreement for scientific collaboration, STRI Tupper Center, Panama City, 1990. Smithsonian Institution Archives, no. SIA2009-1280.

Costa Ricans and several serious administrative and financial crises, OTS survived and grew enormously. With National Science Foundation funds, OTS expanded its facilities and branched out into research as well as teaching. In 1968, OTS bought Holdridge's Finca La Selva to manage it as a research station and site for field courses. The following year, OTS began leasing another U.S.-owned property, Palo Verde, as a contrasting, seasonally dry forest station. In 1973, OTS also took over operations of a station at Las Cruces and the private Wilson Botanical Gardens, developed ten years earlier by a Florida couple.[25] Shifting toward the operation of a constellation of field stations and surrounding reserves, OTS thus converged with STRI in its pattern of land ownership, despite its distinct origins. Having passed through the global postcolonial moment of the 1960s and 1970s, OTS and STRI embodied many of its contradictions. They emerged as tropical biology's premier institutions, larger than ever and more integrated into Latin American society than in the past, yet still overwhelmingly dominated by U.S. scientists.

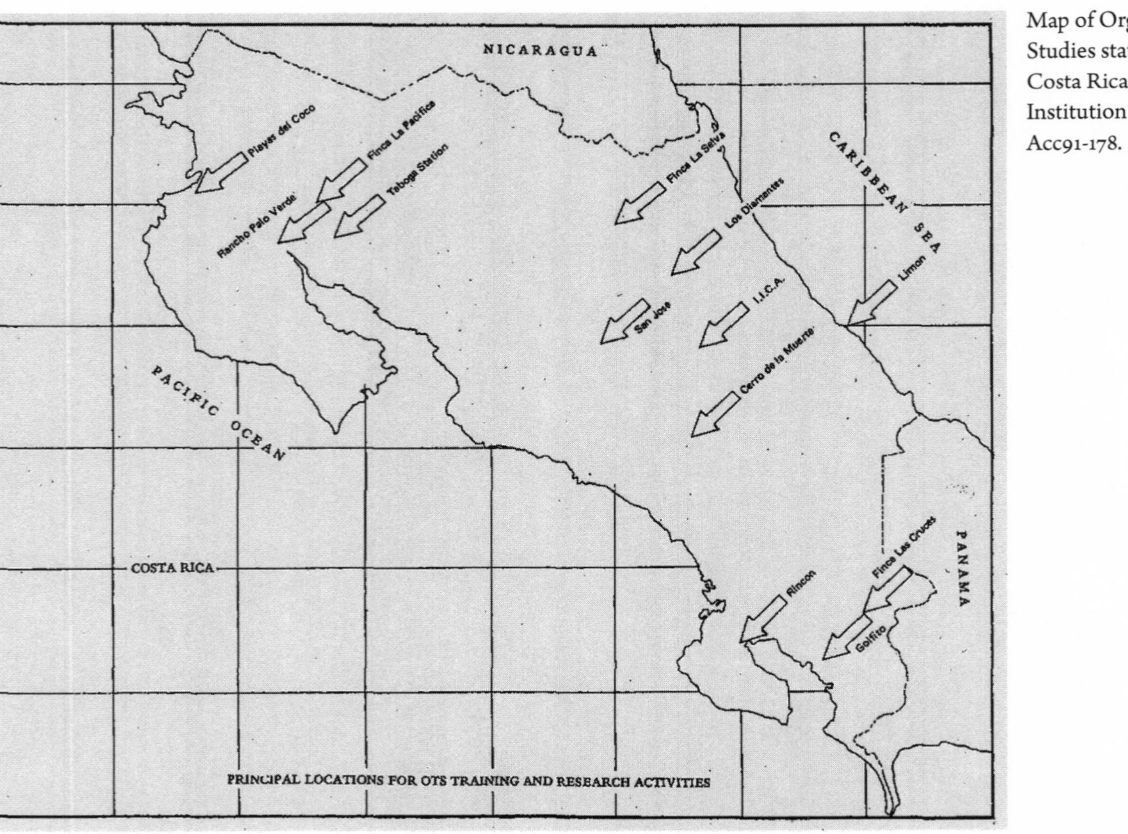

Map of Organization for Tropical Studies stations and field sites, Costa Rica, 1971. Smithsonian Institution Archives, box 6, Acc91-178.

The organizers of the 1986 BioDiversity Forum, then, were certainly not strangers to the international politics of science, shrouded in "leafy obscurity." The majority were not just tropical biologists but had long-standing and deep connections to either OTS or STRI. Indeed, among them were OTS president Peter Raven and STRI director Ira Rubinoff, both well versed in international scientific diplomacy. Yet, up to this point, neither STRI nor OTS had had a serious role in conservation. What pushed them into conservation work were, in fact, efforts to protect their stations. As part of STRI's 1977 canal treaty corollary, Rubinoff secured custodianship of a new Barro Colorado Nature Monument, including BCI and five surrounding mainland peninsulas, contiguous to the east with another new park, the Parque Nacional Soberanía (Sovereignty National Park). This not only provided BCI with a buffer against poaching and deforestation (and thus, in part, its viability for testing island biogeography) but also provided additional international legal standing for STRI's ongoing presence. As a nature monument, Barro Colorado would be protected by the 1940 Convention on Nature Protection and Wildlife Preservation in the Western Hemisphere, ratified by both the United States and Panama.[26] Although public conservation outreach had been declared a major objective of both STRI and OTS at their formation, involvement in such activities was negligible until the late 1970s. OTS sponsored no courses on applied conservation or forestry in its early days, despite requests from the Costa Rican side.[27]

The first generation of OTS students initially had little awareness of conservation issues.[28] At the time of their acquisition, OTS's stations were remote and surrounded by undeveloped land. During the late 1960s and early 1970s, however, Costa Rica's countryside began to experience high rates of deforestation as government policies encouraged "frontier" settlement and the expansion of multinational banana, logging, and ranching ventures.[29] Rural development had provided U.S. ecologists inroads into Costa Rica during the 1950s, but it now proved a threat to their long-term research sites. Daniel Janzen, among OTS's first and most famous graduates, was "flattened" by the clearing of two private field areas in 1972.[30] A Tropical Science Center station used by OTS at Rincón de Osa on the lands of a U.S.-owned timber company was also compelled to shut down in 1973, a victim of an unsavory real estate deal.[31] Logging in Guanacaste likewise forced a bitterly disappointed Stephen Hubbell to relocate his planned long-term tree diversity census project to BCI in 1979.[32] Development on the Osa Peninsula convinced a number of OTS associates to lobby for the reservation of what would become the Parque Nacional Corcovado, but the last straw was a threat to La Selva itself. In 1980,

OTS's neighbors adjacent to La Selva, the Vargas family, were pressured by Costa Rica's Instituto de Tierra y Colonización to develop their land.[33]

With the specter of La Selva becoming a depauperate "island" forest fragment, members of OTS and a variety of high-profile associates were spurred into action. Chief among these were Raven, Ehrlich, and Wilson. They made the rounds of U.S. foundations and published an alert in the pages of *Science* about a "global problem in which . . . biologists have a more than ordinary personal stake," that of extinction and the "reduction of genetic and species diversity by the destruction of tropical habitats." It culminated in a plea for donations to save La Selva.[34] They raised enough money to buy the Vargas ranch, doubling La Selva's size, and secured Costa Rica's establishment of a larger *zona protectora* around the site.[35] It marked a watershed in tropical biologists' involvement in activism, although still framed as "preservation for science."[36]

Raven was at the same time head of a committee preparing an assessment of "Research Priorities in Tropical Biology" for the National Research Council. Its report underlined the connections between ecological knowledge, land conservation, and the solution of economic and population problems in tropical countries. Human welfare in the tropics depended on sound biological research, they argued, but warned, "Scientific knowledge about tropical ecosystems is extremely incomplete." With "only about a sixth" of tropical species known to science, "virtually no knowledge is available to apply to the solution of many ecological problems."[37] To face a crisis of habitat destruction, the committee argued, reserves must be created where they could both "preserve the maximum biological diversity" and provide places where biologists could discover the basic insights needed for "sound development" alongside conservation in the tropics.[38]

Committee members, particularly Janzen and Arturo Gómez-Pompa (who had been instrumental in developing Mexico's Los Tuxtlas station), had previously warned that tropical rainforests were a "nonrenewable resource," facing destruction by global economic pressures and the inappropriate application of temperate-zone-based assumptions.[39] The argument gained traction, however, as fellow committee member Norman Myers brought the group's perspective to his 1979 book *The Sinking Ark: A New Look at the Problem of Disappearing Species*. Whereas Raymond Dasmann's *A Different Kind of Country* had offered a hopeful vision in 1968 of the potential to preserve and restore natural and human diversity, Myers presented an unfolding apocalypse of accelerating global extinction rates, based on new quantitative estimates of species diversity and deforestation. It used Costa Rica's expanding cattle lands as a vivid example of this loss of biological diversity. Before "blaming the slash-and-burn cultiva-

Teleconference on BioDiversity, 24 September 1986. From left: Thomas Lovejoy (World Wildlife Fund and Smithsonian Institution), Joan Martin-Brown (United Nations Environment Programme), Edward O. Wilson (Museum of Comparative Zoology, Harvard), Maureen Bunyan (moderator, WUSA-TV), Paul Ehrlich (Stanford University), Michael Robinson (National Zoological Park, Smithsonian Institution), and Peter Raven (Missouri Botanical Garden and Organization for Tropical Studies). Smithsonian Institution Archives, negative no. 86-12211.12A.

tors," however, Myers asked his temperate-zone readers to consider their own position within the flows of global capitalism: "Which American conservationist has not taken satisfaction at the 'humble hamburger' with its humble price?"[40] Sounding the alarm, it argued that conservationists should shift their priorities south, where most of the planet's species lived. Following Myers's book and the report of Raven's committee, the National Research Council identified the loss of biological diversity as "one of the highest ranking concerns." Walter Rosen suggested a national event to address it.[41]

The Legacies of U.S. Tropical Stations

The stage was set for the National Forum on BioDiversity—all that remained was Rosen's felicitous contraction of *biological diversity* to *biodiversity*. The subsequent career of biodiversity as a global environmentalist cause is well known. The significance of its roots in the twentieth-century history of tropical biology, however, has often been overlooked. Biologists came to new understandings of life's diversity by working in the tropics during the twentieth century. Migrants from the northern temperate zone, U.S. biologists became residents at tropical field stations—some for short periods, some seasonally, some permanently. They came to know these places deeply. In many cases, they also came to care deeply about them. To set down tropical roots, however, U.S. biologists became embedded in the networks of empire, centered in the circum-Caribbean. This came in many forms: interimperial solidarity with the British in the West Indies, ties to U.S. government power structures in Puerto Rico or the Panama Canal Zone, or reliance on land-owning U.S. corporations and private individuals (backed by U.S. economic and military might, either implicitly or, as in Cuba under the Platt Amendment, by law). These relationships of international power shaped the transportation networks and hierarchies of labor that enabled U.S. biologists to do their fieldwork. They also informed the ways in which biologists justified their work in order to secure patronage. Ironically, the corporations and government agencies that U.S. biologists depended on to sustain their access to tropical environments through much of the century were at the same time responsible for the exploitation of those environments.

The twentieth-century history of tropical biology has thus left a complex legacy. Tropical stations fostered deep, place-based knowledge, but they often presented landscapes that were, except for station laborers, unpeopled. Too often, rather than fostering connections, they insulated U.S. biologists from local communities of scientists and nonscientists. In the period from the 1959 Cuban Revolution to the 1979 return of the Canal Zone to Panama, this insulation increasingly became a liability. Tropical biologists repositioned themselves. *Biodiversity* emerged from both the intellectual revolution in species diversity and a necessity for tropical biologists to justify the practical value of their basic research, not only for U.S. interests, but also in the interests of tropical countries and the whole world. Imperial powers had long imagined tropical lands as rich in untapped resources. The concept of biodiversity yoked this biological wealth to a global conservation ethos. As a resource, biodiversity was like the goose that laid golden eggs; it would be worth far more for all if cared for and preserved for the future than if slaughtered in an instant of greed.

Tropical biologists had begun to see the neocolonial extraction of the biological resources of the tropics as a threat to their own profession—the future of tropical biology, as a scientific field, depended on the conservation of tropical life in the field. As Janzen saw it, if biologists wanted a tropics to study in the future they would "have to buy it."[42] The concept of biodiversity offered a way to navigate an ambiguous relationship with global capital—one that critiqued its excesses and yet could harness its patronage. Arguing provocatively that "the developed nations should return some of the resources ripped off the tropics over the past 400 years," Janzen saw the purchase of scientific parks as a weapon against centuries of exploitation by foreign interests and local land-owning elites.[43] Critics like the environmental historian Ramachandra Guha, however, have pointed to Janzen's efforts to control Costa Rican territory as itself a prime example of a new "conservation imperialism."[44] As the 1986 biodiversity blitz secured concessions from the U.S. Agency for International Development and the World Bank, it is easy to see continuity in the relationship of U.S. biologists to the centers of global power over the century. In courting U.S. foreign policy and international development as a patron, Wilson did little that would have been unfamiliar to Thomas Barbour or William Beebe. Global biodiversity conservation might be new, but it had a precedent in tropical biologists' efforts to sustain their stations. It is worth recalling that this patronage was never a simple prospect, always subject to the winds of geopolitical change and fraught with trade-offs.

Beyond the American Tropics

From the very beginning of the twentieth century, U.S. ecologists had lobbied for tropical stations in order to transcend the temperate-zone bias of their science. By the close of the century, it was clear that this bias could never be overcome by northern visitors and transplants alone. Biology and conservation needed voices from within tropical countries. Today, tropical environments still remain far less studied than those of the temperate zones, but U.S. and European scientists no longer have the field to themselves. Mexico and Brazil, where U.S. dominance was always less complete, have since the 1960s supported their own biological stations and growing communities of self-identified tropical biologists.[45] Under the leadership of Puerto Rican ecologist Ariel Lugo (a protégé of Howard Odum who began his career carrying bricks for the Rain Forest Project), El Yunque has been an important Long Term Ecological Research Network site since 1988.[46] STRI and OTS have also become far more inclusive since the 1970s, employing and serving many Latin American scientists and students. Indeed, despite the persisting underrepresentation of tropical field studies in ecology, Panama and Costa Rica are among the most studied areas in the world for their size.[47] A significant portion of this research is authored by Panamanian and Costa Rican scientists, but U.S. researchers still dominate—the patterns of the twentieth century continue to shape the landscape of tropical research.[48]

Sites long cut off from flows of capital have faced more obstacles. Harvard's former Soledad station is now the Jardín Botánico de Cienfuegos. It has persisted, despite scarcities of funds and fuel, through the heroic effort of dedicated Cubans and now recasts itself as a repository of plant biodiversity.[49] In Guyana, the entomologist Margaret S. Collins, an African American student of Alfred Emerson, reopened a station at Kartabo in 1979.[50] It has not been maintained for continuous research, but new stations have emerged, along with a growing recognition of the need for meaningful collaboration with indigenous people.[51] In Jamaica, Cinchona remains a public garden, cash-strapped but beloved by foreign ecotourists and local hikers. The long history of data collected there remains crucial to understanding the area, and UNESCO's 2015 declaration of the surrounding Blue and John Crow Mountains as a World Heritage Site may bring much-needed funding.[52]

The histories of these former sites of U.S. science underline the fragility of tropical field stations as institutions but also suggest avenues of hope for the future. New stations and other forms of long-term environmental research sites and projects are being established throughout the tropical world—increasingly with impetus from within.[53] Few new stations endure for many years, but our ability to comprehend a changing world depends on their survival. That survival depends not only on funds but also on the cultivation of relationships among people and the strengthening of institutions. Tropical stations must become sites for creating not only knowledge of biodiversity but also new, equitable local and international scientific relations. The future of research in the tropics demands a full recognition of its history.

Notes

Abbreviations Used in the Notes

Acc	Accession
AHS	Arizona Historical Society
DU	Duke University David M. Rubenstein Rare Book and Manuscript Library
FTG	Fairchild Tropical Botanic Garden Special Collections
HUA	Harvard University Archives
IRTA	Institute for Research in Tropical America
JHU	Johns Hopkins University, Ferdinand Hamburger Archives
NARA	National Archives and Records Administration
NAS	National Academy of Sciences Archives
NYBG	New York Botanical Garden Mertz Library Archives
OTS	Organization for Tropical Studies
PUA	Princeton University Library's Manuscripts Division
RAC	Rockefeller Archive Center
RG	Record Group
RU	Record Unit
SIA	Smithsonian Institution Archives
STRI	Smithsonian Tropical Research Institute, Earl S. Tupper Tropical Sciences Library
WCS	Wildlife Conservation Society Library and Archives

Introduction

1. Conference details come from Acc92-030, SIA. See also Wilson, *Naturalist*, 359–60.

2. On the concept of biodiversity and the emergence of conservation biology, see Takacs, *Idea of Biodiversity*; Lévêque and Mounolou, *Biodiversity*; Oksanen and Pietarinen, *Philosophy and Biodiversity*; Sarkar, *Biodiversity and Environmental Philosophy*; Meine, Soulé, and Noss, "'Mission-Driven Discipline'"; Farnham, *Saving Nature's Legacy*; Barrow, *Nature's Ghosts*, 353–59; Robin, "Rise of the Idea of Biodiversity"; Vadrot, *Politics of Knowledge and Global Biodiversity*.

3. Wilson, *BioDiversity*, 8.

4. Ibid., 3.

5. Ibid., 8.

6. Ibid., 14.

7. Coiner Walter Rosen states that he intended to suggest a less cumbersome word for a well-established scientific concept. Rosen, "What's in a Name?" On the first appearances of *biological diversity* and *biodiversity*, see Farnham, *Saving Nature's Legacy*.

8. Libby Robin has critiqued scholars' concentration on this constricted time frame for "promulgat[ing] an 'origins' story rather than a nuanced sense of what drives the global concept of biodiversity, and its uses today. While there is no doubt that a 'global environmentalist story' accelerates from the 1980s, it is useful to document its roots in actual places, in case histories." Robin, "Biological Diversity as a Political Force," 40.

9. Ecology is so central to the study of biology in the tropics that *tropical ecology* is often used interchangeably with *tropical biology*. I use *tropical biology* to refer to the general field of study and *tropical ecology* only when referring specifically to ecological research.

10. For example, see Zuk, "Temperate Assumptions"; Robinson, "Is Tropical Biology Real?"

11. On tropicality, see Arnold, *Problem of Nature*, 141–68; Slater, "Amazonia as Edenic Narrative"; Arnold, " 'Illusory Riches' "; Stepan, *Picturing Tropical Nature*; Sheller, *Consuming the Caribbean*; Driver, "Imagining the Tropics"; Driver and Martins, *Tropical Visions in an Age of Empire*; Tobin, *Colonizing Nature*; Thompson, *Eye for the Tropics*; Carey, "Inventing Caribbean Climates"; Sutter, "The Tropics."

12. Voices from within nations in the tropics have historically challenged aspects of this discourse but have also deployed it in their own campaigns for internal reforms. For example, see Peard, *Race, Place, and Medicine*.

13. Pratt, *Imperial Eyes*.

14. Darwin to J. S. Henslow, 18 May 1832, in Burkhardt and Smith, *Correspondence of Charles Darwin*, 237.

15. Enright, *Maximum of Wilderness*, 22. See also Slater, *In Search of the Rain Forest*, 288–89. Nancy Stepan addresses this phenomenon of disappointment in *Picturing Tropical Nature*, 45–66.

16. Wallace, *Tropical Nature*, vii.

17. Ibid., 65. Wallace here refers only to plants, but he uses nearly identical language in his chapter on tropical animals.

18. Ibid., 69–70.

19. Ibid., 27, 65.

20. Ibid., 65.

21. Ibid., 122–23.

22. Ibid., 123.

23. On station science compared with expeditionary science, see Vetter, *Field Life*; de Bont, *Stations in the Field*; Kingsland, "Role of Place in the History of Ecology"; Kohler, *All Creatures*.

24. On this literature, see Wright and Finnegan, *Spaces of Global Knowledge*; McCook, "Global Currents in National Histories of Science"; Rosenberg, "Transnational Currents in a Shrinking World"; Safier, "Global Knowledge on the Move"; Roberts, "Situating Science in Global History"; Raj, *Relocating Modern Science*.

25. Pauly, "Summer Resort and Scientific Discipline"; Benson, "From Museum Research to Laboratory Research"; Benson, "Naples Stazione Zoologica"; Maienschein, *One Hundred Years Exploring Life*; Maienschein, *Transforming Traditions*; Pauly, *Biologists and the Promise of American Life*, 94–96, 145–64.

26. Kohler, *Landscapes and Labscapes*. See also Gieryn, "City as Truth-Spot"; Kingsland, "Frits Went's Atomic Age Greenhouse."

27. Vetter, *Field Life*; Muka, "Right Tool and the Right Place"; de Bont, *Stations in the Field*; Raby, "Ark and Archive"; Wille, "Coproduction of Station Morphology"; Alagona, "Sanctuary for Science."

28. De Bont, *Stations in the Field*, 3.

29. Mickulas, *Britton's Botanical Empire*, 130–33; Kohler, *Landscapes and Labscapes*, 128–29; Kingsland, *Evolution of American Ecology*, 82–83.

30. Hagen, "Problems in the Institutionalization of Tropical Biology."

31. Cittadino, *Nature as the Laboratory*. Robert-Jan Wille in "De Stationisten" also argues for the station's importance in expanding European botanists' professional prospects.

32. Jack, "Biological Field Stations"; Brockway, *Science and Colonial Expansion*; Drayton, *Nature's Government*; Osborne, "Acclimatizing the World"; Schiebinger and Swan, *Colonial Botany*; Schiebinger, *Plants and Empire*; McCook, "Neo-Columbian Exchange."

33. De Bont notes comparable heterogeneity, excepting gardens, in European zoological stations. De Bont, *Stations in the Field*, 4.

34. For example, see Goetzmann, *Exploration and Empire*; Carter, *Surveying the Record*; Vetter, "Science along the Railroad"; Kohler, *All Creatures*.

35. As Catherine Christen notes, tropical biologists themselves remark on the importance of this shift. Christen, "At Home in the Field," 558–59. Kohler and de Bont have also explored phenomena of "residential" field science. Kohler, "Paul Errington, Aldo Leopold, and Wildlife Ecology"; de Bont, *Stations in the Field*, 199–208.

36. Leigh, "Barro Colorado," 90–91.

37. Billick and Price, "Ecology of Place," 5.

38. Ibid., 5–6.

39. On the underrepresentation of tropical research, see Amano and Sutherland, "Four Barriers to the Global Understanding of Biodiversity Conservation"; Martin, Blossey, and Ellis, "Mapping Where Ecologists Work"; Collen et al., "Tropical Biodiversity Data Gap"; Kier et al., "Global Patterns of Plant Diversity and Floristic Knowledge"; Janzen, "Impact of Tropical Studies."

40. Surveying publications between 2004 and 2009, Laura Martin and colleagues found that nine of the seventy-three countries represented had significantly more sites than expected based on size. Only two were tropical: Costa Rica (with forty-nine times more studies than expected) and Panama (thirty-three times more studies). This made Central America—despite the general underrepresentation of the global tropics—the region "most overstudied by a factor of 8." Martin, Blossey, and Ellis, "Mapping Where Ecologists Work," 198, web tables 7 and 8.

41. In a study by Gabriela Stocks and colleagues, U.S. scientists comprised 35–45 percent of authors of papers on tropical ecology. All other countries were represented by 12 percent or fewer authors. However, scientists from Brazil and Mexico had a much larger share of lead authorship compared with scientists from other developing countries. Stocks et al., "Geographical and Institutional Distribution of Ecological Research." See also Wilson et al., "Conservation Research Is Not Happening Where It Is Most Needed"; Livingston et al., "Perspectives on the Global Disparity"; Malhado et al., "Geographic and Temporal Trends in Amazonian Knowledge Production"; Braker, "Changing Face of Tropical Biology?"; Gálvez et al., "Scientific Publication Trends and the Developing World."

42. Chazdon and Whitmore, *Foundations of Tropical Forest Biology*, 2–3.

43. See, for example, Watkins and Donnelly, "Biodiversity Research in the Neotropics"; Malhado, "Amazon Science Needs Brazilian Leadership"; Amano and Sutherland, "Four Barriers to the Global Understanding of Biodiversity Conservation"; Theobald et al., "Global Change and Local Solutions."

44. See Bemis, "'America' and Americans"; Renda, *Taking Haiti*, xvii. Philip J. Pauly specifically discusses the expansionist use of *American* in nineteenth-century natural history in *Biologists and the Promise of American Life*, 7–8.

45. This region is also referred to as the Greater Caribbean or Caribbean Basin. For historical and geographical perspectives on this region, see Mintz and Price, *Caribbean Contours*; Knight and Palmer, *Modern Caribbean*; Gaspar and Geggus, *Turbulent Time*; Randall and Mount, *Caribbean Basin*; Hillman and D'Agostino, *Understanding the Contemporary Caribbean*.

46. For perspectives on this imperial transition, see the essays in part 1 of McCoy and Scarano, *Colonial Crucible*.

47. See, for example, Colby, *Business of Empire*; Bucheli, *Bananas and Business*; Soluri, *Banana Cultures*; Tucker, *Insatiable Appetite*; Striffler, *Shadows of State and Capital*; Harrison, *King Sugar*; Ayala, *American Sugar Kingdom*; LeGrand, "Living in Macondo."

48. Overfield, "Agricultural Experiment Station and Americanization"; Overfield, "Science Follows the Flag"; Kramer, *Blood of Government*; Anderson, *Colonial Pathologies*; Okihiro, *Pineapple Culture*; Bankoff, "First Impressions"; Anderson, "Science in the Philippines"; Manganaro, "Assimilating Hawai'i."

49. Garfield, *In Search of the Amazon*.

50. On such collaborations and colonial power dynamics, see Henson, "Invading Arcadia"; Quintero Toro, "¿En Qué Anda la Historia de la Ciencia y el Imperialismo?"; Quintero Toro, *Birds of Empire, Birds of Nation*; Rodriguez, "Beyond Prejudice and Pride."

51. Sutter, "Nature's Agents or Agents of Empire?"; McNeill, *Mosquito Empires*; Espinosa, *Epidemic Invasions*.

52. Overfield, "Science Follows the Flag"; Bankoff, "Breaking New Ground?"; Fitzgerald, "Exporting American Agriculture"; McCook, *States of Nature*; Soluri, *Banana Cultures*.

53. Mintz and Price, *Caribbean Contours*, 5. See also Gowricharn, *Caribbean Transnationalism*.

54. Heggie, "Why Isn't Exploration a Science?"

55. On biodiversity as a globalizing discourse, see Taylor, "How Do We Know We Have Global Environmental Problems?"; Castree, "Geopolitics of Nature"; Zimmerer, "Biodiversity."

Chapter One

1. "American Tropical Laboratory," 415.

2. Kingsland, "Battling Botanist"; Magnus, "Down the Primrose Path"; Kohler, *Landscapes and Labscapes*, 53, 88, 93, 206, 221–24.

3. "American Tropical Laboratory," 415.

4. On laboratory "placelessness," see Livingstone, *Putting Science in Its Place*, 3; Kohler, *Landscapes and Labscapes*, 7–11, 213; Gieryn, "City as Truth-Spot"; Gooday, "Placing or Replacing the Laboratory."

5. "American Tropical Laboratory," 415.

6. McCook, " 'World Was My Garden,' " 499.

7. Coulter's close friendship with Arthur likely later helped bend his ear to MacDougal's proposal for a tropical station. Cittadino, "Ecology and the Professionalization of Botany"; Rodgers, *John Merle Coulter*.

8. Darwin and Darwin, *Power of Movement in Plants*; de Chadarevian, "Laboratory Science versus Country-House Experiments"; Ayres, *Aliveness of Plants*; Whippo and Hangarter, " 'Sensational' Power of Movement."

9. MacDougal, "Tendrils of Passiflora caerulea."

10. MacDougal, "Mechanism of Movement and Transmission of Impulses."

11. MacDougal, "Mimosa," 48.

12. Cittadino, *Nature as the Laboratory*.

13. Nicolson, "Humboldtian Plant Geography after Humboldt."

14. MacDougal, "Plant Zones"; MacDougal, "Recent Botanical Explorations." On C. Hart Merriam's life zones, see Kohler, *All Creatures*, 97–98, 151–54; Nicolson, "Humboldtian Plant Geography after Humboldt."

15. Worster, *Nature's Economy*, 190–93, 198; Coleman, "Evolution into Ecology?"

16. Warming and Knoblauch, *Ökologischen Pflanzengeographie*; Warming, *Plantesamfund*.

17. Schimper, *Pflanzen-Geographie*; Schimper, *Plant-Geography*.

18. See, for example, Cittadino, *Nature as the Laboratory*; Goodland, "Tropical Origin of Ecology."

19. The Indonesian government continues to operate this garden, now known as the Bogor Botanical Gardens.

20. MacDougal, "Botanic Gardens I," 181; "American Tropical Laboratory," 415.

21. Dammerman, "History of the Visitors' Laboratory." For the site's significance in German, Indonesian, and Dutch science, see Pyenson, *Empire of Reason*, 10–13; Cittadino, *Nature as the Laboratory*; Moon, *Technology and Ethical Idealism*, 29, 32–34; Goss, *Floracrats*; Wille, "Coproduction of Station Morphology."

22. De Bont, *Stations in the Field*, 6–7; Wille, "Coproduction of Station Morphology."

23. Rowland, "Plea for Pure Science." See Lucier, "Origins of Pure and Applied Science."

24. MacDougal, "Botanic Gardens I," 180–81.

25. Cowles, "New Treatise on Ecology," 214. This was also an impetus behind the establishment of mountain stations, where a gradation of climates at different elevations were at hand for comparison; Vetter, "Rocky Mountain High Science"; Vetter, "Labs in the Field?"

26. Schimper, *Pflanzen-Geographie*, iv; Schimper, *Plant-Geography*, vi.

27. Quoted in translation in Cittadino, *Nature as the Laboratory*, 140.

28. See, for example, Wallace, *Tropical Nature*, 31.

29. Cryptogams are today considered paraphyletic. Fungi and algae are no longer classified as plants. On the growing interest in cryptogams within the new botany, see Cittadino, "Ecology and the Professionalization of Botany," 176.

30. It appeared as *Regenwald* in German and *regnskov* in Danish. Schimper is often incorrectly cited as coining the term, likely because English speakers first encountered it through translation of his work.

31. Warming and Vahl, *Oecology of Plants*, 13.

32. Ibid., 93. See similar statements in Warming, *Lagoa Santa*.

33. Warming, "Vegetation of Tropical America," 2.

34. Quoted in translation in Cittadino, *Nature as the Laboratory*, 140.

35. MacDougal, "Mimosa," 47–48.

36. See, for example, Haberlandt, *Botanische Tropenreise*, 36; Fairchild, "Sensitive Plant as a Weed."

37. MacDougal, "Mimosa," 61.

38. MacDougal, "Mechanism of Movement and Transmission of Impulses," 293.

39. "American Tropical Laboratory," 415.

40. He clarified, "You understood of course that the matter you so kindly sent was to be used editorially." Coulter to MacDougal, 12 October 1896, box 1, Daniel T. MacDougal Papers, NYBG. See also Coulter to MacDougal, 24 October 1896, ibid.

41. MacDougal, "A Tropical Laboratory," 496.

42. MacDougal, "Botanic Gardens I," 186.

43. See Raby, "Laboratory for Tropical Ecology."

44. "American Tropical Laboratory," 415.

45. Fairchild was at Buitenzorg from April to December of 1896, while MacDougal's editorial was published. Fairchild, *World Was My Garden*. See also Galloway, "Buitenzorg Gardens"; Dammerman, "History of the Visitors' Laboratory"; Goss, *Floracrats*, 71–72. Fairchild would play a key role in the establishment of future tropical stations (see chapters 2 and 3 in this volume) and the introduction of new plants into the U.S. landscape. Douglas, *Adventures in a Green World*; Pauly, *Fruits and Plains*, 126.

46. "American Tropical Laboratory," 416.

47. Ibid.

48. Pérez, *Cuba in the American Imagination*, 32–38. See also Craib and Burnett, "Insular Visions." On broader shifts in the geographical discourse of American empire, see Smith, *American Empire*.

49. Pérez argues that, in the language of nineteenth-century geopolitics, proximity and "neighborliness implied duty" and "conferred rights." Pérez, *Cuba in the American Imagination*, 34, 110.

50. "American Tropical Laboratory," 416.

51. MacDougal, "Botanic Gardens I," 186.

52. MacDougal, "Tropical Laboratory Commission," *Botanical Gazette* 23, no. 2 (1897): 129; MacDougal, "Tropical Laboratory Commission," *Botanical Gazette* 23, no. 3 (1897): 126–27.

53. Schimper to MacDougal, 9 February 1897, and Goebel to MacDougal, 9 February 1897, box 1, and George Murray to MacDougal, 7 October 1897, box 2, MacDougal Papers.

54. Brooks, "Notes from the Biological Laboratory," 132. Humphrey's seminar was reportedly on "Warming's Ecological Plant Geography." Warming's *Plantesamfund* would not be translated into English until 1909, however, and then as the essentially new book, *Oecology of Plants*.

55. Humphrey, "Tropical Laboratory," 50.

56. The members included among their ranks a young Thomas Hunt Morgan. Brooks, "Johns Hopkins Marine Laboratory"; Swindle, "Notes from a Marine Biological Laboratory." See also Muka, "Working at Water's Edge," 28–82.

57. Coulter to MacDougal, 4 May 1897, folder 2, box 1, MacDougal Papers.

58. MacDougal, "Tropical Laboratory Commission," *Botanical Gazette* 23, no. 3 (1897): 208.

59. Karl Goebel, quoted in ibid., 207–8; Macdougal, "New Botanical Laboratory," 395.

60. "Tropical Laboratory," 202–3. Italics added.

61. Swingle to MacDougal, 19 February 1897, folder 1, box 1, MacDougal Papers.

62. Jared G. Smith to MacDougal, 11 February 1897, folder 1, box 1, MacDougal Papers.

63. Humphrey, "Tropical Laboratory," 50.

64. Swingle to MacDougal, 27 December 1896, folder 1, box 1, MacDougal Papers. See also Farlow to MacDougal, 15 February 1897, folder 2, box 1, MacDougal Papers.

65. Swingle to MacDougal, 19 February 1897, and J. T. Scovill to MacDougal, 25 November 1896, folder 1, box 1, MacDougal Papers.

66. Smith to MacDougal, 11 February 1897, folder 1, box 1, MacDougal Papers. Smith would later take charge of the USDA Hawai'i Agricultural Experiment Station. See Overfield, "Science Follows the Flag."

67. Humphrey, "Tropical Laboratory," 50.

68. Swingle to MacDougal, 27 December 1896, folder 1, box 1, MacDougal Papers.

69. Humphrey, "Botany in Jamaica," 85.

70. See, for example, Brockway, *Science and Colonial Expansion*; Drayton, *Nature's Government*; Schiebinger, *Plants and Empire*; Schiebinger and Swan, *Colonial Botany*.

71. Karl Goebel, quoted in MacDougal, "Tropical Laboratory Commission," 207–8.

72. Farlow to MacDougal, 15 February 1897, folder 2, box 1, MacDougal Papers.

73. Humphrey, "Botany in Jamaica." In this *Science* article, Humphrey suggested the feasibility of a laboratory for "the unsolved problems of tropical vegetation," but the point was somewhat buried in the text, and the suggestion did not receive response in the pages of *Science*.

74. Kramer, "Empires, Exceptions, and Anglo-Saxons."

75. See correspondence in folder 2, box 1, MacDougal Papers.

76. MacDougal to William Fawcett, 9 January 1897, and Fawcett to Sir William Thiselton-Dyer, 15 February 1897, Directors' Correspondence, Royal Botanic Gardens, Kew, JSTOR Global Plants, http://plants.jstor.org/stable/10.5555/al.ap.visual.kldc13681 and http://plants.jstor.org/stable/10.5555/al.ap.visual.kldc13680 (accessed 9 April 2015).

77. Campbell, "Botanical Aspects," 34.

78. Ibid., 35.

79. Hogue, "Cruise Ship Diplomacy," 33–34.

80. Stark, *Jamaica Guide*, 42–44. See also the advertisement in Stark, *Bermuda Guide*, 184.

81. Campbell, "Botanical Aspects," 34.

82. Ibid., 35.

83. Ibid., 37.

84. MacDougal, "Movements of Plants"; MacDougal, "Mimosa."

85. See, for example, Brockway, *Science and Colonial Expansion*; Drayton, *Nature's Government*; Schiebinger, *Plants and Empire*; Schiebinger and Swan, *Colonial Botany*.

86. *Daily Gleaner*, 20 November 1885, 1.

87. On the Boston Fruit Company lease, see Fawcett to Daniel Morris, 26 June 1891, Directors' Correspondence, Royal Botanic Gardens, Kew, JSTOR Global Plants, http://plants.jstor.org/stable/10.5555/al.ap.visual.kldc13561 (accessed 12 April 2015). See also Stark, *Jamaica Guide*, 75.

88. See Fawcett to Sir William Thiselton-Dyer, 15 May 1893, Directors' Correspondence, Royal Botanic Gardens, Kew, JSTOR Global Plants, http://plants.jstor.org/stable/10.5555/al.ap.visual.kldc13628 (accessed 12 April 2015). This letter shows that Fawcett was interested to "see what help can be obtained from the Yankees" as early as Humphrey's 1893 visit. On the origins of the garden, see Edwards, "Historiography of the Hill Gardens."

89. Humphrey, "Botany in Jamaica," 85.

90. Campbell, "Botanical Aspects," 37–38.

91. See, for example, Brockway, *Science and Colonial Expansion*; Drayton, *Nature's Government*; Schiebinger, *Plants and Empire*.

92. On British hill stations, see Kenny, "Climate, Race, and Imperial Authority"; Kenny, "Claiming the High Ground"; Kennedy, *Magic Mountains*; Livingstone, "Tropical Climate and Moral Hygiene."

93. See, for example, "Botanist in Jamaica."

94. Campbell, "Botanical Aspects," 40.

95. Humphrey, "Tropical Laboratory," 50.

96. MacDougal, "Mimosa," 49.

97. Fawcett, "Public Gardens and Plantations," 367. Fawcett was ultimately unable to obtain leave to attend the meeting. "Botanical Society of America," 185.

98. Humphrey to MacDougal, 19 October 1896, folder 1, box 1, MacDougal Papers.

99. "Botanical Society of America," 180, 185.

100. Humphrey, "Tropical Laboratory," 50.

101. A third, H. L. Clark, caught a mild version and survived. See Brooks, "Notes from the Biological Laboratory," 2. The initial uncertainty around the cause of Humphrey's and Conant's deaths are likely due to an unwillingness on the part of local officials to declare the disease yellow fever and thus risk a port closure. See Espinosa, *Epidemic Invasions*, 25.

102. Ibid., 24–30.

103. Those interested primarily in a site for marine research shifted their attention to subtropical Florida; Muka, "Working at Water's Edge," 41.

104. On the lead-up to the U.S. intervention, see Espinosa, *Epidemic Invasions*; Pérez, *War of 1898*; Pérez, *Cuba and the United States*.

105. Coulter to Farlow, 17 May 1898, cited in Rodgers, *John Merle Coulter*, 170.

106. Coulter to MacDougal, 22 September 1898, folder 2, box 1, MacDougal Papers.

107. Coulter to MacDougal, 5 October 1898, folder 2, box 1, MacDougal Papers; Coulter to Farlow, 1898, cited in Rodgers, *John Merle Coulter*, 172.

108. Overfield, "Agricultural Experiment Station and Americanization"; Overfield, "Science Follows the Flag"; McCook, " 'World Was My Garden.' "

109. Mickulas, *Britton's Botanical Empire*, 155.

110. Boom, "Botanical Expeditions of the New York Botanical Garden"; Kingsland, *Evolution of American Ecology*, 79. See also Mickulas, *Britton's Botanical Empire*.

111. Sastre-D. J. and Santiago-Valentín, "Botanical Explorations of Puerto Rico"; Baatz, "Imperial Science and Metropolitan Ambition"; Brock, "American Empire and the Scientific Survey of Puerto Rico."

112. Coulter to MacDougal, 5 October 1898.

113. MacDougal to Britton, 7 October 1898, cited in Mickulas, *Britton's Botanical Empire*, 129.

114. Farlow to MacDougal, 3 February 1897, folder 1, box 1, MacDougal Papers.

115. Ames, "George Lincoln Goodale," 641.

116. Harvard University, *Reports, 1898–1899*, 242–43.

117. See chapter 2 and McCook, *States of Nature*, 56–60.

118. Coulter to Farlow, 1897, cited in Rodgers, *John Merle Coulter*, 168.

119. Carnegie Institution of Washington, *Year Book, 1902*, 8.

120. Ibid.

121. See Craig, *Centennial History of the Carnegie Institution*, 253n15. On the Carnegie Institution's early support of plant ecology and biological laboratories, see McIntosh, "Pioneer Support for Ecology"; Ebert, "Carnegie Institution of Washington and Marine Biology." A tropical laboratory was also estimated to cost $39,500 for five years,

nearly twice the desert lab estimate. Carnegie Institution of Washington, *Year Book, 1902*, 11.

122. Carnegie Institution of Washington, *Year Book, 1902*, 8–9 (italics added).

123. Ibid., 9.

124. Ibid., 6.

125. Britton, "Tropical Station," 2.

126. Ibid., 251; Robinson, "Visit to the Botanical Laboratory," 191.

127. "Laboratories," 284.

128. "Notes, News and Comment" (1903), 139; "Notes, News and Comment" (1904), 152.

129. Robinson, "Visit to the Botanical Laboratory," 187.

130. Ibid., 194.

131. Brackett, "Summer in the Tropics," 7.

132. Ibid., 6. *Plant World* would transform into *Ecology* in 1920 under the editorship of another Cinchona visitor, Forrest Shreve. Bowers, "Plant World and Its Metamorphosis."

133. Brackett, "Summer in the Tropics," 6.

134. Robinson, "Visit to the Botanical Laboratory," 193–94.

135. On the social relations of field scientists and assistants, see Jacobs, "Intimate Politics of Ornithology"; Fan, *British Naturalists in Qing China*; Raffles, "Intimate Knowledge"; Camerini, "Wallace in the Field."

136. Brackett, "Summer in the Tropics," 10.

137. Board of Agriculture and Department of Public Gardens and Plantations, *Annual Report* for 1905, 16.

138. See Seifriz, "Length of the Life Cycle of a Climbing Bamboo."

139. Brackett, "Summer in the Tropics," 11.

140. He is recorded as the collector of a type specimen in 1891; see JSTOR Global Plants, http://plants.jstor.org/stable/10.5555/al.ap.specimen.ny00073509 (accessed 11 December 2016). The last reference to his observations of plants is in Seifriz, "Length of the Life Cycle of a Climbing Bamboo," 86, 91. See also Underwood, "Explorations in Jamaica," 114; Robinson, "Visit to the Botanical Laboratory," 193–94; Brackett, "Summer in the Tropics," 10–11. See Board of Agriculture and Department of Public Gardens and Plantations, *Annual Reports* for 1901–6.

141. Quoted in Johnson et al., "Cinchona as a Tropical Station," 918.

142. Ibid., 919.

143. Ibid., 918.

144. Fawcett, "Public Gardens and Plantations," 367; Asprey and Robbins, "Vegetation of Jamaica," 363. See also Frodin, *Guide to Standard Floras*, 289–90.

145. Fuertes, "Voices of Tropical Birds," 96, 342.

146. Ibid., 100. See, for example, Kohler, "Paul Errington, Aldo Leopold, and Wildlife Ecology"; de Bont, *Stations in the Field*; Raby, "Ark and Archive."

147. Johnson, "Invasion of Virgin Soil," 305.

148. Johnson, "Origin and Development of a Tropical Forest," 274.

149. Coker, "Professor Duncan Starr Johnson," 510.

150. Johnson, "Origin and Development of a Tropical Forest," 275.

151. Johnson, "Invasion of Virgin Soil," 311.

152. Johnson, "Revegetation of a Denuded Tropical Valley," 305.

153. Johnson, "Origin and Development of a Tropical Forest," 276.

154. Shreve, "Collecting Trip"; Shreve, "Winter at the Tropical Station."

155. Shreve to MacDougal, 15 April 1907, MS 0452, folder 90, box 8, Daniel Trembly MacDougal Papers, AHS. See also Bowers, *Sense of Place*, 17.

156. Shreve to MacDougal, 15 April 1907, MS 0452, folder 90, box 8, MacDougal Papers, AHS. Robert E. Kohler suggests that Shreve was attracted to Cinchona for its laboratory but that the distinctive forest conditions eventually "enticed him into the field." In fact, Shreve's correspondence and publications demonstrate that, from the start, he saw fieldwork as his primary objective, with laboratory work as a necessary adjunct to any investigation of organism-environment relationships. There is little evidence that he spent "most of his time in the station's physiological laboratory"; his descriptions of the forest types of the Blue Mountains entailed significant trekking through difficult terrain, with repeated returns to observe seasonal change. Kohler, *Landscapes and Labscapes*, 128.

157. Shreve to MacDougal, 6 September 1909, MS 0452, folder 144, box 11, MacDougal Papers, AHS.

158. Shreve, *Montane Rain-Forest*, 107.

159. Shreve, "Direct Effects of Rainfall."

160. Shreve, *Montane Rain-Forest*, 106.

161. This made Shreve among the earliest critics of applying the plant association concept to rainforests as a whole. This critique is usually identified as originating with A. Aubréville in 1938; see, for example, Chazdon and Denslow, "Floristic Composition and Species Richness."

162. Shreve, *Montane Rain-Forest*, 109–10.

163. Ibid., 110.

164. "Review: *Montane Rain-Forest*," 245.

165. Fuller, "Montane Rain-Forest," 239. Kohler argues that the fusion of experimentalism and fieldwork remained incomplete in Shreve's work at Cinchona. Reviews such as Fuller's, however, indicate that contemporaries saw Shreve as making significant headway in this direction; Kohler, *Landscapes and Labscapes*, 155.

166. Johnson, "Cinchona Botanical Station," 525.

167. Tilley, *Africa as a Living Laboratory*, 10–11.

168. I use the term *colonial science* (1) to designate science in a colonial territory; (2) to emphasize how science was embedded in the power structures, infrastructures, economies, and politics of the territories in which it took place; and (3) to counter the exceptionalism of much of the literature on "American science," which has rarely recognized science in U.S. colonies and neocolonies. This usage serves to highlight transimperial linkages between science in U.S. and European colonial territories. On this final point, see Kramer, "Power and Connection"; McCoy and Scarano, *Colonial Crucible*.

169. Johnson, "Cinchona Botanical Station," 523.

170. Ibid.

171. Britton, "Report, 1907," 13.

172. Underwood, "Explorations in Jamaica," 118.

173. "Laboratories for Botanical Research," 538–39; Johnson, "Cinchona Botanical Station," 523.

174. Lease between the Colonial Secretary of Jamaica and Charles D. Walcott, Secretary of the Smithsonian Institution, 28 December 1916, folder 2, box 107, RU45, SIA; Johnson to Walcott, 15 February 1921, folder 8, box 11, RU45, SIA; Board of Regents of the Smithsonian Institution, *Annual Report, 1921*, 12.

175. See, for example, Skutch, *Bird Watcher's Adventures*, 29; Tanner, "Four Montane Rain Forests"; Mordecai, "Cinchona"; Cooke, "Great House, Not So Great."

Chapter Two

1. See Langley, *Banana Wars*; Striffler and Moberg, *Banana Wars*; Colby, *Business of Empire*.

2. Barbour, *Naturalist at Large*, 300.

3. Ibid., 16.

4. Barbour, "Zoögeography of the East Indian Islands," 6–9; Barbour and Barbour, *Letters*, 102, 112–19, 138–63.

5. Barbour, *Naturalist at Large*, 42–44. See also, Barbour and Barbour, *Letters*. On Ali's relationship with Wallace and colonial social connections to fieldwork in Malaysia, see Camerini, "Wallace in the Field."

6. See, for example, Kohler, *All Creatures*, 10–13, 94–98; Vuilleumier and Andors, "Origins and Development of North American Avian Biogeography."

7. See, for example, Heggie, "Why Isn't Exploration a Science?"

8. Barbour, "Zoögeography of the East Indian Islands."

9. On Wallace's line, see Camerini, "Evolution, Biogeography, and Maps"; Vetter, "Wallace's Other Line"; Quammen, *Song of the Dodo*, 25–27.

10. Barbour, "Zoögeography of the East Indian Islands," 166.

11. Wallace, *Geographical Distribution of Animals*, 78–80.

12. Winsor, *Reading the Shape of Nature*, 252.

13. On the congress, see Fernós, *Amistad y Progreso*.

14. Barbour, *Naturalist at Large*, 97, 119; Barbour, "Herpetology of Jamaica," 271.

15. Today known as the Cayman Trough, this feature was discovered during Agassiz's 1880 expedition on the U.S. Coast Survey Steamer *Blake*.

16. Barbour, "Zoögeography of the West Indies," 227.

17. Barbour and Noble, "Revision of the Lizards of the Genus Ameiva," 417–18.

18. Barbour, "Matthew's 'Climate and Evolution,'" 8.

19. Barbour, "Herpetology of Jamaica," 273, 293–99; McCook, "Neo-Columbian Exchange," 26–27.

20. On this phenomenon, see Barrow, *Nature's Ghosts*, 117–18, 133, 262.

21. In emphasizing ancient life forms, Barbour followed, among others, the German-Brazilian biogeographer Hermann von Ihering's "History of the Neotropical Region." See also Lomolino, Sax, and Brown, *Foundations of Biogeography*, 11. Incidentally, von Ihering established his own tropical station at Alto da Serra in Brazil, donated to the state in 1909. It remained little known to U.S. researchers. Drummond, Franco, and Ninis, "Brazilian Federal Conservation Units," 467; Dean, *With Broadax and Firebrand*, 261; Blakeslee, "Paradise for Plant Lovers."

22. Barbour, "Herpetology of Jamaica," 282–85.

23. Their presence in the West Indies also suggested that the islands had never been entirely submerged.

24. Barbour, "Matthew's 'Climate and Evolution'"; Barbour, "For Zoogeographers Only"; Winsor, *Reading the Shape of Nature*, 245–66.

25. See also Rainger, *Agenda for Antiquity*, chap. 8.

26. Barbour, "Matthew's 'Climate and Evolution,'" 1. Barbour did not attack Matthew's assumptions about the relative fitness of northern species, let alone the racism and imperialism that scholars have noted in the theory's discussion of humans. Maasen, Mendelsohn, and Weingart, *Biology as Society*, 117–19; Bowler, *Earth Encompassed*, 345–46; Quintero Toro, *Birds of Empire, Birds of Nation*.

27. Barbour, "Matthew's 'Climate and Evolution,'" 8.

28. Ibid., 2, 6.

29. Bigelow, "Thomas Barbour," 18.

30. Barbour to David Fairchild, 24 December 1930, "Fairchild, David, 1929–30, 31," box 14, UAV 298.20, HUA.

31. Allen, *Growth of E. Atkins & Co.*, 18–21; Atkins, *Sixty Years in Cuba*, 30–137; Ayala, *American Sugar Kingdom*, 89–94; Pérez, *Cuba and the United States*, 56–58.

32. For more on Atkins, science, and Soledad, see Raby, "Making Biology Tropical," 72–110; Singerman, "Inventing Purity," 52–105.

33. McCook, *States of Nature*, 56–60. Loss of productivity over time was more likely due to soil exhaustion. Funes Monzote, *From Rainforest to Cane Field*; Tucker, *Insatiable Appetite*, 7–42.

34. Fernández Prieto, "Islands of Knowledge"; Fernández Prieto, *Espacio de Poder*; Fernández Prieto, "Azúcar y Ciencia."

35. Overfield, "Science Follows the Flag"; Fernández Prieto, "Saberes Híbridos."

36. Ames and Plimpton, *Oakes Ames*, 145–46.

37. Edgar Anderson, quoted in Ames and Plimpton, *Oakes Ames*, 389. The course inspired Richard Evans Schultes, Ames's most famous student, to study medicinal and hallucinogenic plants in Mexico and the Amazon. Ames and Schultes's relationship is recounted in Davis, *One River*.

38. See correspondence in UAV 231, HUA, and McCook, *States of Nature*, 58. On the water lily *Victoria regia*'s symbolism of tropical exuberance and colonial power, see Stepan, *Picturing Tropical Nature*, 33.

39. Ames to Blanche Ames, 18 January 1903, "Cuban Letters," box 1, HUG 4139.5, HUA. See also Ames and Plimpton, *Oakes Ames*, 145.

40. On how station science expanded into terrestrial environments around the turn of the twentieth century, see de Bont, *Stations in the Field*; Vetter, "Labs in the Field?"

41. Barbour, "Bermudan Fishes." The station, renamed the Bermuda Institute of Ocean Sciences in 2007, was established in its current permanent location on Ferry Reach, St George's, in 1931. See Muka, "Working at Water's Edge"; Bermuda Biological Station for Research, *First Century*, xii, 8.

42. Ebert, "Carnegie Institution of Washington and Marine Biology," 180–82; Kohler, *Landscapes and Labscapes*; Muka, "Working at Water's Edge," 7–9.

43. Schneider, "Local Knowledge, Environmental Politics"; de Bont, *Stations in the Field*.

44. Kohler, *All Creatures*, 12–13; Vetter, *Field Life*, 138–202.

45. On hotels and colonial circulation, see Peleggi, "Social and Material Life of Colonial Hotels."

46. Henderson and Powell, "Thomas Barbour and the Utuwana Voyages," 303–4.

47. Barbour, *Naturalist at Large*, 63.

48. Pruna, "National Science in a Colonial Context"; Pruna, *Ciencia y Científicos en Cuba Colonial*.

49. Barbour, *Naturalist in Cuba*, 88.

50. Barbour, "Zoögeography of the West Indies," 255; Barbour, *Naturalist in Cuba*, 88–89.

51. Barbour, *Naturalist at Large*, 81–85; Barbour and Ramsden, "Herpetology of Cuba," 81.

52. Barbour and Ramsden, "Herpetology of Cuba."

53. Pérez, *Intervention, Revolution, and Politics in Cuba*, 88–103.

54. Thomas Barbour, 29 March 1917, 837.00/1279; 20 January 1918, 123B23/134; 18 January 1919, 837.124/11, State Department Central File, RG59; and Thomas Barbour to Major Sherman Miles, 2 December 1920, Military Intelligence Division File 51-312, RG165, NARA.

55. Harris, "Edwin F. Atkins."

56. Barbour, *Naturalist in Cuba*, 5.

57. Barbour, *Naturalist at Large*, 97.

58. Ames and Plimpton, *Oakes Ames*, 72.

59. See Barbour's correspondence with Atkins and his statement about his friendship with Atkins in "Reports," box Q-Z, UAV 231.5, HUA.

60. Oakes Ames, "The Atkins Garden and Research Laboratory," n.d., box 2, HUG 4139.5, HUA; East and Weston, *Report on the Sugar Cane Mosaic Situation*; Harvard University, *Report, 1919–1920*, 240–41.

61. Harvard University, *Report, 1919–1920*, 203–4; "Atkins Fund," box 2, Atkins Institution Papers, Arnold Arboretum Archives of Harvard University.

62. Edwin F. Atkins to the President and Fellows of Harvard College, 2 January 1920, "Mrs. Edwin F. Atkins," box A-P, UAV 231.5, HUA.

63. New York Zoological Society, *Annual Report, 1916*, 53–54. On the establishment of the station, see also Bridges, *Gathering of Animals*, 364–67; Gould, *Remarkable Life of William Beebe*, 188–202; Enright, *Maximum of Wilderness*, 88–97.

64. Roosevelt, "Naturalists' Tropical Laboratory," 46.

65. Beebe, Hartley, and Howes, *Tropical Wild Life*, ix. On the tour, see Roosevelt, "Where the Steady Trade-Winds Blow"; Canfield, *Roosevelt in the Field*, 364–65.

66. Gould, *Remarkable Life of William Beebe*, 35, 86.

67. See, for example, Niles and Beebe, *Our Search for a Wilderness*, 209–10, 314.

68. Barrow, *Passion for Birds*, 27–30; Kohler, *All Creatures*, 231–39; Johnson, *Ordering Life*, 62–67.

69. "Handbook of the Ecological Society of America," 11; Beebe, *Geographic Variation*, 15.

70. Niles and Beebe, *Our Search for a Wilderness*, 280–81, 336–37.

71. Ibid., 313.

72. "Pará trip," box 1, 1005B, WCS. Today the forest is the Parque Estadual do Utinga.

73. Beebe, *Notes on the Birds of Pará*, 55.

74. Beebe, "Yard of Jungle," 44.

75. Beebe, "Fauna of Four Square Feet," 107–8.

76. Ibid.; Beebe, "Yard of Jungle."

77. Tobey, *Saving the Prairies*, 48–75; Kohler, *Landscapes and Labscapes*, 100–107.

78. Beebe, "Fauna of Four Square Feet," 112.

79. Ibid., 117.

80. Ibid., 115.

81. Ibid., 116.

82. Warming, "Vegetation of Tropical America."

83. Beebe, "Fauna of Four Square Feet," 118.

84. Ibid., 119.

85. Beebe, Hartley, and Howes, *Tropical Wild Life*, 23.

86. Beebe Journal, p. 105, 23 March 1909, folder 5, box 4, William Beebe Papers, 1830–1961: C0661, PUA; "Tropical Research Station field reports, 1916–1922, 1924," box 1, 1005A, WCS.

87. Bridges, *Gathering of Animals*, 292–94, 301; Gould, *Remarkable Life of William Beebe*, 189.

88. New York Zoological Society, *Annual Report, 1916*, 82.

89. New York Zoological Society, *Annual Report, 1915*, 57.

90. Beebe, Hartley, and Howes, *Tropical Wild Life*, 27.

91. On early twentieth-century ideas about *wild life*, as distinct from today's *wildlife*, see Benson, "From Wild Lives to Wildlife and Back."

92. Ishmael, *Guyana Story*, 52; Hyles, *Guiana and the Shadows of Empire*, 4.

93. On Beebe's ideas about wilderness, see Enright, *Maximum of Wilderness*, 73–110; Kroll, *America's Ocean Wilderness*, 65–94.

94. Beebe, *Jungle Peace*, 141.

95. Ibid., 142.

96. See, for example, McConnell, *Land of Waters*, 101; Ishmael, *Guyana Story*, 251.

97. Hunter, "Tropical Wild Life," 21.

98. Niles and Beebe, *Our Search for a Wilderness*, 237. On how this myth shaped the colony's geographical construction, see Burnett, *Masters of All They Surveyed*.

99. Beebe, *Jungle Peace,* 142.

100. Ibid., 144.

101. Enright, *Maximum of Wilderness,* 106.

102. Beebe, quoted in Gould, *Remarkable Life of William Beebe,* 195.

103. Beebe, *Jungle Peace,* 89–90.

104. See, for example, ibid., 140, 146; Beebe, *Edge of the Jungle,* 114, 231.

105. Beebe, Hartley, and Howes, *Tropical Wild Life,* 27.

106. Ibid.

107. Beebe, *Jungle Peace,* 147.

108. See Canfield, *Roosevelt in the Field,* 327–55; Millard, *River of Doubt.*

109. Beebe, *Jungle Peace,* 134, 140; Gould, *Remarkable Life of William Beebe,* 194.

110. See, for example, Beebe, *Jungle Peace,* 78, 106, 163, 173, 175.

111. Gould, *Remarkable Life of William Beebe,* 195, 212.

112. Beebe, *Jungle Peace,* 142.

113. Beebe, Hartley, and Howes, *Tropical Wild Life,* 44, 75.

114. "Tropical Research Station field reports, 1916–1922, 1924"; Beebe, quoted in Gould, *Remarkable Life of William Beebe,* 194. See also Taylor, Taylor, and Moore, *Selected Letters of Anna Heyward Taylor,* 97–109, 156–74.

115. Beebe, Hartley, and Howes, *Tropical Wild Life,* 174.

116. Ibid., 167.

117. Beebe, *Jungle Peace,* 147.

118. Ibid., 4.

119. Ibid., 3.

120. "Sterling Debenture Arrests"; "Raid Men Who Sold $10,000,000 in Stock"; "Bartica Rubber Company." It became a textbook case of such fraud: "Bartica Company Annual Report, 1912," 390.

121. New York Zoological Society, *Annual Report, 1916,* 53.

122. Beebe, Hartley, and Howes, *Tropical Wild Life,* xiii.

123. "Tropical Research Station field reports, 1916–1922, 1924."

124. See Gould, *Remarkable Life of William Beebe,* 204–6.

125. Beebe to Madison Grant (secretary of the New York Zoological Society), 22 September 1916, box 1, folder "Tropical Research Station Field Reports 1916–1924," 1005A; see also "History of the Department of Tropical Research: assorted memoranda, lists, and drafts, 1900–1961," box 1, 1005A, WCS; Weed, *Mines Handbook,* 2002; Josiah, *Migration, Mining, and the African Diaspora,* 49–50.

126. New York Zoological Society, *Annual Report, 1920,* 52; New York Zoological Society, *Annual Report, 1921,* 104. See also "History of the Department of Tropical Research"; "Tropical Research Station financial statements and fundraising, 1919–1922," box 1, 1005A, WCS.

127. "Tropical Research Station field reports, 1916–1922, 1924."

128. Beebe, Hartley, and Howes, *Tropical Wild Life*; Beebe, *Jungle Peace*; Beebe, *Edge of the Jungle.*

129. New York Zoological Society, *Annual Report, 1919,* 117.

130. The Bronx Zoo briefly maintained a handful in the late 1990s. "Census of Rare Animals in Captivity," 575. No major zoos currently hold *Opisthocomus hoazin*. International Species Information System Member Support to author, email, 7 October 2015.

131. Beebe to William Hornaday, 10 July 1920, folder 14, box 15, Beebe Papers.

132. "Tropical Research Station field reports, 1916–1922, 1924"; "Emerson, Alfred," box 4, and "Kartabo, 1916–1922" and "British Guiana, 1916–1922" (guest books), box 5, 1005A, WCS.

133. Barbour, "Recent Check-List," 284.

134. Ibid., 284–85.

135. Beebe, Hartley, and Howes, *Tropical Wild Life*, 255.

136. Penard, "Remarks on Beebe's 'Tropical Wild Life,'" 224; Beebe, Hartley, and Howes, *Tropical Wild Life*, 255–57.

137. Barbour, "Recent Check-List," 286.

138. Beebe, "Higher Vertebrates," 205.

139. Barbour, "Recent Check-List," 286.

140. Ibid., 288 (italics added). Modern workers have also criticized Beebe's lack of voucher material. Because "few surveys have been done" in Guyana, this has serious conservation consequences. Donnelly, Chen, and Watkins, "Sampling Amphibians and Reptiles," 55–56.

141. Beebe to William Hornaday, 16 September 1920, folder 14, box 15, Beebe Papers.

142. Barbour, "Recent Check-List," 286.

143. Ibid., 288.

144. Beebe to Hornaday, 16 September 1920, folder 14, box 15, Beebe Papers.

145. See Barbour and Noble, "Revision of the Lizards of the Genus Ameiva"; Beebe, "Lizards of the Genus Ameiva."

146. Dupree, *Science in the Federal Government*, 309–11, 326–27.

147. The idea of an institute dedicated to research in the tropics was first suggested at the 1919 annual meeting of the NRC's Division of Biology and Agriculture. The organization followed up by forming a Committee on Phytopathology in the Tropics to address problems of concern to U.S. fruit, sugar, coffee, and rubber interests. Then, citing the unique political conditions of each area, it held two conferences in 1920, one on scientific research in the Philippines and another on research in tropical America. "Proceedings: November, 1920," Conference on Scientific Research in Philippines and Other Tropical Countries, and "National Research Council Division of Biology and Agriculture Conference on Research in Tropical America," 12 June 1920, Conference on Research in Tropical America, NAS.

148. When the group decided against dual memberships, Alexander Ruthven, and later Stephen Forbes, served as representatives of the Ecological Society of America. Forbes to Ruthven, 21 July 1921, folder 1, box 1, RU134, and "Circular of Information," 16 March 1922, folder 1, box 25, RU135, SIA.

149. "Outline of a Plan for the Work of the Institute," 1, 13 April 1922, folder 1, box 25, RU135, SIA.

150. Ibid., 1.

151. Ibid., 2.

152. "Summary of the Situation as Expressed in preliminary Announcements," IRTA, NAS; "Outline of a Plan for the Work of the Institute," 2, 3.

153. "Outline of a Plan for the Work of the Institute," 3.

154. On the ideology of basic science after World War II, see Appel, *Shaping Biology*, 2, 18–19.

155. "Outline of a Plan for the Work of the Institute," 2.

156. "Plan for Comprehensive Investigations of the Animal and Plant Life of South and Central America," [1920?], folder 2, box 25, RU135, SIA.

157. Wheeler, "Organization of Research," 54. See also Manning, *Black Apollo of Science*, 85.

158. Wheeler, "Organization of Research," 65.

159. Barbour to Hitchcock, 24 February 1922, folder 2, box 1, RU134, SIA. See also Osborn to Beebe, 3 February 1921, folder 5, box 16, PU.

160. "Outline of a Plan for the Work of the Institute," 3.

161. Ibid. See also "Correspondence in re Desirable Location for a Tropical Laboratory," "Duncan Starr Johnson Correspondence, Institute for Research in Tropical America, 1920–21," box 1, RG04.070, JHU.

162. "Plan for Comprehensive Investigations."

163. Hitchcock later reflected, "I am not sure that this was a wise thing to do. We probably must look to interested industrial institutions for the bulk of our funds." Hitchcock to Fairchild, 27 March 1924, folder "Barro Colorado Laboratory," David Fairchild Collection, FTG.

164. "IRTA Constitution," 1921, folder 1, box 25, RU135, SIA. Franklin Sumner Earle, of the Central Aguirre sugar company, Puerto Rico, and John R. Johnston of the United Fruit Company were the first such associate members.

165. "Plan for Comprehensive Investigations."

166. Hitchcock was a man of liberal social views, including international cooperation and support of women's rights, which often put him at odds with colleagues. For example, see Henson, "Invading Arcadia," 582; Henson, "'What Holds the Earth Together.'" Earle's opinion was shaped by his experience as director of Cuba's Estación Experimental Agronómica de Santiago de las Vegas. See McCook, *States of Nature*, 50–53.

167. Hitchcock to "the members of the Institute for Research in Tropical America," 16 April 1921, folder 1, box 1, RU134, SIA.

168. "National Research Council Division of Biology and Agriculture Committee on Organization and Incorporation for the Institute for Research in Tropical America," 15 January 1921, folder 1, box 25, RU135, SIA.

169. "Plan for Comprehensive Investigations."

170. Johnson to Earle, 23 December 1921, "Duncan Starr Johnson Correspondence."

171. Ibid.

172. Barbour to Frank R. Lillie (NRC), 26 September 1922, folder 2, box 25, RU135, SIA.

173. "National Research Council Division of Biology and Agriculture Committee." The National Science Foundation would play a similar role in facilitating tropical biology later in the century; see chapter 5 and Appel, *Shaping Biology*, 197–205.

174. Such an association could have helped to maintain Cinchona, or possibly the Carnegie Institution's foundering marine station at Dry Tortugas. Ebert, "Carnegie Institution of Washington and Marine Biology." The possibility of IRTA taking over the Tortugas station is discussed in Hitchcock to Members of the Executive Committee, IRTA, 15 January 1923, folder 2, box 1, RU134, SIA.

175. "History of the Department of Tropical Research"; New York Zoological Society, *Annual Report, 1921*, 105.

176. See, for example, the selection by John R. Johnston in "Correspondence in re Desirable Location for a Tropical Laboratory" and I. W. Bailey to Duncan Johnson, 9 February 1921, "Duncan Starr Johnson Correspondence."

177. Barbour, *Naturalist at Large*, 193.

178. Barbour to Lillie, 26 September 1922.

179. In the Philippines, Strong had overseen the government biological laboratory (focusing on medical diagnoses and research). Anderson, *Colonial Pathologies*, 111–12; Anderson, "Science in the Philippines," 299–300.

180. "Fairchild Panama Trip Aug.–Sept, 1921," Expedition Reports, FTG; Fairchild, "Jungles of Panama," 145.

181. Barbour to Ruthven, 14 July 1923, folder 1, box 1, RU134, SIA.

182. Barbour, "More About Harvard Biology," 212.

183. Ibid.

184. Emerson, "Jungle Laboratory of Tropical Biology."

Chapter Three

1. On the canal's construction, the transformation of the Chagres Valley, and the formation of Barro Colorado Island, see Carse, *Beyond the Big Ditch*; Castro Herrera, "Isthmus in the World"; Harmon, *Río Chagres*; Foster and Brokaw, "Structure and History of the Vegetation of Barro Colorado Island"; Leigh, *Tropical Forest Ecology*, 3–12.

2. Board of Regents of the Smithsonian Institution, *Annual Report, 1921*, 12. See also Duncan Johnson to Charles Walcott (Smithsonian secretary), 15 February 1921, folder 8, box 11, RU45, SIA. Duncan Johnson increasingly took advantage of the United Fruit Company's Jamaican landholdings as field sites. Among his students was the influential ornithologist and environmental philosopher Alexander Skutch. Skutch, *Bird Watcher's Adventures*, 29; Abarca Jiménez, *Alexander Skutch*.

3. On Beebe's transition to marine biology, see Kroll, *America's Ocean Wilderness*, 68–70; Gould, *Remarkable Life of William Beebe*, 221–342.

4. Fairchild, *World Was My Garden*, 468; Chapman, *My Tropical Air Castle*, 3–4.

5. "Mr. James Zetek"; "Zetek"; Snyder, Wetmore, and Porter, "James Zetek"; Heckadon-Moreno, *Naturalistas de Istmo de Panamá*, 139–49.

6. In this, Zetek, like other Canal Zone entomologists, divided his loyalties between science and the U.S. mission in Panama. See Sutter, "Nature's Agents or Agents of Empire?"

7. Zetek to Altus Quaintance (Bureau of Entomology), 25 March 1923, folder 11, box 1, RU134, SIA.

8. Zetek to Albert Spear Hitchcock, 4 March 1923, folder 2, box 25, RU135, SIA.

9. On the land lease program, see Carse, *Beyond the Big Ditch*, 131–56. On labor and race in the Canal Zone, see Greene, *Canal Builders*; Conniff, *Black Labor on a White Canal*.

10. Zetek to Governor Jay J. Morrow, 19 April 1923, folder 2, box 25, RU135, SIA.

11. Zetek to Victor Shelford, 22 April 1923, folder 2, box 25, RU135, SIA.

12. Schneider, "Local Knowledge, Environmental Politics"; McIntosh, *Background of Ecology*, 58–59.

13. "Handbook of the Ecological Society of America."

14. Kinchy, "Borders of Post-war Ecology," 32. See also Tjossem, "Preservation of Nature and Academic Respectability."

15. Rumore, "Preservation for Science."

16. Shelford, *Naturalist's Guide*, 3.

17. Their effort was able to extend only as far south as the Amazon River. The inclusion of the Philippines in this guide to the "Americas" perhaps best reveals how the project depended more on networks of U.S. science and empire than on physical geography. Tropical stations already played an important role in this scientific geography: William Beebe authored the chapter on the Guianas; Forrest Shreve authored the Jamaican chapter. The only Latin American authors on the project were the European-born botanists Henri Pittier for Venezuela and Brother León (Joseph Sylvestre Sauget, a frequenter of the Soledad station) for Cuba. Shelford's committee unsuccessfully requested contact information from IRTA for a wider array of Latin American naturalists: Shelford to J. R. Schramm (National Research Council), 6 May 1923, folder "Proposed: Committee on Preservation of Natural Conditions of Ecological Society of America: 1920–1921, Funds for Naturalist's Guide," Projects, Division of Biology and Agriculture Collection, NAS.

18. Barbour to members of the Institute for Research in Tropical America, 14 July 1923, folder 1, box 25; Hitchcock to Zetek, 19 February 1923, folder 2, box 25; Zetek, "Report of the Canal Zone Field Station, July 1, 1922–June 30, 1923," folder 2, box 11; Zetek, "The History of Barro Colorado, Part One: The Early Years, 1911–1923," 1949, folder 4, box 21, RU135, SIA.

19. Zetek to Colonel William C. Erwin (land inspector), 20 April 1923, folder 2, box 25, and Zetek, "The History of Barro Colorado," undated, folder 5, box 10, RU135, SIA. To pay tribute to Erwin's assistance, Zetek worked hard but unsuccessfully to rename (and Anglicize) BCI to "Erwin Island."

20. Smithsonian Institution, "Biological Survey," 15. The survey was proposed as early as 1904. Despite an appeal to President Roosevelt casting the project in the imperial mold of the U.S. Great Western Surveys and European colonial surveys, federal

funding was not immediately forthcoming. See, for example, S. P. Langley (Smithsonian secretary) to President Roosevelt, 21 March 1904, folder 14, box 42, RU45, SIA. On the survey itself, see Heckadon-Moreno, *Naturalistas de Istmo de Panamá*, 73–78; Carse, *Beyond the Big Ditch*, 209–11; Henson, "Baseline Environmental Survey."

21. "Plan Zoological Park on Island in Gatun Lake," *Panama Star and Herald*, 16 April 1923, folder 2, box 25, RU135, SIA. On the naturalization of Lake Gatun, see Carse, "'Like a Work of Nature.'"

22. Zetek to Morrow, 19 April 1923, box 25, folder 2, RU135, SIA.

23. Zetek to Shelford, 22 April 1923, box 25, folder 2, RU135, SIA.

24. Whether he believed this exaggerated claim or made it up to drum up press for the proposed park and station is unclear. "Plan Zoological Park on Island in Gatun Lake"; "'Bug Preserve' for the Canal Zone Now Planned," *Panama Star and Herald*, 28 March 1923, box 25, folder 2, RU135, SIA.

25. Recounted in Fairchild to Barbour, 10 September 1924, folder 11, box 1, RU134, SIA.

26. Zetek to Morrow, 27 March 1923, folder 33-B-51(5), entry 34, RG185, NARA. See also Zetek to Morrow, 19 April 1923, folder 2, box 25, RU135, SIA. The former is the missing letter referred to in Sapp, *Coexistence*, 21.

27. Morrow, "The Panama Canal, Canal Zone, Executive Office," 17 April 1923, and Morrow to Zetek, 16 April 1923, folder 2, box 25, RU135, SIA.

28. Zetek, quoted in Governor Morrow to Fairchild, 14 December 1923, folder 2, box 25, RU135, SIA.

29. Allee and Allee, *Jungle Island*, 17–19 (italics added).

30. Starry, "Jungle Island in Panama," 15.

31. Chapman, *My Tropical Air Castle*, 27–28. On Chapman's relationships with Colombian scientists, see Quintero Toro, "Trading in Birds"; Quintero Toro, *Birds of Empire, Birds of Nation*.

32. Van Dyke, *In the West Indies*, 210–11.

33. See, for example, Fairchild to Barbour, 13 August 1923, "1923–1928," box 14, UAV 298.20, HUA. For Fairchild's role in the patronage networks of tropical biology, see also chapters 1 and 2 in this volume.

34. Hitchcock, "Biological Station," x.

35. Kohler, *All Creatures*, 18. Focusing his study on the U.S. West, Kohler has called such areas "inner frontiers." The concept of the "frontier," however, has long been contentious among U.S. Western historians engaging with the colonial history of U.S. expansion, as well as among environmental historians critical of wilderness concepts. (See, e.g., Limerick, *Legacy of Conquest*; Klein, "Reclaiming the 'F' Word"; Cronon, "Trouble with Wilderness.") As a ten-mile strip of U.S. territory carved out of the heart of the Republic of Panama, the Canal Zone also renders discussion of frontiers problematic. (A "borderlands" approach is more appropriate; see Donoghue, *Borderland on the Isthmus*.) I appreciate Kohler's effort to understand the development of intensive field practices in relation to settlement patterns but eschew "inner frontiers" to recognize the human presence on both sides of this boundary, in Panama as in the U.S. West.

36. Chapman, *Life in an Air Castle*, 207–8; Enders, "Mammalian Life Histories," 388.

37. Kenoyer, "Botanizing," 322.

38. Allee, "Barro Colorado Laboratory," 521.

39. Standley, "Flora of Barro Colorado Island," 2. The plant ecologist Leslie Kenoyer, in contrast, suggested that BCI's forest was "probably . . . a little less typical and verges more on the monsoon type than is true over much of the tropical American rain-forest area." He also compared BCI's forest to conditions reported at other forests, largely at stations including Cinchona, Kartabo, and Buitenzorg. Critical evaluations of BCI's typicality were available but rare during the station's early years. Kenoyer, "General and Successional Ecology," 206.

40. James Zetek, "Tropical Biological Station," 1923, folder 3, box 4, RU134, SIA; Chapman, *My Tropical Air Castle*, 10; Leigh, Rand, and Windsor, *Ecology of a Tropical Forest*, 79; Carse, *Beyond the Big Ditch*, 145–54.

41. Statement relinquishing leases, signed by Alman Challa, Manuel Ortega, Temas Polo, and Sixto Rivera, 18 February 1924, folder 3, box 25, RU135, SIA. These were the four leaseholders; seven individuals were reportedly living on the island in 1923. Barbour, "The Tropical Biological Station on Barro Colorado Island in the Panama Canal Zone," 1923, folder "Barro Colorado Biological Laboratory: Beginning of Program: 1922–1923," Institute for Research in Tropical America, Institutions Associations Individuals, Division of Biology and Agriculture Collection, NAS.

42. Lasso, "From Citizens to 'Natives' "; Maurer and Yu, *Big Ditch*, 110–13; Conniff, *Black Labor on a White Canal*, 47.

43. Lasso, "From Citizens to 'Natives.' "

44. Warren, *Hunter's Game*; Jacoby, *Crimes against Nature*. The U.S. approach stood in stark contrast to the approach that would be developed in Mexico, where campesinos continued to inhabit the national parks created in the wake of the Mexican Revolution. Wakild, *Revolutionary Parks*.

45. Goldman and Zetek, "Panama," 620.

46. See, for example, "Steamer Routes, Isthmus to other parts of the world," in Hitchcock, "Report on Panama," 1923, box 4, folder 3, RU134, SIA.

47. Harvard University, *Report, 1923–1924*, 68. See also Katherine Atkins to Barbour, 21 December 1924 (letter and diagram), "Mrs. Edwin F. Atkins," box A-P, UAV 231.5, HUA.

48. "Second Annual Report of the Barro Colorado Biological Station," folder 2, box 11, RU135, SIA.

49. On the pitfalls of this system, see Hagen, "Problems in the Institutionalization of Tropical Biology."

50. See, for example, Barbour to Hartshorn, 24 November 1937, folder 2, box 17, RU135, SIA.

51. Barbour, *Naturalist at Large*, 193–96.

52. Harvard University, *Report, 1924–1925*, 164.

53. Annual reports of the Barro Colorado Biological Station, folder 2, box 11, RU135, SIA.

54. Barbour collaborated with the Brazilian zoologist Afranio do Amaral to develop the company's snake antivenin program there. Ditmars, "Antivenin Institute"; Clark, "United Fruit Company's Cooperative Action."

55. Harvard University, *Report, 1924–1925*, 164; Salt and Myers, *Report on Sugar-Cane Borers*; Harvard University, *Report, 1925–1926*, 171; Harvard University, *Report, 1926–1927*, 174; Harvard University, *Report, 1927–1928*, 183. Tropical scientists became a "reserve labor force" for the company, akin to the science-state relationship explored in Mukerji, *Fragile Power*.

56. See, for example, the correspondence in "United Fruit," box Q–Z, UAV 231.5, HUA.

57. He was employed by the Republic of Panama on several occasions, organizing an exhibition for the republic's 1915–16 National Exposition in Panama City, and from 1916 to 1918 he served as professor of biology and hygiene at the National Institute of Panama. "Mr. James Zetek"; "Zetek"; Snyder, Wetmore, and Porter, "James Zetek."

58. Graham Bell Fairchild, oral history interview by Joel B. Hagen, 7 June 1989, RU009559, SIA.

59. Annual reports of the Barro Colorado Biological Station.

60. See ibid.; "Buildings and Grounds: General Folder 1," box 1, RU135, SIA.

61. Zetek to Barbour, 15 May 1925, folder 3, box 25; Zetek to Barbour, 22 May 1935, and telegraphed reply, 27 May 1935, folder 6, box 22, RU135, SIA. Barbour agreed, despite his eagerness to cooperate with companies in other ways.

62. See reports and correspondence in box 11, RU134, SIA. See also Snyder and Zetek, "Damage by Termites"; Snyder and Zetek, "Test of Wood Preservatives"; Snyder and Zetek, "Test House"; Snyder and Zetek, "Inspection of Test Building."

63. Robert K. Enders, oral history interview by Neal Griffith Smith, 13 April 1976, RU009562, SIA.

64. These figures are estimates based on the visitors reported in the station's annual reports and guest books: folders 6 and 7, box 4, RU134, and box 11, RU135, SIA.

65. Barbour to Katherine Atkins, 30 January 1935, box 1A, UAV 298.20, HUA.

66. Bisby, "Barro Colorado Island Laboratory"; Townsend, "Island in the Tropics."

67. Van Tyne, "Barro Colorado Laboratory as a Station for Ornithological Research"; Mayfield, "In Memoriam."

68. Frost, "Animal Life"; Frost, "Collecting Leaf-Miners."

69. Barbour, *Naturalist at Large*, 197. The number of high school teachers remained low compared to biological stations within the United States, which often placed a much higher value on education. See Benson, "Why American Marine Stations?"

70. The "air castle" of Chapman's titles references Thoreau's experiment in living in the woods near Walden Pond. Thoreau exhorted readers to follow him in "put[ting] the foundations under" their "castles in the air." Thoreau, *Walden*, 346; Chapman thus cast BCI as Walden's tropical counterpart.

71. "Sixty Birds in New Group Exhibited at Museum."

72. Wilcox, "Paradise for Zoologists"; "Noah's Ark in Jungle Isle."

73. Barbour to Zetek, 18 July 1931, box 26, folder 1, RU135, SIA.

74. "Visitors' register 1928–1940," folder 6, box 4, RU134, SIA.

75. See folder 6, box 4, RU134, SIA.

76. Annual reports of the Barro Colorado Biological Station; Enders interview.

77. Hitchcock, "The advantages of a research station in Panama," folder 1, box 25, RU135, SIA.

78. On Chardón and his travels, see Fernández Prieto, "Islands of Knowledge," 796; Brock, "American Empire and the Scientific Survey of Puerto Rico"; McCook, *States of Nature*, 6–10, 105–27.

79. McCook, *States of Nature*, 60.

80. The others included Gonzalo Martínez Fortún y Foyo, Alberto J. Fors y Reyes, and Jorge Dechapelle. Harvard University, *Report, 1937–38*, 369; Barbour, *Naturalist in Cuba*, 304. See also Alvarez Conde, *Historia de la Botanica*, 167, 256, 315.

81. Pruna, *Historia de la Ciencia y la Tecnología*; Pruna, "National Science in a Colonial Context"; Glick, "Science and Society in Twentieth-Century Latin America"; Vessuri, "Academic Science." Panama's Instituto Nacional dates to 1907, and the University of Panama was not founded until 1935. On Panama's continued cultural and intellectual ties to Colombia, and especially Cartagena, see Ropp, "Beyond United States Hegemony."

82. Fernós, *Amistad y Progreso*, 144–47. On ideas of the practical in science in Latin America, see McCook, *States of Nature*, 3; Vessuri, "Academic Science"; Medina et al., *Beyond Imported Magic*.

83. Gould, *Remarkable Life of William Beebe*, 180–84.

84. Barbour to Fairchild, 2 March 1924, folder 4, box l, RU134, SIA.

85. Fairchild to Barbour, 10 September 1924, folder 11, box 1, RU134, SIA.

86. Ibid.

87. Henson, "Invading Arcadia," 587.

88. Carpenter to Barbour, 5 November 1932, box 15, folder 3, RU135, SIA. Jan Sapp has downplayed the barriers to women at BCI, suggesting that overnight restrictions were lifted by the 1930s. Sapp, *Coexistence*, 33. Carpenter's and many other letters, however, demonstrate that housing was indeed not routinely available for families or single women before World War II. Women continued to face many formal and informal obstacles at the station. Into the 1940s and 1950s widespread confusion remained over whether women were banned from overnight stays. See, for example, Alexander Wetmore to Barbour, 4 September 1945, box 4, RU7006; Carl Koford to Wetmore, n.d. December 1950, folder 4, box 8, RU135, SIA.

89. Zetek to Barbour, 6 October 1931, including excerpt of letter from Duncan to Zetek, folder 1, box 26, RU135, SIA. Barbour drew on his scientific network to vet potential BCI visitors. He forwarded this letter to Chapman with the note "What do you know about this bird?" Chapman was an ornithologist, but the question was about Duncan's legitimacy as a researcher.

90. Rau, *Jungle Bees and Wasps*, 5.

91. Ibid., 6.

92. Fairchild interview.

93. Fausto Bocanegra, oral history interviews by Giselle Mora, August 1988, RU009561, SIA. On segregation in the Canal Zone, see Conniff, *Black Labor on a White Canal.*

94. Zetek to Tom Clark, 30 January 1955, folder 4, box 7, RU135, SIA. Privately, workers said they found Zetek's Spanish *bruta*. Enders interview.

95. These distinctions were intended to prevent laborers from organizing. Greene, *Canal Builders*, 123–58.

96. See, for example, Oscar Dean Kidd, oral history interviews by Pamela M. Henson, Patricia Escobar Paramo, Elizabeth Stockwell, Mitch Aide, and Carolien Haverkate, 1987, 1990, RU009574, SIA.

97. Zetek to Barbour, 15 August 1931, folder 1, box 26, RU135, SIA.

98. Chapman, *My Tropical Air Castle*, 212.

99. Zetek to Barbour, 17 March 1935, box 26, folder 3, RU135, SIA.

100. Zetek to Barbour, 8 January 1934, box 26, folder 3, RU135, SIA.

101. See, for example, Zetek to Barbour, 12 May 1927, box 25, folder 4; Zetek to Barbour, 18 September 1932, box 26, folder 2, RU135, SIA.

102. Sutter, " 'First Mountain to Be Removed.' "

103. Enders interview.

104. See correspondence in folder 5, box 25, RU135, SIA.

105. *Petit marronage* describes the actions of slaves who ran away temporarily, a resistance strategy for securing better treatment. See Thompson, *Flight to Freedom*, 53–90.

106. Greene, *Canal Builders*.

107. Beebe to William Temple Hornaday, 10 July 1920, folder 14, box 15, William Beebe Papers, PUA. See also Beebe to Madison Grant, 10 July 1920, folder 12, box 15, Beebe Papers. On the strikes, see Randall and Mount, *Caribbean Basin*, 66.

108. Frank Gorden Walsingham to Barbour, 19 September 1933, "F. G. Walsingham," box Q–Z, UAV 231.5, HUA.

109. McCook, *States of Nature*, 137. See also Pérez, *Cuba: Between Reform and Revolution*, 204–5, 212.

110. Zetek to Chapman, 11 October 1928, folder 4, box 25, RU135, SIA.

111. Bocanegra interviews.

112. Allee and Allee, *Jungle Island*, 39–42.

113. For criticism of the concept of local knowledge, see Raffles, "Intimate Knowledge"; Jacobs, "Intimate Politics of Ornithology"; Tilley, "Global Histories." See also the discussion of Creole science in McCook, "Global Currents in National Histories of Science." At U.S. stations, the intimacies of local and scientific labor sometimes fostered shared environmental politics. Schneider, "Local Knowledge, Environmental Politics." The Canal Zone's segregation seems to have prevented this at BCI.

114. Bocanegra interviews.

115. Ibid.; Pamela M. Henson, oral history interview by Pablo Rodriguez-Martinez, June 1990, RU009581, SIA.

116. Weston, "Biological Station," 381.

117. Ibid., 382.

118. Ibid., 381.

119. Chapman, *Life in an Air Castle,* vii.

120. For example, Gross, "Jungle Laboratory"; Coursen, "Jungle Laboratory."

121. Royte, *Tapir's Morning Bath,* 36. See also Hagen, "Problems in the Institutionalization of Tropical Biology," 230; Sapp, *Coexistence,* 25.

122. Chapman, *Life in an Air Castle,* vii.

123. These can be seen as "archival" field practices. Raby, "Ark and Archive."

124. On Chapman, the emergence of bird watching, and its relationship with professional ornithology, see Barrow, *Passion for Birds,* 156–60; Dunlap, *In the Field, among the Feathered,* 13–36; Vuilleumier, "Dean of American Ornithologists," 389–402.

125. Chapman, *My Tropical Air Castle,* 82. See also Van Tyne, "Barro Colorado Laboratory."

126. Chapman, "Nesting Habits of Wagler's Oropendola," 124.

127. Ibid.; Chapman, *My Tropical Air Castle,* 4, 78–146.

128. Chapman, *My Tropical Air Castle,* 82–83. His long-term study ended with the fall of the oropendulas' nesting tree, weakened by the clearing of adjacent forest. Chapman's observation—via the clearing that made the birds visible to him—did affect the animals' lives.

129. Chapman adapted this apparatus, developed by sportsman photographers, for an explicitly scientific purpose. See Kucera and Barrett, "History of Camera Trapping," 11.

130. Etienne Benson documents the midcentury controversy over the ethics of using such technologies to monitor ostensibly wild animals. On BCI, comparable techniques were uncontroversial. Chapman admitted dismay that camera flashes startled BCI's shy mammals but felt satisfied that camera trapping posed no threat to the "balance of nature" or the authenticity of observations. Compared to killing or live traps, camera traps were less invasive. Perhaps the island's status as a scientific, rather than public, park also bolstered the legitimacy of such techniques. Benson, *Wired Wilderness.*

131. Chapman, "Who Treads Our Trails?," 342; Chapman, *My Tropical Air Castle,* 196.

132. "Seventeen Years of Camera-Trapping."

133. Enders, "Some Mammals from Barro Colorado Island," 281.

134. Enders, "Mammalian Life Histories," 391.

135. Enders, "Some Mammals from Barro Colorado Island," 281.

136. STRI, "BCI Species Database," http://biogeodb.stri.si.edu/biodiversity/bci/ (accessed 2 October 2014); Allen, "Varied Bats of Barro Colorado Island"; Kalko and Handley, "Neotropical Bats."

137. Carpenter, "Behavior and Social Relations of Howling Monkeys," 2; Haraway, *Primate Visions,* 88–90.

138. Montgomery, *Primates in the Real World,* 51–70.

139. See Carpenter, "Behavior and Social Relations of Howling Monkeys"; Carpenter, "Behavior of Red Spider Monkeys"; Carpenter, "Behavior and Social Relations of Free-Ranging Primates."

140. Carpenter to Barbour, 9 February 1933 and Carpenter to Barbour, 15 December 1935, box 15, folder 3, RU135, SIA; Carpenter, "Behavior of Red Spider Monkeys," 175. See also Haraway, *Primate Visions,* 91.

141. Montgomery, *Primates in the Real World*, 57, 66.

142. Enders, "Mammalian Life Histories," 389.

143. Montgomery, "Place, Practice and Primatology"; Montgomery, "Primates in the Real World," 85–86; Rees, *Infanticide Controversy*, 27–46.

144. Rees, "Place That Answers Questions"; Rees, *Infanticide Controversy*, 27–46.

145. Chapman, *My Tropical Air Castle*, 168.

146. Kenoyer, "Fern Ecology of Barro Colorado Island," 7; Croat, *Flora of Barro Colorado Island*, 15.

147. Barbour, "Sixth Annual Report," 18 March 1930, box 11, folder 2, RU135, SIA.

148. Scott, *Seeing Like a State*, 11–22.

149. James Zetek, June 1951 (copy), "Information for Scientists," p. 24, box 4, folder 5, RU134, SIA.

150. Nils Lindahl Elliot also notes this hierarchy at work in "Memory of Nature," 251–52.

151. J. P. McEvoy, "Barro Colorado, Tropical Noah's Ark," *Reader's Digest*, May 1955, box 21, folder 4, RU135, SIA.

152. Zetek, "Information for Scientists," 24. This card cabinet still exists, located in the herbarium on BCI. It was last updated during the directorship of Carl Koford.

153. Calhoun, "Jungle Folk Sanctuary."

154. Calhoun, "Panama's Island Ark."

155. See also Raby, " 'Jungle at Our Door.' "

156. Chapman, *My Tropical Air Castle*, 2–4. A nearly identical statement without qualifications appears in Chapman, "In an Eden Where Man Befriends Beast."

157. Allee and Allee, *Jungle Island*, 2.

158. Zetek to Shelford, 22 April 1923, box 25, folder 2, RU135, SIA.

159. Chapman, "Island Ark," 61; Chapman, *My Tropical Air Castle*, 5.

160. Calhoun, "Jungle Folk Sanctuary," 13; Chapman, *My Tropical Air Castle*, 5.

161. Chapman, *My Tropical Air Castle*, 5–6.

162. Ibid., 4.

163. Weston, "Biological Station," 380.

164. Fairchild, *World Was My Garden*, 468.

165. Chapman, *My Tropical Air Castle*, 11.

166. Enders, "Mammalian Life Histories," 389.

167. Enders, "Changes Observed," 105.

168. Annual reports of the Barro Colorado Biological Station, folder 2, box 11, RU135; Enders interview.

169. Chapman, *Life in an Air Castle*, 218–19.

170. Zetek, "First Annual Report," 1941, folder 4, box 11, RU135, SIA.

171. Hagen, "Problems in the Institutionalization of Tropical Biology," 238.

172. Leigh, "Barro Colorado," 91.

173. Standley, *Flora of the Panama Canal Zone*, 3–4. By today's standards, BCI's flora was barely known in those early years of the station, but Standley's work provided an important foundation for the more intensive plant taxonomy and ecology completed in the 1960s and 1970s; see the epilogue in this volume.

174. Park, "Studies in Nocturnal Ecology," 210.

175. See Billick and Price, "Ecology of Place"; de Bont, *Stations in the Field.*

176. Beebe, "Fauna of Four Square Feet"; Beebe, *Studies of a Tropical Jungle.*

177. Emerson to Barbour, 20 February 1929, folder 1, box 2, RU134, SIA. This was his second request for cooperation; his first was Emerson to Barbour, 9 February 1928, "Alfred Emerson," box 13, UAV 298.20, HUA. Emerson emphasized the value of maintaining multiple tropical stations for ecological comparison. He had tried to keep Kartabo going as a "summer station" of the University of Pittsburgh after Beebe's departure but no longer had the support of his department's chair. Warder Allee unsuccessfully revisited the issue in the early 1930s. See correspondence between Allee and Barbour in folder 1, box 15, RU135, SIA.

178. Chapman, *My Tropical Air Castle,* 1.

Chapter Four

1. Allee, Emerson, and Thomas Park worked at the University of Chicago, Orlando Park at Northwestern, and Schmidt at the Field Museum.

2. Victor Dropkin to James Zetek, 7 December 1939, and Ralph Buchsbaum to James Zetek, 18 November 1939, folder 7, box 2, RU134, SIA. On the Ecology Group, see Mitman, *State of Nature,* 134.

3. Dropkin, "Host Specificity Relation of Termite Protozoa"; Dropkin, "Use of Mixed Colonies of Termites." Emerson also cited Dropkin's unpublished work as evidence of how termites use chemical markers to distinguish colony members from outsiders; Emerson, "Social Coordination and the Superorganism," 191.

4. Ralph Buchsbaum, best known for his 1938 book *Animals without Backbones,* exposed a broad set of audiences to BCI. He presented footage of BCI at various venues, including the 1941 meeting of the Illinois State Academy of Science, where almost a thousand members attended his talk "A Summer in a Tropical Rain Forest." Paton, "The Illinois State Academy of Science." In the 1960s, as a producer of films for *Encyclopædia Britannica,* Buchsbaum featured BCI among his subjects.

5. Allee et al., *Principles of Animal Ecology,* 237. Buchsbaum's participation would be terminated by World War II. Schmidt, "Warder Clyde Allee," 17.

6. Buchsbaum to Zetek, 18 November 1939, folder 7, box 2, RU134, SIA.

7. Ibid.; Dropkin to Zetek, 7 December 1939, folder 7, box 2, RU134, SIA. Dropkin stated that more attendees had been to BCI than to the Marine Biological Laboratory, while Ralph Buchsbaum claimed there were "exactly the same number" who had visited each station. According to Buchsbaum, these included Allee, the Buchsbaums, Dropkin, Emerson, Orlando Park, his students Albert Barden and Eliot Williams Jr., and Paul Voth. Karl Schmidt and Nicholas Collias were also regulars at the Ecology Group and BCI visitors.

8. Dropkin to Zetek, 7 December 1939, folder 7, box 2, RU134, SIA.

9. Buchsbaum to Zetek, 18 November 1939, folder 7, box 2, RU134, SIA.

10. McCook, " 'World Was My Garden.' "

11. Robert K. Enders, oral history interview by Neal Griffith Smith, 13 April 1976, RU009562, SIA.

12. Mitman, "When Nature *Is* the Zoo," 129. See also Mitman and Burkhardt, "Struggling for Identity"; Collias, "Role of American Zoologists and Behavioural Ecologists."

13. Buchsbaum to Zetek, 18 November 1939, folder 7, box 2, RU134, SIA.

14. Mertz, "Thomas Park."

15. Buchsbaum to Zetek, 18 November 1939, folder 7, box 2, RU134, SIA.

16. Ibid.

17. Mertz, "Thomas Park"; Kingsland, *Modeling Nature*, 183. For his views on the quantitative study of populations and communities, see Park, "Analytical Population Studies in Relation to General Ecology"; Park, "History and Scope of Population Ecology."

18. See Erickson, *World the Game Theorists Made*; Bocking, *Ecologists and Environmental Politics*; Edwards, *Closed World*; Kwa, *Mimicking Nature*.

19. Harvard University, *Report, 1942–1943*.

20. Howard, "Role of Botanists during World War II," 1994, 234–35.

21. Barbour, "Marine Biological Laboratories"; Ebert, "Carnegie Institution of Washington and Marine Biology."

22. Reports 1–5, box 1, 1005A, WCS. See also Gould, *Remarkable Life of William Beebe*, 347–49. On the Rockefellers' ventures in Venezuela, see Rivas, *Missionary Capitalist*.

23. Beebe, *High Jungle*, 60. Pittier in fact originally conceived of the idea of a station in this park but had been unsuccessful in establishing one. On his difficult work under Vincente Gómez, see McCook, *States of Nature*, 34–45.

24. "Expedition of the Department of Tropical Research," 422.

25. The Committee for Inter-American Artistic and Intellectual Relations was a nongovernmental organization run by representatives of the Guggenheim, Carnegie, and Rockefeller Foundations but funded by the U.S. government's Office of the Coordinator of Inter-American Affairs, a propaganda department headed by Nelson Rockefeller. *History of the Office of the Coordinator of Inter-American Affairs*, 94.

26. "Brazil-US Agreement," 8; "Science News"; Wadsworth, "Forest Research Institution"; West, "USDA Forest Service."

27. Howard, "Role of Botanists during World War II," 2000, 83.

28. Judith A. Warnement, "An Interview with Richard A. Howard at the Harvard University Herbaria on April 15, 1993," Papers of Richard Howard, Arnold Arboretum Archives of Harvard University.

29. Howard, "Role of Botanists during World War II," 1994, 207.

30. Montgomery, *Primates in the Real World*, 65.

31. Wilson, "Philip Jackson Darlington, Jr"; Darlington, "Experiments on Mimicry"; Darlington, "Origin of the Fauna of the Greater Antilles."

32. Skutch, *Bird Watcher's Adventures*, 109–44.

33. On U.S. and Latin American networks of science in the Pacific and the exploitation of a key resource, see Cushman, *Guano and the Opening of the Pacific World*.

34. Annual reports of the Barro Colorado Biological Station, box 11, RU135, SIA.

35. Harvard University, *Report, 1941–42,* 411; Harvard University, *Report, 1942–43,* 294. See also Finlay, *Growing American Rubber.*

36. Harvard University, *Report, 1942–43,* 238; Harvard University, *Report, 1947–1948,* 251.

37. Barbour to William Crocker (National Research Council), undated, folder 2, RU135, SIA.

38. The island retained its name. *BCI* and *CZBA* were used interchangeably during this period. I use *BCI* when referring to the island or station site itself and *CZBA* when referring to this managing board.

39. First annual report of the Canal Zone Biological Area, folder 4, box 11, RU135, SIA.

40. Even before the U.S. declaration of war, the number of soldiers in the Canal Zone doubled to more than 22,000 between 1939 and 1940. By 1943 there were 63,000. Lindsay-Poland, *Emperors in the Jungle,* 45.

41. Testing continued into the 1970s; the full extent was not made clear to Panamanians until the 1990s, during preparations for the handover of the canal to Panama. Environmental remediation is still ongoing. Ibid., 72–73; Lindsay-Poland and Fellowship of Reconciliation (U.S.) Chemical Weapons Working Group, *Test Tube Republic*; Pugliese, "Panama."

42. John Lindsay-Poland notes that tests focused on racial differences in sensitivity to chemical weapons. Lindsay-Poland and Fellowship of Reconciliation (U.S.) Chemical Weapons Working Group, *Test Tube Republic,* 1; Lindsay-Poland, *Emperors in the Jungle,* 57–58.

43. Wetmore to E. F. Bullene, 30 November 1943, folder 1, box 153, RU7006, SIA. See also Pamela M. Henson, oral history interview by Neal G. Smith, 6 June 1990, RU9584, SIA; Lindsay-Poland, *Emperors in the Jungle,* 44–73. Robert MacArthur and E. O. Wilson would later use Wetmore's censuses in formulating their theory of island biogeography.

44. Board of Regents of the Smithsonian Institution, *Annual Report, 1947,* 4.

45. Loveridge, "Thomas Barbour, Herpetologist," 39.

46. Barbour, *Naturalist at Large,* 193.

47. "Canal Zone Biological Area Board of Directors Annual Meeting," 21 June 1946, and Wetmore to Zetek, 16 July 1946, folder 2, box 7, RU135, SIA.

48. In his optimism about international unity and cooperation, as in his ecological ideas, Park drew from his training at Chicago. Mitman, *State of Nature.*

49. Park, "Future of Ecology," 6.

50. Ibid., 8.

51. Ibid.

52. Ibid., 7 (italics added).

53. Ibid., 6–7.

54. Elton, *Animal Ecology,* xiii.

55. Ibid., 28–29.

56. Ibid., xiv–xvii. On Amani, see Conte, "Imperial Science, Tropical Ecology, and Indigenous History"; Geissler and Kelly, "Field Station as Stage." On Huxley, Elton, and ecology in Africa, see Anker, *Imperial Ecology.*

57. Park, "History and Scope of Population Ecology."

58. Bates, "Advantage of the Tropical Environment for Studies on the Species Problem," 235. It is notable that Bates's chapter appeared in the Dutch botanist and historian of science Frans Verdoorn's *Plants and Plant Science in Latin America*. This volume was intended to foster international cooperation in the face of the divisions in the plant science community caused by World War II and was spurred by a Rockefeller Foundation grant "to prepare a list of Latin American investigators in pure and applied plant sciences." See Diary of Harry M. Miller, 20 April 1942, box 329, RG12, M–R FA393, RAC.

59. Bates, "Advantage of the Tropical Environment for Studies on the Species Problem," 235.

60. On Dobzhansky's interest in wild populations, see Kohler, *Lords of the Fly*, 250–93.

61. Provine, *Sewall Wright and Evolutionary Biology*, 386–88; Glick, "Rockefeller Foundation and the Emergence of Genetics in Brazil"; Araújo, "Spreading the Evolutionary Synthesis," 469–73.

62. Dobzhansky, quoted in Provine, *Sewall Wright and Evolutionary Biology*, 387.

63. Miller diary, 24 February 1941.

64. The Rockefeller Foundation had long supported Dobzhansky's work, originally bringing him from Russia to T. H. Morgan's "fly lab" in 1927 and funding his previous expeditions. Kay, *Molecular Vision of Life*, 79.

65. See Garfield, *In Search of the Amazon*, 49–85; Wilkinson, "Tapping the Amazon for Victory." The initial plan was to attach Dobzhansky to a "war job" as an entomologist with the Section on Health and Sanitation, but instead he was asked to substitute temporarily for Rockefeller fellow Andre Dreyfus, who would be absent from São Paulo. See Miller diary, 18 November 1942. Like Beebe, Dobzhansky was also supported by the Committee for Inter-American Artistic and Intellectual Relations.

66. Provine, *Sewall Wright and Evolutionary Biology*, 386–88. See also Glick, "Rockefeller Foundation and the Emergence of Genetics in Brazil"; Araújo, "Spreading the Evolutionary Synthesis."

67. Dobzhansky to Wright, 28 September 1943, cited in Araújo, "Spreading the Evolutionary Synthesis."

68. Dobzhansky and Pavan, "Local and Seasonal Variations."

69. Ibid.; da Cunha, Dobzhansky, and Sokoloff, "On Food Preferences of Sympatric Species of Drosophila"; Dobzhansky, "Genetics of Natural Populations."

70. Pavan and da Cunha, "Dobzhansky and the Development of Genetics."

71. Da Cunha, "Dobzhansky and His Evolution," 297. On aviation networks in Brazil, see Cruz, "Flight of the Steel Toucans."

72. Miller diary, 27 May 1942; Dobzhansky, *Roving Naturalist*, 110. On Camargo and the Instituto Agronomico do Norte, see Dean, *Brazil and the Struggle for Rubber*, 88–89; Grandin, *Fordlandia*, 333–35. Camargo's success would be, for political reasons, relatively short lived.

73. Dobzhansky and the Rockefeller Foundation undeniably stimulated Brazil's *Drosophila* research program. Some point out, however, that genetics was already an active

field in Brazil before his arrival and that it continued to develop independently (especially in plant and human genetics) after his final visit in 1955–56. See, for example, Araújo, "Spreading the Evolutionary Synthesis," 473. For a variety of perspectives, see Glick, "Rockefeller Foundation and the Emergence of Genetics in Brazil"; da Cunha, "Dobzhansky and His Evolution"; Pavan and da Cunha, "Dobzhansky and the Development of Genetics"; Schwartzman, *Space for Science*, 143–44, 163; de Souza and Santos, "Emergence of Human Population Genetics."

74. Black, Dobzhansky, and Pavan, "Some Attempts to Estimate Species Diversity"; Murça Pires, Dobzhansky, and Black, "Estimate of the Number of Species of Trees." On Murça Pires, see Prance, "João Murça Pires"; Lisboa and Almeida, "Vida e Obra de João Murça Pires."

75. Black, Dobzhansky, and Pavan, "Some Attempts to Estimate Species Diversity," 413.

76. Davis and Richards, "Vegetation of Moraballi Creek," 108.

77. Dobzhansky, *Roving Naturalist*, 88, letter dated 14 August 1949.

78. Black, Dobzhansky, and Pavan, "Some Attempts to Estimate Species Diversity"; Murça Pires, Dobzhansky, and Black, "Estimate of the Number of Species of Trees." Although their plots were not far from where Beebe had sampled his four square feet of forest floor (see chapter 2)—they were essentially undertaking a similar census at an enormous scale—they made no reference to his work.

79. Beard, "Some Ecological Work."

80. His facility with languages aided him in this endeavor. See Stanley, Argent, and Whitehouse, "Botanical Biography of Professor Paul Richards"; Willis, "Paul Westamacott Richards."

81. Eggeling, "Review: The Tropical Rain Forest"; Tansley, "Review: The Tropical Rain Forest"; Hodge, "Review: The Tropical Rain Forest"; Champion, "The Tropical Rain Forest."

82. Murça Pires, Dobzhansky, and Black, "Estimate of the Number of Species of Trees," 475.

83. See Raunkiær, "Formationsundersøgelse Og Formationsstatistik"; Kenoyer, "Raunkaier's Law of Frequence"; McIntosh, *Background of Ecology*, 135–37.

84. Fisher, Corbet, and Williams, "Relation between the Number of Species and the Number of Individuals." For a historical overview of diversity indices, see Sarkar, "From Ecological Diversity to Biodiversity."

85. Preston, "Commonness, and Rarity, of Species."

86. Murça Pires, Dobzhansky, and Black, "Estimate of the Number of Species of Trees," 477.

87. Black, Dobzhansky, and Pavan, "Some Attempts to Estimate Species Diversity," 425.

88. Dobzhansky, "Evolution in the Tropics," 212.

89. Dobzhansky did note irregularities in the data, stemming from differences in topography, the extent of scientific study, and so on. Ibid., 214; Fischer, "Latitudinal Variations in Organic Diversity."

90. Dobzhansky, "Evolution in the Tropics," 209–10.

91. Ibid., 210.

92. He credited the idea of "challenge and response" to the historian Arnold Toynbee. Ibid., 220.

93. Ibid., 221.

94. Worster, *Nature's Economy*; Kingsland, *Modeling Nature*; McIntosh, *Background of Ecology*.

95. Slack, *Hutchinson and the Invention of Modern Ecology*.

96. McIntosh, *Background of Ecology*, 225; Taylor, *Unruly Complexity*, 56.

97. Howard T. Odum and Cynthia Barnett, "Howard T. Odum Interview" (2001), http://ufdc.ufl.edu/AA00004025/00001 (accessed 10 December 2016), 9, 16. Despite his contributions, Howard Odum was not given credit as an author on the first editions, causing tension between the brothers. Odum, *Fundamentals of Ecology*; Odum and Odum, *Fundamentals of Ecology*.

98. On tropes about coral specifically and changing conceptions of coral reefs as durable or fragile, threatening or threatened, see Sponsel, "Coral Reefs."

99. Odum and Odum, "Trophic Structure and Productivity of a Windward Coral Reef Community," 291.

100. Ibid.

101. Ibid., 302.

102. Ibid., 291.

103. Richards, *Tropical Rain Forest*, 16, 219–21, 401. Similar language is used throughout.

104. Odum and Odum, "Trophic Structure and Productivity of a Windward Coral Reef Community," 292, 319. See also Odum and Pinkerton, "Time's Speed Regulator."

105. Odum, "Curriculum Vitae. University of Florida, George A. Smathers Libraries," http://ufdc.ufl.edu/UF00101105/00001 (accessed 18 December 2016).

106. Odum, "Possible Effect of Cloud Cover." Based on his observations and the historical records available to him in the Canal Zone, Odum instead correlated northward migration with seasonal changes in cloud cover and their attendant shift in the predominant light wavelengths.

107. Lugo, "Contributions of H. T. Odum to Tropical Ecology," 23. The Institute of Tropical Meteorology was a collaboration of the U.S. military, the University of Chicago, and the University of Puerto Rico in Rio Piedras and was responsible for the development of early models of tropical circulation. For the approaches this course introduced Odum to, see Riehl, *Tropical Meteorology*; Anthes, "Hot Towers and Hurricanes."

108. Since 1968 it has been called the Marine Science Institute.

109. Excerpt from the Diary of J. G. Harrar, 28 December 1956, and Grant in Aid, 22 October 1958, folder 2306, box 238, subseries 200.D, RG1.2, RAC.

110. "A Comparison between Tropical Rainforest Community and a Tropical Plankton Community," 2, stamped 25 January 1957, folder 2306, box 238, subseries 200.D, RG1.2, RAC.

111. Golley, Odum, and Wilson, "Structure and Metabolism," 9.

112. "Comparison between Tropical Rainforest Community and a Tropical Plankton Community," 2.

113. Ibid.

114. Odum, McConnell, and Abbott, "Chlorophyll 'A' of Communities," 72.

115. "Comparison between Tropical Rainforest Community and a Tropical Plankton Community," 3. The "availability of cheap labor" also shaped the International Biological Program's choice of more complex sites for ecosystem studies in tropical Africa. See Lachenal, "At Home in the Postcolony," 4.

116. "Comparison between Tropical Rainforest Community and a Tropical Plankton Community," 3.

117. HMM [Harry M. Miller] note, 22 January 1957; see also R. S. Morison diary, 1 May 1957, folder 2306, box 238, subseries 200.D, RG1.2, RAC.

118. "Plan for Making Holistic Measurement of Rainforest," January 1958, folder 2306, box 238, subseries 200.D, RG1.2, RAC.

119. Vernberg, "Field Stations," 303–4; "Training Reactor in Puerto Rico."

120. Odum, McConnell, and Abbott, "Chlorophyll 'A' of Communities"; Odum, Abbott, and Selander, "Studies on the Productivity of the Lower Montane Rain Forest"; Odum, Burkholder, and Rivero, "Measurements of Productivity"; Odum, "Ecological Tools and Their Use"; Golley, Odum, and Wilson, "Structure and Metabolism"; Odum, Copeland, and Brown, "Direct and Optical Assay of Leaf Mass."

121. Brock, "American Empire and the Scientific Survey of Puerto Rico"; Mickulas, *Britton's Botanical Empire*.

122. The Spanish Crown Reserve had been created in 1876. Roosevelt declared the area U.S. public land in 1903 and it was designated the Luquillo National Forest in 1907. Its name changed to the Caribbean National Forest in 1935, and it was renamed El Yunque National Forest in 2007 to better reflect long-standing local name usage. See Wadsworth, "Review of Past Research"; Brokaw et al., *Caribbean Forest Tapestry*.

123. Holdridge, "Forestry in Puerto Rico"; Holdridge, "Determination of World Plant Formations"; Holdridge, *Life Zone Ecology*.

124. Wadsworth, "Forest Research Institution," 45; Steen, *Interview with Frank H. Wadsworth*.

125. Odum, "Ecological Tools and Their Use," 66.

126. Ibid.

127. Odum, Copeland, and Brown, "Direct and Optical Assay of Leaf Mass," 433.

128. Odum, "Ecological Tools and Their Use," 66–67.

129. "Comparison between Tropical Rainforest Community and a Tropical Plankton Community," 1.

130. Taylor, "Technocratic Optimism"; Taylor, *Unruly Complexity*.

131. Odum and Pigeon, *Tropical Rain Forest*, C-23.

132. See Briggs, *Reproducing Empire*; Lapp, "Rise and Fall of Puerto Rico as a Social Laboratory."

133. Kannowski, "Tropical Radioecology," 1131. See also Hagen, *Entangled Bank*, 167–68.

134. Bocking, *Ecologists and Environmental Politics*, 66–67.

135. Keiner, "Panatomic Canal and the US Environmental Management State."

136. Odum and Pigeon, *Tropical Rain Forest*, A-11.

137. Ibid., A-5.

138. Ibid., A-9.

139. Young, *History of the US Department of Defense Programs*, 46–47; Orians and Pfeiffer, "Ecological Effects of the War in Vietnam," 547; Wadsworth, "Review of Past Research," B-44; Dowler and Tschirley, "Effect of Herbicides."

140. Odum, "Rain Forest and Man," A-5.

141. Odum, "Summary," I-276.

142. On MacArthur's approach, see Odenbaugh, "Searching for Patterns, Hunting for Causes"; Kaspari, "Knowing Your Warblers"; Ishida, "Patterns, Models, and Predictions"; Pianka and Horn, "Ecology's Legacy from Robert MacArthur"; Dritschilo, "Mathematician Who Knew His Warblers"; Swanson, "Bridging the Gap"; Brown, "Legacy of Robert MacArthur"; Fretwell, "Impact of Robert MacArthur."

143. Wilson, "Eminent Ecologist," 12.

144. MacArthur, *Geographical Ecology*, 1.

145. MacArthur, "Fluctuations of Animal Populations and a Measure of Community Stability"; Pianka and Horn, "Ecology's Legacy from Robert MacArthur," 218; "Robert H. MacArthur," 995.

146. Odum, *Fundamentals of Ecology*.

147. Sarkar, "From Ecological Diversity to Biodiversity," 392–93.

148. MacArthur, "Fluctuations of Animal Populations and a Measure of Community Stability," 535.

149. MacArthur, "Population Ecology of Some Warblers"; "George Mercer Award." See also Kaspari, "Knowing Your Warblers."

150. Klopfer, *Politics and People in Ethology*, 39.

151. Bond, *How 007 Got His Name*; Parkes, "In Memoriam"; Lewis, "Scientists or Spies?"

152. See Colby, *Business of Empire*; Palmer, *Launching Global Health*.

153. Jones, "Colonization in Central America"; Sandoval-García, *Shattering Myths on Immigration and Emigration*; Rankin, *History of Costa Rica*.

154. MacArthur, "Population Ecology of Some Warblers," 618.

155. "Inter-American Institute of Agricultural Sciences." Today known as the Instituto Interamericano de Cooperación para la Agricultura, it sponsored the formation of the Centro Agronómico Tropical de Investigación y Enseñanza in 1973.

156. Pierce, "Environmental History of La Selva," 47.

157. Jimenez Saa, "Dr. Leslie R. Holdridge."

158. Holdridge, quoted in Pierce, "Environmental History of La Selva," 48.

159. Skutch, *Bird Watcher's Adventures*; Stiles, "In Memoriam"; Abarca Jiménez, *Alexander Skutch*; *Naturalist in the Rainforest*.

160. Stiles, "In Memoriam."

161. Skutch, "Helpers at the Nest."

162. Stiles, "In Memoriam," 710.

163. MacArthur, "Population Ecology of Some Warblers," 613–15.

164. Skutch, "Life Histories of Central American Birds." See also Cody, "General Theory of Clutch Size"; Pianka, "Latitudinal Gradients in Species Diversity."

165. With Edward O. Wilson, he would address these problems through the concept of r/K-selection. MacArthur and Wilson, *Theory of Island Biogeography*, 151.

166. MacArthur, "On the Relative Abundance of Bird Species," 293; Kingsland, *Modeling Nature*, 183–87.

167. Kingsland, *Modeling Nature*, 186–87.

168. Recher, "Population Density and Seasonal Changes"; MacArthur, Recher, and Cody, "Relation between Habitat Selection and Species Diversity"; Recher and Recher, "Avifauna of the Sierra de Luquillo."

169. MacArthur and Levins, "Competition, Habitat Selection, and Character Displacement"; Levins and MacArthur, "Maintenance of Genetic Polymorphism"; MacArthur and Levins, "Limiting Similarity." Levins was critical of the ongoing exploitation of Puerto Rico's environment by the United States. His involvement with the Puerto Rican Independence movement during this period would lead to his denial of tenure in 1966. Richard Levins, "A Permanent and Personal Commitment." http://www.peacehost.net/WhiteStar/Voices/eng-levins.html (accessed 6 March 2016).

170. MacArthur and Wilson, "Equilibrium Theory of Insular Zoogeography"; MacArthur and Wilson, *Theory of Island Biogeography*.

171. See folder: MacArthur Robert 1969–1972, box 9, Acc10-060, SIA. See also the annual reports of the Smithsonian Institution, 1961–72; Klopfer and MacArthur, "Causes of Tropical Species Diversity"; Klopfer and MacArthur, "Niche Size and Faunal Diversity"; MacArthur and MacArthur, "Bird Species Diversity"; MacArthur, MacArthur, and Preer, "Bird Species Diversity II"; MacArthur and Connell, *Biology of Populations*; MacArthur, Recher, and Cody, "Relation between Habitat Selection and Species Diversity."

172. See, for example, Henson interview.

173. Smithsonian Institution, *Smithsonian Year, 1965*, 170.

174. Their graphical model assumed ideal conditions; the text of *The Theory of Island Biogeography* did consider additional environmental factors.

175. See folder: MacArthur Robert 1969–1972.

176. MacArthur, *Geographical Ecology*, 223.

177. Ibid., 216.

178. Ibid., 199.

179. Odum, "Rain Forest and Man," A-7.

180. Against this view, see Pianka and Horn, "Ecology's Legacy from Robert MacArthur." For "colonization" language, see McIntosh, *Background of Ecology*, 280; Slack, *Hutchinson and the Invention of Modern Ecology*, 275.

181. Worster, *Nature's Economy*.

Chapter Five

1. Edward O. Wilson interview, 20 January 2014, OTS History Videos, OTS.
2. MacArthur and Wilson, "Equilibrium Theory of Insular Zoogeography."
3. Wilson, *Naturalist*, 152.
4. Wilson interview.

5. Wilson, *Naturalist*, 152.
6. Wilson interview.
7. Wilson, *Naturalist*, 147.
8. Wilson, *Diversity of Life*, 408.
9. Ibid.
10. Wilson, *BioDiversity*, 3.
11. Harvard University, *Report, 1960–1961*, 476; Smith, "Bundy Denies Reports of Cuban Interference." See also Cahan, "Harvard Garden in Cuba"; Hazen, "Cienfuegos Botanical Garden."
12. Harvard University, *Report, 1960–1961*, 477.
13. Ibid.
14. See McCook, *States of Nature*, 60, 136–37.
15. Harvard University, *Report, 1949–1950*, 289.
16. Harvard University, *Report, 1948–1949*, 269. The United States also recognized Philippine independence in 1946. Its state scientific agencies, which had already largely been "Filipinized" by the 1930s, had to be rebuilt after World War II and the Japanese invasion. Anderson, "Science in the Philippines."
17. Clement, quoted in Smith, "Clement Tells of Cuban Research."
18. Harvard University, *Report, 1954–1955*, 363–64.
19. Ibid., 289.
20. Both institutions were founded by Wilson Popenoe, protégé of David Fairchild.
21. McCook, *States of Nature*, 47–76. On regional exchange relationships, see Fernández Prieto, "Islands of Knowledge."
22. See, for example, Navia et al., "Nutrient Composition of Cuban Foods"; Harvard University, *Report, 1957–1958*.
23. Harvard University, *Report, 1955–1956*, 596.
24. Folder: Cartoon, Atkins Institution Papers, Arnold Arboretum Archives of Harvard University.
25. Smith, "Clement Tells of Cuban Research."
26. Harvard University, *Report, 1954–1955*, 365. On Acuña Galé, see Alvarez Conde, *Historia de la Botanica*.
27. Harvard University, *Annual Report, 1958–1959*, 456.
28. Pite, "Force of Food," 73–79.
29. Harvard University, *Annual Report, 1958–1959*, 456.
30. Ibid., 455.
31. Ibid., 456.
32. Harvard University, *Annual Report, 1959–1960*, 576.
33. See Pérez, *Cuba and the United States*, 241–62.
34. "Harvard, out of Cuban Garden," 49.
35. It was soon handed over to the Universidad Central de Las Villas and by 1962 would pass into the hands of the Academia de Ciencias. Contreras and Albuerne, "Jardín Botánico de Cienfuegos," 152; Ojeda Quintana et al., "Jardín Botánico de Cienfuegos," 52–53.
36. Smith, "Bundy Denies Reports of Cuban Interference."

37. In 1966, Esperanza Vega would marry Ian Duncan Clement after his first wife's death in Costa Rica the previous year. "Reports of Deaths of American Citizens Abroad, 1835–1974," box 21, RG59, entry 5166, NARA; Massachusetts, Marriage Index, 1901–55 and 1966–70 [online database], Department of Public Health, Registry of Vital Records and Statistics, and Massachusetts Vital Records Index to Marriages [1916–70], vols. 76–166, 192–207 (facs. ed.; Boston, MA: New England Historic Genealogical Society), www.ancestry.com (accessed 18 December 2016).

38. Harvard University, *Report, 1961–1962*, 565–66; "Harvard, out of Cuban Garden." Gonzalez worked at the Fairchild Tropical Botanical Garden until his retirement in 1980. Popenoe, "News of the Garden," 4.

39. Harvard University, *Report, 1961–1962*, 565–66; Smith, "Corporation Vote Ends Cuban Research Project." The Atkins Fund was later redirected to graduate training and research in the tropics, supporting the work of P. Barry Tomlinson, for example. Harvard University, *Harvard Annual Report, 1970–1971*, 393.

40. Theodore H. Hubbell to R. L. Wilbur, 10 October 1958, "1. 1957 May—Report on proposed center for tropical studies, issued by the University of Michigan. Letter by Dr. Theodore H. Hubbell, Chairman," OTS Historical Documents.

41. Hubbell studied under William Morton Wheeler, but he first developed his interest in tropical natural history while living in the Philippines as a child, when his father was a civil engineer in Manila. (See the finding aid to the Theodore H. Hubbell Papers, Bentley Historical Library, University of Michigan.) Theodore Hubbell's son, Stephen P. Hubbell, followed in his footsteps to become an important tropical ecologist in his own right. He dedicated his *Unified Neutral Theory of Biodiversity and Biogeography* to his father, who "first introduced me to 'biodiversity' before it was called such, on many field trips in the United States, Mexico, and Central America. He is a hard act to follow." Hubbell, *Unified Neutral Theory*, xii.

42. On the role of Miranda, Beltrán, and Hernández-X in Mexican biology and conservation, see Simonian, *Defending the Land of the Jaguar*; Soto Laveaga, *Jungle Laboratories*, 33, 99–103; Wakild, *Revolutionary Parks*, 155–58; Mancilla, *Faustino Miranda*; Garrison Wilkes, "Efraim Hernández Xolocotzi-Guzman." On the national and transnational development of biological science and conservation in Mexico more broadly, see Ledesma-Mateos and Barahona, "Institutionalization of Biology in Mexico"; Barahona, "Transnational Science and Collaborative Networks."

43. Miranda had founded an important botanical garden in Tuxtla Gutierrez, as director of the Instituto Botánico before coming to UNAM. Committee on the Proposed Center for Tropical Studies, report of the Subcommittee on Site and Lands, 5 August 1958, "1. 1957 May—Report on proposed center for tropical studies, issued by the University of Michigan. Letter by Dr. Theodore H. Hubbell, Chairman," OTS Historical Documents.

44. Raymond Fosberg to William H. Hatheway, 5 June 1962, folder: Trinidad Conference, box 6, Acc91-178, SIA.

45. Another new biological institute had been proposed for British Guiana, but it fell through amid the political turmoil of independence efforts and U.S. and British

interventions against the leftist government of Cheddi Jagan. Hodge and Keck, "Biological Research Centres in Tropical America," 115–16. See Rabe, *US Intervention in British Guiana*; Palmer, *Cheddi Jagan and the Politics of Power*.

46. See Appel, *Shaping Biology*, 186–89; National Science Foundation, *Eleventh Annual Report, 1961*, 4.

47. "Editorial: Loss of Cuban Botanical Garden," 49. See also Appel, *Shaping Biology*, 199.

48. Appel, *Shaping Biology*, 199. On Keck, see Mayr and Provine, *Evolutionary Synthesis*, 142–46.

49. National Science Foundation, *Tenth Annual Report*, 78. Its report's published title, "Conference on Tropical Botany," dropped the reference to U.S. interests.

50. Fischer, "Latitudinal Variations in Organic Diversity"; Klopfer and MacArthur, "Niche Size and Faunal Diversity"; MacArthur, "On the Relative Abundance of Species"; Preston, "Time and Space and the Variation of Species"; Whittaker, "Vegetation of the Siskiyou Mountains."

51. While the authorship of the Fairchild conference report is not clear, the language of tropical problems appears broadly in the writings of several participants.

52. *Conference on Tropical Botany*, 1.

53. Ibid., 5–6.

54. Ibid., 1.

55. Ibid., 2–3.

56. Ibid., 5.

57. Ibid., 4, 3.

58. Ibid., 3.

59. Ibid., 15.

60. On "one world" see Sluga, "UNESCO and the (One) World of Julian Huxley"; Jordan, "Small World of Little Americans."

61. *Conference on Tropical Botany*, 1.

62. Ibid., 13.

63. Ibid., 1.

64. Fosberg to Hatheway, 5 June 1962, folder: Trinidad Conference, box 6, Acc91-178, SIA.

65. Fosberg to Keck, 14 November 1961, folder: Institute for Tropical Botany, box 5, Acc91-178, SIA.

66. Hatheway to Fosberg, 25 May 1962, folder: Trinidad Conference, box 6, Acc91-178, SIA. Hernández-X had been trained in the United States but spent his career in Mexico and was committed to the development of Mexican agroecology. See Garrison Wilkes, "Efraim Hernández Xolocotzi-Guzman."

67. Hodge and Keck, "Biological Research Centres in Tropical America," 107.

68. Ibid., 110. Miranda would also establish the Estación de Biología Tropical "Los Tuxtlas" in 1967; see Gómez-Pompa, *Mi Vida en Las Selvas Tropicales*, 77–82; Ruiz Cedillo and Durand, "Estación Biológica Tropical 'Los Tuxtlas'"; Estrada, Coates-Estrada, and Martínez-Ramos, "Estación Biológica Tropical Los Tuxtlas"; Lot-Helgueras, "Estación Biológica Tropical Los Tuxtlas."

69. See 1005C, and annual reports in box 1, 1005A, WCS. See also Gould, *Remarkable Life of William Beebe*, 367–68. Venezuela's Department of Agriculture continued to administer the Rancho Grande site. Since 1966, the Universidad Central de Venezuela has operated it as the Estación Biológica Fernández Yépez. See Arnal, "Beginnings of Modern Ornithology in Venezuela," 612–21.

70. On U.S. relations with the British West Indies during the Cold War and decolonization, see Parker, "Remapping the Cold War"; Parker, *Brother's Keeper*; Fraser, "Ambivalent Anti-colonialism." Simla was then being operated for the New York Zoological Society by Jocelyn Crane—the first woman director of a tropical research station. See 1005D, and "Simla, 1967–1971," box 5, 1005A, WCS.

71. Thirteen out of nineteen U.S. participants had attended the 1960 Fairchild conference. Neotropical Botany Conference, Imperial College of Tropical Agriculture, Trinidad, 1–7 July 1962, folder: Tropical Biology 1962–1963, box 216, RU50, SIA.

72. This was Giris "George" Hanna Sidrak, a recent immigrant to Jamaica. "Personalities, 'S'," http://discoverjamaica.com/snames.htm (accessed 18 December 2016).

73. The exception was Paul Allen, director of United Fruit's Lancetilla experiment station, Honduras, and former director of the Fairchild Tropical Garden and, in Panama, the Summit Garden (Experiment Garden of the Canal Zone) and the Missouri Botanical Garden's orchid station. Hodge, "Paul Hamilton Allen."

74. A. C. Smith to Leonard Carmichael, 10 July 1962, folder: Tropical Biology 1962–1963, box 216, RU50, SIA.

75. In fact, the International Society for Tropical Ecology had already been founded in 1956, based in India and comprising a membership primarily from the Old World tropics and Europe. The only ATB member to comment on the society at the time was Fosberg: he felt it was having difficulty taking off. He would later serve as its president but continued to struggle to bring the society and the ATB into closer relation. (See, e.g., Fosberg to Robert Dressler, 12 May 1967, folder: CZBA-STRI, box 5, Acc91-178, SIA.) The disconnect between the two fledgling organizations suggests the ongoing divide between tropical research in the two hemispheres and former colonial domains.

76. Harvard was also about to begin an interdisciplinary program in evolutionary biology that sought tropical field opportunities. Norman Hartweg, "Summary of Coral Gables Meeting of Institutional Representatives," 8 March 1963, 2, folder 176, box 5, OTS Records, DU.

77. "Tropical Biology Program, University of Southern California," 1962, folder 176, box 5, OTS Records, 1962–1974, DU; Jay Savage interview, 25 September 2012, OTS History Videos, OTS; Jay M. Savage, "The Founding of the Organization for Tropical Studies, Inc.," http://www.ots.ac.cr/images/downloads/historicaldocuments/2008%20article%20on%20founding%20of%20ots%2C%20present%20at%20the%20pre-conception%2C%20conception%2C%20and%20birth%20a%20personal%20perspective%20by%20jay%20savage.pdf (accessed 26 October 2012).

78. See Levy, *To Export Progress*; Bu, "Educational Exchange and Cultural Diplomacy." See also Miller, "'Effective Instrument of Peace'"; Doel and Harper, "Prometheus Unleashed."

79. Jay Savage, "Final Report on Advanced Biology Institute Held at the University of Costa Rica during July and August, 1961," box 1, OTS Reference Collection, DU; Harvard University, *Report, 1961–1962*, 565. See also Stone, "OTS: A Success Story," 145–47.

80. "Final Report on the Conference on Problems in Education and Research in Tropical Biology Held at the Universidad de Costa Rica, April 23–27, 1962," folder 176, box 5, OTS Records, DU.

81. Representing Harvard, E. O. Wilson and Reed Rollins took the place of Clement, who went on to work on Latin American exchange projects at the NSF.

82. Savage interview; Savage, "Founding of the Organization for Tropical Studies." See also folder 176, box 5, OTS Records, DU.

83. "Final Report on the Conference on Problems."

84. Hartweg, "Summary of Coral Gables Meeting."

85. Hubbell, "Organization for Tropical Studies."

86. Hartweg, "Summary of Coral Gables Meeting."

87. "Final Report on the Conference on Problems."

88. Committee on a Proposed Center for Tropical Studies, First Report, 15 May 1958, "1. 1957 May—Report on proposed center for tropical studies, issued by the University of Michigan. Letter by Dr. Theodore H. Hubbell, Chairman," OTS Historical Documents.

89. Hubbell, "Organization for Tropical Studies," 239.

90. Hartweg, "Summary of Coral Gables Meeting."

91. "Final Report on the Conference on Problems."

92. Quoted in Palmer and Molina Jiménez, *Costa Rica Reader*, 1; Bowman, "New Scholarship on Costa Rican Exceptionalism," 123.

93. See Evans, *Green Republic*; Allen, *Green Phoenix*; Christen, "Development and Conservation on Costa Rica's Osa Peninsula"; Christen et al., "Latin American Environmentalism."

94. On Costa Rican exceptionalism, see Bowman, "New Scholarship on Costa Rican Exceptionalism"; Booth, *Costa Rica*, 29; Rankin, *History of Costa Rica*; Christian, "'Latin America without the Downside.'"

95. Longley, *Sparrow and the Hawk.*

96. Rankin, *History of Costa Rica*, 117–18.

97. Hopkins, *Policymaking for Conservation in Latin America*, 42; Levy, *To Export Progress*, 94, 162–63.

98. Hodge and Keck, "Biological Research Centres in Tropical America," 107.

99. Savage interview; Savage, "Founding of the Organization for Tropical Studies." See also folder 106, box 3, OTS Records, DU.

100. See Gómez, Savage, and Janzen, "Searchers on That Rich Coast." The most famous case in point is the Swiss-born naturalist Henri Pittier's work for the government of Costa Rica around the turn of the twentieth century; McCook, *States of Nature*, 28–34.

101. Following migrating sea turtles, Carr had worked in Costa Rica periodically for years. His mentor Barbour's connections with United Fruit were instrumental to his initial fieldwork in Honduras and Costa Rica. Davis, *Man Who Saved Sea Turtles*, 61.

102. De Girolami, "Reseña Histórica," 346 (my translation).

103. Burlingame, "Evolution of the Organization for Tropical Studies," 454. See also Monge-Nájera and Diaz, "Thirty-Five Years of Tropical Biology."

104. An emphasis on scientific and national value of Costa Rica's richness in species went back at least to the work of Pittier. McCook, *States of Nature*, 33. For an analysis of similar dynamics in Colombia, see Quintero Toro, "Trading in Birds"; Quintero Toro, *Birds of Empire, Birds of Nation.*

105. Savage interview; Savage, "Founding of the Organization for Tropical Studies." See also folder 91, box 3, OTS Records, DU.

106. These projects included courses for the Associated Colleges of the Midwest. See folder 211, box 6, OTS Records, DU.

107. "Organization for Tropical Studies: A Descriptive Statement," 1963, folder 176, box 5, OTS Records, DU.

108. For examples of OTS concerns about scientific imperialism and funding tied to host country benefits, see Savage to Leslie A. Chambers, 27 March 1963, folder 87, box 3, OTS Records, DU; Tangley, "Studying (and Saving) the Tropics," 381.

109. See Conniff, *Panama and the United States*, 116–39; LaFeber, *Panama Canal*, 133–41.

110. Moynihan to Carmichael, 26 February 1963, folder 2, box 23, RU135, SIA.

111. Hodge and Keck, "Biological Research Centres in Tropical America," 110.

112. Ibid.

113. "Minutes of the Barro Colorado Island Meetings," 31 October 1953, folder 1, box 7, RU135, SIA.

114. Bates to Carmichael, 2 November 1953, folder 3, box 23, RU135, SIA.

115. Allee to Carmichael, 28 October 1953, folder 3, box 23, RU135, SIA; Emerson, quoted in "Minutes of the Barro Colorado Island Meetings."

116. Eisenmann, quoted in "Minutes of the Barro Colorado Island Meetings." Eisenmann was a close associate of Ernst Mayr. He did not have a doctorate in science but came to ornithology after a career in law. Bull and Amadon, "In Memoriam: Eugene Eisenmann"; Heckadon-Moreno, *Naturalistas de Istmo de Panamá*, 253–60.

117. Zetek to John E. Graf (assistant secretary of the Smithsonian), 24 August 1956, folder 4, box 23, RU135, SIA.

118. Walter Clark to John E. Graf, 30 June 1954, folder 3, box 7, RU135, SIA; Graf to Cleaveland C. Soper, 22 December 1953, folder 6, box 9, RU135, SIA.

119. On Canal Zone school segregation, see Donoghue, *Borderland on the Isthmus*, 64. Koford had extensive field experience in Mexico and Peru, and his research was instrumental to the conservation of the California condor. Barrow, *Nature's Ghosts*, 290–300.

120. Fausto Bocanegra, oral history interviews by Giselle Mora, August 1988, RU009561, SIA.

121. Commissary prices were rising during this period, but assistant secretary Graf placed great stock in "the fear of fresh vegetables that exists in the Zone and the prejudice against 'native' beef" held by white Zonians. Graf to Koford, 9 September 1956, folder 4, box 8, RU135, SIA. See also related correspondence in this folder.

122. Bocanegra interviews. See also folder 5, box 22, RU135, SIA.

123. Zetek to John Graf, 21 March 1955, folder 4, box 16, RU135, SIA.

124. See folder 6, box 22, and folder 1, box 23, RU135, SIA.

125. Smithsonian Institution, *Smithsonian Year, 1965*, 309; Appel, *Shaping Biology*, 202.

126. See Raby, "'Jungle at Our Door.'"

127. Smithsonian Institution, *Smithsonian Year, 1969*, 226.

128. Moynihan, "Report, 1958," 185.

129. See chapter 4.

130. Moynihan to Remington Kellogg, 6 March 1958, folder 4, box 23, RU135, SIA.

131. Moynihan, "Report, 1960," 172; Smithsonian Institution, *Smithsonian Year, 1970*, 50.

132. Christen, "At Home in the Field," 573–74.

133. Moynihan to Carmichael, 26 February 1963.

134. Moynihan to Kellogg, 14 August 1962, folder 4, box 32a, RU50, SIA.

135. Moynihan to Carmichael, 26 February 1963.

136. Ibid.

137. Moynihan to Carmichael, 16 April 1963, folder 2, box 23, RU135, SIA.

138. Moynihan, "Report, 1964." See also folder 5, box 32a, RU50, SIA.

139. Donoghue, *Borderland on the Isthmus*, 18–19; LaFeber, *Panama Canal*, 138–39.

140. Moynihan to Carmichael, 17 January 1964, folder 1, box 2, RU135, SIA.

141. Moynihan, "Report, 1964," 232.

142. Moynihan to Carmichael, 5 July 1963, folder 3, box 23, RU135, SIA.

143. Ibid. His position echoes that of Thomas Barbour, who sheltered his stations from the interwar trend toward "organized" research (see chapter 2). On the IBP, see Aronova, Baker, and Oreskes, "Big Science and Big Data"; Coleman, *Big Ecology*; Lewis, *Inventing Global Ecology*, 151–52.

144. As historian Joel Hagen has argued, this also helped to bring new shared intellectual perspectives; Hagen, "Problems in the Institutionalization of Tropical Biology," 243, 247.

145. Lee Talbot to S. Dillon Ripley, 31 August 1967, folder: STRI, box 5, Acc91-178, SIA. Not all new staff scientists felt comfortably integrated and at home on BCI. Robert L. Dressler grew particularly frustrated with life in "the monkey park"—meaning the human rather than the animal community. His frustration led him to travel to work more closely with the ATB and to attempt to support the Rancho Grande station in Venezuela. Dressler to Richard S. Cowan, 17 November 1966, folder: CZBA-STRI, box 5, Acc91-178, SIA. See also Robert L. Dressler, oral history interview by Catherine A. Christen, 1 July 1997, History of Tropical Biology Interviews, RU009606, SIA.

146. On Ripley, see Lewis, *Inventing Global Ecology*.

147. Fosberg was in charge of this effort. See folder: Tropical Biology Task Force, box 1, Acc91-178, SIA.

148. A modernized canal would be a boon to Panama, while a new canal at an alternative site in Nicaragua, Costa Rica, or Colombia would destroy Panama's economy. Lawrence, "Exception to the Rule?," 42; Keiner, "Panatomic Canal and the US Environmental Management State." See also Doel and Harper, "Prometheus Unleashed."

149. Indeed, Howard Odum's project was in part justified as an environmental impact assessment in the case nuclear technology would be used in that construction. On Odum's Rain Forest Project as big science, see chapter 4 and Hagen, *Entangled Bank*, 167–68. On the "Pan-Atomic Canal," see Keiner, "Panatomic Canal and the US Environmental Management State."

150. Moynihan, "Report, 1965," 14.

151. Moynihan, "A Preliminary Outline of the Objectives and Future Programs of the Smithsonian Tropical Research Institute," 1 May 1966, folder 1, box 37, RU254, SIA.

152. See Edward H. Kohn to Leonard Pouliot, 4 July 1969, box 4, Acc10-060, SIA; Smithsonian Institution, *Smithsonian Year, 1965*, 171. Moynihan also attempted to attract Daniel Simberloff, Wilson's collaborator in a famous experimental test of island biogeography. Moynihan to Simberloff, 22 January 1968, folder: Mayr, Ernst, box 9, Acc10-060, SIA. See Simberloff and Wilson, "Experimental Zoogeography of Islands"; Quammen, *Song of the Dodo*, 427–30.

153. Moynihan, "Smithsonian Tropical Research Institute," 178.

154. "Objectives for the Panama Conference on Tropical Biology," folder: Panama Conference, box 6, Acc91-178, SIA. The Gorgas Memorial Laboratory also took part. It was represented by Graham B. "Sandy" Fairchild, David Fairchild's son, who had been present at the construction of the first BCI laboratory.

155. Ripley, "Conference in Panama to Plan an Ecological Program in that Area," 26 August 1966, folder: Panama Conference, box 6, Acc91-178, SIA.

156. See Keiner, "Two-Ocean Bouillabaisse"; Keiner, "Panatomic Canal and the US Environmental Management State"; Hagen, "Problems in the Institutionalization of Tropical Biology," 243–47; Rubinoff, "Sea-Level Canal Controversy"; Rubinoff, "Mixing Oceans and Species."

157. Buechner and Fosberg, "Contribution toward a World Program in Tropical Biology," 532–33.

158. Paul Ehrlich was at the time giving public interviews ahead of the publication of his landmark book on global population, *Population Bomb*.

159. Hubbell, "Organization for Tropical Studies," 236.

160. See Dasmann, *Called by the Wild*, 114–15.

161. Dasmann, *Different Kind of Country*, 35.

162. Ibid., 38.

163. Woodwell and Smith, *Diversity and Stability*; Matthews and Matthews, "Adaptive Aspects of Insular Evolution"; Prance, *Biological Diversification*.

164. Wilson has stated that such applications were far from their minds during the theory's formulation. However, the idea that large reserves were inherently more effective at preserving a variety of species was already widespread; it was brought up at the 1960 Fairchild conference as a truism, for example. See the section titled "Tropical Problems" in this chapter.

165. On the diversity/stability debate, see Ives, "Community Diversity and Stability"; Justus, "Stability-Diversity-Complexity Debate"; Nikisianis and Stamou, "Quantifying Nature." On debates over the application of island biogeography theory to

reserve design, see Quammen, *Song of the Dodo*, 438–540; Kingsland, "Creating a Science of Nature Reserve Design"; Lewis, *Inventing Global Ecology*, 141–235; Sarkar, "Complementarity and the Selection of Nature Reserves."

166. Sarkar, "From Ecological Diversity to Biodiversity"; Sarkar, *Biodiversity and Environmental Philosophy*; Magurran, *Measuring Biological Diversity*; Pielou, *Ecological Diversity*; McIntosh, "Index of Diversity."

167. Meine, "Conservation Biology"; Meine, Soulé, and Noss, " 'Mission-Driven Discipline.' "

168. Hubbell, "Organization for Tropical Studies," 238.

169. Ripley, "Perspectives in Tropical Biology," 539.

170. Tropical biologists were thus particularly anxious to avoid entanglements with covert U.S. operations during this period. Moynihan warned Smithsonian secretary and former spy Ripley, for example, "I do hope that all the existing SI arrangements with AID are 'clean.' STRI is presently engaged in rather delicate negotiations. . . . We cannot afford any unfortunate revelations in the immediate future." Moynihan to Ripley, 29 December 1972, folder: Ripley, S. Dillon, 1967–1974, box 1, Acc10-060, SIA. Likewise, Amy Gilmartin, a botanist and ATB committee member whose tireless work kept the *ATB Bulletin* running, tentatively expressed her relief at working for a neutral, nonpolitical scientific organization, "Thank god, here's one organization with no tie-in with the CIA etc. (I hope!)." Gilmartin to Jean H. Langenheim, 30 October 1968, folder: ATB Committee Members, Correspondence, 1968–1970, box 1, Acc06-131, SIA.

171. Lucier, "Origins of Pure and Applied Science," 536.

Epilogue

1. See Tangley, "Biological Diversity Goes Public"; Joyce, "Species Are the Spice of Life."

2. See Mazur and Lee, "Sounding the Global Alarm"; Takacs, *Idea of Biodiversity*.

3. Joyce, "Species Are the Spice of Life," 20.

4. Ehrlich and Peter Raven's concept of coevolution had also gained currency in environmentalist circles with the initiation of Stewart Brand's *CoEvolution Quarterly* in 1974. They had posited coevolution as a mechanism for explaining the greater species diversity of tropical communities. Ehrlich and Raven, "Butterflies and Plants," 606.

5. Egbert Leigh, verbal personal communication with the author, 8 January 2011. See also Croat, *Flora of Barro Colorado Island*.

6. Rubinoff and Leigh, "Dealing with Diversity," 116.

7. Chapman, *My Tropical Air Castle*, 6. See chapter 3.

8. Willis, "Populations and Local Extinctions." See also Karr, "Avian Extinction," 344–47.

9. Of course, data from BCI had also been used to construct the theory being tested; see chapter 5.

10. Terborgh, "Preservation of Natural Diversity"; Diamond, "Island Dilemma"; Wilson and Willis, "Applied Biogeography."

11. Wilson, *BioDiversity*, 12.

12. The project was renamed the Biological Dynamics of Forest Fragments Project in 1987, and the U.S. side of the partnership transferred, with Lovejoy, from the World Wildlife Fund to the Smithsonian Institution's National Museum of Natural History. In 2000, it moved from the museum to Smithsonian Tropical Research Institute administration and is part of the Center for Tropical Forest Science. The Brazilian partner is the Instituto Nacional de Pesquisas da Amazônia (National Institute for Research in the Amazon), which had been founded in 1954 and had operated a silviculture station at Manaus since 1972. See annual reports in box 3, Acc08-098, SIA.

13. On the SLOSS (single-large or several-small) reserves debate, see Sarkar, "Complementarity and the Selection of Nature Reserves"; Lewis, *Inventing Global Ecology*, 141–235; Kingsland, "Creating a Science of Nature Reserve Design"; Quammen, *Song of the Dodo*, 438–540.

14. Alwyn H. Gentry's classic volume *Four Neotropical Rainforests* compares BCI, La Selva, Cocha Cashu, and Manaus, cementing their role as model sites.

15. On the failure of this round of treaties, see Conniff, *Panama and the United States*, 116–39; LaFeber, *Panama Canal.*

16. See, for example, Pennisi, "Panama Lab Overcomes Political Turmoil."

17. For more on this arrangement, see Christen, "At Home in the Field," 555–56.

18. Croat to Raven, 30 June 1972, folder 27, collection 1, RG4/3/2/11, Missouri Botanical Garden Archives; Thomas Croat, verbal personal communication with the author, 30 October 2013.

19. Leigh, personal communication.

20. Donoghue, *Borderland on the Isthmus*, 90.

21. Amy Jean Gilmartin (chairman, Association for Tropical Biology Education and Research Committee) to Thomas R. Soderstrom (Association for Tropical Biology executive director), 23 September 1968, folder: Alphabetical Correspondence, 1968–1969, box 1, Acc06-131, SIA.

22. Ibid. OTS did not have a facility on the Universidad de Costa Rica campus until 2004. Jay Savage interview, 25 September 2012, OTS History Videos, OTS.

23. Gilmartin to Soderstrom, 23 September 1968.

24. Burlingame, "Evolution of the Organization for Tropical Studies."

25. See especially minutes of the meeting of the board of directors of the OTS, 10–11 November 1967, folder 1, box 1; annual OTS board of directors meeting, 12–13 January 1973, folder 3, box 1; OTS board of directors, meeting minutes, 8–9 November 1974, folder 2, box 1, OTS Reference Collection, DU; Stone, "OTS: A Success Story," 166–67. OTS even considered acquiring the Lancetilla station, as the United Fruit Company found it increasingly expensive to operate. "Minutes of the Executive Committee Meeting of the Organization for Tropical Studies," 28 April 1967, folder 174, box 5, OTS Records, DU; "Lancetilla," box 5, Acc91-178, SIA.

26. Folder: Correspondence with Department of State 1974–1982, folder: Correspondence with Department of State 1983–1984, box 1, Acc12-055, SIA; Christen, "At Home in the Field," 566.

27. Burlingame, "Evolution of the Organization for Tropical Studies."

28. Lawrence E. Gilbert interview, 29 July 2014; Peter H. Raven interview, 29 July 2014, OTS History Videos, OTS. Paul W. Richards, an early voice of warning about the fate of global rainforests (see chapter 4), was an OTS instructor in 1965, however.

29. Raven interview; Pierce, "Environmental History of La Selva"; Locher, "Migration and the Costa Rican Environment," 287–89.

30. Allen, *Green Phoenix*, 33. Janzen's significant contributions to the question of species diversity include Janzen, "Why Mountain Passes Are Higher in the Tropics"; Janzen, "Herbivores and the Number of Tree Species."

31. Christen, "Tropical Field Ecology and Conservation Initiatives"; Christen, "Development and Conservation on Costa Rica's Osa Peninsula."

32. Working with Robin Foster, he established his fifty-hectare Forest Dynamics Plot there in 1980. Hubbell, "To Know a Tropical Forest," 328. See also Raby, "Ark and Archive."

33. Pierce, "Environmental History of La Selva," 50–51; Stone, "OTS: A Success Story," 178.

34. Eisner et al., "Conservation of Tropical Forests," 1314.

35. Raven interview; Pierce, "Environmental History of La Selva"; Evans, *Green Republic*, 29.

36. This idea went back to the beginning of the Ecological Society of America and the founding of BCI. See chapter 3 and Rumore, "Preservation for Science."

37. Raven et al., *Research Priorities in Tropical Biology*, 1.

38. Ibid., 34–35. See also Myers, *Conversion of Tropical Moist Forests*.

39. Janzen, "Uncertain Future of Tropics"; Gómez-Pompa, Vazquez-Yanes, and Guevara, "Tropical Rain Forest"; Gómez-Pompa, *Mi Vida en las Selvas Tropicales*, 77–82.

40. Myers, *Sinking Ark*, 147–48.

41. Rosen, "What's in a Name?," 710.

42. Janzen, "Future of Tropical Ecology," 306. This was not a new strategy: U.S. scientists had been buying Costa Rican land for study since Alexander Skutch in the 1940s (see chapters 4 and 5).

43. Janzen, "Uncertain Future of Tropics," 89. Rubinoff made a similar argument in "Tax Rich Nations, Save the Jungle."

44. Guha, "Authoritarian Biologist," 19–20.

45. Yantko and Golley, "Census of Tropical Ecologists"; Gómez-Pompa and Butanda, *Index of Current Tropical Ecology Research*.

46. Ariel Lugo interview, 6 January 2014, OTS History Videos, OTS; Waide and Lugo, *NSF Proposal*; Campbell and Lugo, "Conversation with Ariel Lugo." Lugo also spoke at the 1986 BioDiversity Forum.

47. Martin, Blossey, and Ellis, "Mapping Where Ecologists Work." See also Baker, "Way Forward for Biological Field Stations"; Deikumah, McAlpine, and Maron, "Biogeographical and Taxonomic Biases"; Pitman, "Research in Biodiversity Hotspots Should Be Free"; Clark, "Ecological Field Studies in the Tropics."

48. Stocks et al., "Geographical and Institutional Distribution of Ecological Research," 402. See also Livingston et al., "Perspectives on the Global Disparity"; Braker, "Changing Face of Tropical Biology?"

49. Peter Del Tredici, verbal personal communication with author, 13 July 2009; Ojeda Quintana et al., "Jardín Botánico de Cienfuegos"; Domínguez Soto, Ojeda Quintana, and Moreno, "Jardín Botánico de Cienfuegos y el Sistema Agrario Municipal."

50. Warren, "Black Women Scientists," 59–62.

51. Watkins and Donnelly, "Biodiversity Research in the Neotropics"; Kalamandeen, "Indigenous Rights, Conservation, and Climate Change."

52. See Edwards, "Historiography of the Hill Gardens," 82–84; Grubb and Tanner, "Montane Forests and Soils of Jamaica"; Chai and Tanner, "One-Hundred-Fifty-Year Legacy of Land Use"; Graves, "Historical Decline and Probable Extinction of the Jamaican Golden Swallow"; World Heritage Committee, *Decisions Adopted, 2015*.

53. No list of tropical stations is complete or up-to-date, because of the instability of many institutions and communication difficulties. The best currently available are Leibniz-Institute of Freshwater Ecology and Inland Fisheries, "Global Map of Biological Field Stations," http://bfs.igb-berlin.de/index.php/biological-field-station-map.html (accessed 18 December 2016); Leite Pitman, "Field Sites in Tropical Countries," Association for Tropical Biology and Conservation, http://tropicalbiology.org/field-sites/ (accessed 18 December 2016). See also Tydecks et al., "Biological Field Stations"; Organization of Biological Field Stations, "Station Directory," https://obfst.memberclicks.net/directories (accessed 18 December 2016); Huettmann, "Field Schools and Research Stations"; Wyman, Wallensky, and Baine, "Activities and Importance of International Field Stations"; Whitesell, Lilieholm, and Sharik, "Global Survey of Tropical Biological Field Stations." Through the Center for Tropical Forest Science, STRI partners with local researchers and institutions in twenty-four countries who maintain long-term forest census plots around the tropical world. Smithsonian Tropical Research Institute, "Center for Tropical Forest Science—Forest Global Earth Observatory (CTFS-Forestgeo)," http://www.forestgeo.si.edu/ (accessed 18 December 2016).

Bibliography

Manuscript Collections

Baltimore, MD
  Johns Hopkins University, Ferdinand Hamburger Archives
    RG04.070: Records of the Department of Biology
Bronx, NY
  New York Botanical Garden Mertz Library Archives
    Daniel T. MacDougal Papers
  Wildlife Conservation Society Library and Archives
    1005A, Department of Tropical Research, 1900–1971
    1005B, British Guiana and Venezuela terrestrial "When Nature *Is* the Zoo," expeditions, 1916–1924
    1005C, Rancho Grande, Venezuela Expeditions, 1945–1949
    1005D, Simla Station, Trinidad Expeditions, 1949–1961
    WCS Library Photograph Collection
Cambridge, MA
  Harvard University Archives
    HUG 4319, Papers of Oakes Ames, 1897–1967
    UAV 231, Records of the Atkins Garden and Research Laboratory, 1898–1946
    UAV 298, Records of the Museum of Comparative Zoology, 1860–1985
College Park, MD
  National Archives and Records Administration
    RG59, State Department Central File
    RG165, Military Intelligence Division File 51-312
    RG185, Records of the Panama Canal
Coral Gables, FL
  Fairchild Tropical Botanic Garden Special Collections
    Expedition Reports
    David Fairchild Collection
    Fairchild Tropical Botanic Garden Records
    Field Books
Durham, NC
  Duke University David M. Rubenstein Rare Book and Manuscript Library
    Organization for Tropical Studies Records, 1962–1974, UA.26.03.0029
    Organization for Tropical Studies Reference Collection, 1979–1988, UA.01.11.0018

Organization for Tropical Studies
OTS Historical Documents: http://www.ots.ac.cr/index.php?option=com_content&task=view&id=703&Itemid=785 (accessed 12 December 2016)
OTS History Videos: https://www.youtube.com/playlist?list=PLz7w6OjwkQJbKwIS4Np1vePBEhRoeTjEi (accessed 12 December 2016)
Jamaica Plain, MA
Arnold Arboretum Archives of Harvard University
Papers of Richard A. Howard, IB RAH
Records of the Atkins Institution of the Arnold Arboretum, VI AT
New York, NY
American Museum of Natural History, Department of Ornithology
Papers of E. T. Gilliard
Papers of Frank M. Chapman
American Museum of Natural History, Research Library
Photographs, 164 Expeditions: Central America, Panama, Chapman Expedition
Panama City, Panama
Smithsonian Tropical Research Institute, Earl S. Tupper Tropical Sciences Library
Photograph Collection: http://biogeodb.stri.si.edu/bioinformatics/dfm/ (accessed 12 December 2016)
Princeton, NJ
Princeton University Library's Manuscripts Division
William Beebe Papers, 1830–1961, C0661
St. Louis, MO
Missouri Botanical Garden Archives
RG4/1/1/5, Croat, Thomas Bernard, 1938–, Collection 1
Sleepy Hollow, NY
Rockefeller Archive Center
RG1.2, Rockefeller Foundation Records, Projects
RG12, Rockefeller Foundation Records, Officers' Diaries
Tucson, AZ
Arizona Historical Society
Daniel Trembly MacDougal Papers
Washington, DC
National Academy of Sciences Archives
Division of Biology and Agriculture Collection, 1919–1939
Smithsonian Institution Archives
Acc06-131, Association for Tropical Biology, Records, 1966–1970
Acc08-098, Smithsonian Tropical Research Institute Biological Dynamics of Forest Fragments Project, Program Records, 1978–2002
Acc10-060, Smithsonian Tropical Research Institute, Administrative Records, 1959–1990
Acc12-055, Smithsonian Tropical Research Institute, Administrative Records, 1970–1984, 1990–1991

Acc91-178, F. Raymond Fosberg Papers, c. 1946–1984
Acc92-030, National Forum on BioDiversity
RU45, Office of the Secretary, Records, 1890–1929
RU50, Office of the Secretary, Records, 1949–1964
RU134, Canal Zone Biological Area, Records, 1918–1964
RU135, Office of the Secretary, Canal Zone Biological Area Records, 1912–1965
RU254, Assistant Secretary for Science, Records, 1963–1978
RU7006, Alexander Wetmore Papers, circa 1848–1983 and undated
Smithsonian Oral Histories
RU9559, Graham Bell Fairchild, 1989
RU9560, George C. Wheeler, 1989
RU9561, Fausto Bocanegra, 1988
RU9562, Robert K. Enders, 1976
RU9563, Charles F. Bennett Jr. and Anna Carole Bennett, 1975
RU9574, Oscar Dean Kidd, 1987, 1989, 1990
RU9576, Gilberto Ocana, 1989
RU9579, Stanley Rand, 1986, 1989–1990
RU9580, Smithsonian Tropical Research Institute Group Interview, 1990
RU9581, Pablo Rodriguez-Martinez, 1990
RU9582, Ira Rubinoff, 1989–1990
RU9584, Neal Griffith Smith, 1990
RU9585, Nicholas David Edward Smythe, 1990
RU9606, History of Tropical Biology Interviews: Robert L. Dressler, 1997

Books, Articles, Essays, and Other Sources

Abarca Jiménez, Carlos Luis. *Alexander Skutch: La Voz de la Naturaleza*. Santo Domingo de Heredia, Costa Rica: Editorial INBio, 2004.

Alagona, Peter. "A Sanctuary for Science: The Hastings Natural History Reservation and the Origins of the University of California's Natural Reserve System." *Journal of the History of Biology* 45, no. 4 (2012): 651–80.

Allee, Warder Clyde. "The Barro Colorado Laboratory." *Science* 59, no. 1537 (1924): 521–22.

Allee, Warder Clyde, and Marjorie Hill Allee. *Jungle Island*. Chicago: Rand McNally, 1925.

Allee, Warder Clyde, Orlando Park, Alfred Edwards Emerson, Thomas Park, and Karl P. Schmidt. *Principles of Animal Ecology*. Philadelphia, PA: W. B. Saunders, 1949.

Allen, Benjamin. *A Story of the Growth of E. Atkins and Co. and the Sugar Industry in Cuba*. New York: E. Atkins, 1925.

Allen, William. *Green Phoenix: Restoring the Tropical Forests of Guanacaste, Costa Rica*. Oxford: Oxford University Press, 2001.

———. "The Varied Bats of Barro Colorado Island." *BioScience* 46, no. 9 (1996): 639–42.

Alvarez Conde, José. *Historia de la Botanica en Cuba*. La Habana: Junta Nacional de Arqueología y Etnología, 1958.

Amano, Tatsuya, and William J. Sutherland. "Four Barriers to the Global Understanding of Biodiversity Conservation: Wealth, Language, Geographical Location and Security." *Proceedings of the Royal Society of London B: Biological Sciences* 280, no. 1756 (2013): 20122649.

Ames, Oakes. "George Lincoln Goodale (1839–1923)." *Proceedings of the American Academy of Arts and Sciences* 59, no. 17 (1925): 640–44.

Ames, Oakes, and Pauline Ames Plimpton. *Oakes Ames, Jottings of a Harvard Botanist, 1874–1950*. Cambridge, MA: Botanical Museum of Harvard University, 1979.

Anderson, Warwick. *Colonial Pathologies: American Tropical Medicine, Race, and Hygiene in the Philippines*. Durham, NC: Duke University Press, 2006.

———. "Science in the Philippines." *Philippine Studies* 55, no. 3 (2007): 287–318.

Anker, Peder. *Imperial Ecology: Environmental Order in the British Empire, 1895–1945*. Cambridge, MA: Harvard University Press, 2001.

Anthes, Richard A. "Hot Towers and Hurricanes: Early Observations, Theories, and Models." In *Cloud Systems, Hurricanes, and the Tropical Rainfall Measuring Mission (TRMM)*, 139–48. Boston, MA: American Meteorological Society, 2003.

Appel, Toby A. *Shaping Biology: The National Science Foundation and American Biological Research, 1945–1975*. Baltimore, MD: Johns Hopkins University Press, 2000.

Araújo, Aldo Mellender de. "Spreading the Evolutionary Synthesis: Theodosius Dobzhansky and Genetics in Brazil." *Genetics and Molecular Biology* 27, no. 3 (2004): 467–75.

Arnal, Yolanda Texera. "The Beginnings of Modern Ornithology in Venezuela." *Americas* 58, no. 4 (2002): 601–22.

Arnold, David. "'Illusory Riches': Representations of the Tropical World, 1840–1950." *Singapore Journal of Tropical Geography* 21, no. 1 (2000): 6–18.

———. *The Problem of Nature: Environment, Culture and European Expansion*. Cambridge: Blackwell, 1996.

Aronova, Elena, Karen S. Baker, and Naomi Oreskes. "Big Science and Big Data in Biology: From the International Geophysical Year through the International Biological Program to the Long Term Ecological Research (LTER) Network, 1957–Present." *Historical Studies in the Natural Sciences* 40, no. 2 (2010): 183–224.

Asprey, G. F., and R. G. Robbins. "The Vegetation of Jamaica." *Ecological Monographs* 23, no. 4 (1953): 359–412.

Atkins, Edwin F. *Sixty Years in Cuba: Reminiscences*. Cambridge, MA: Riverside Press, 1926.

Ayala, César J. *American Sugar Kingdom: The Plantation Economy of the Spanish Caribbean, 1898–1934*. Chapel Hill: University of North Carolina Press, 1999.

Ayres, P. G. *The Aliveness of Plants: The Darwins at the Dawn of Plant Science*. London: Pickering and Chatto, 2008.

Baatz, Simon. "Imperial Science and Metropolitan Ambition: The Scientific Survey of Puerto Rico, 1913–1934." *Annals of the New York Academy of Sciences* 776, no. 1 (1996): 1–16.

Baker, Beth. "The Way Forward for Biological Field Stations: Change Needed to Ensure Survival and Scientific Relevance." *BioScience* 65, no. 2 (2015): 123–29.

Bankoff, Greg. "Breaking New Ground? Gifford Pinchot and the Birth of 'Empire Forestry' in the Philippines, 1900–1905." *Environment and History* 15, no. 3 (2009): 369–93.

———. "First Impressions: Diarists, Scientists, Imperialists and the Management of the Environment in the American Pacific, 1899–1902." *Journal of Pacific History* 44, no. 3 (2009): 261–80.

Barahona, Ana. "Transnational Science and Collaborative Networks: The Case of Genetics and Radiobiology in Mexico, 1950–1970." *Dynamis* 35, no. 2 (2015): 333–58.

Barbour, Thomas. "Comments on a Recent Check-list." *American Naturalist* 54, no. 632 (1920): 284–88.

———. "A Contribution to the Zoögeography of the East Indian Islands." *Memoirs of the Museum of Comparative Zoölogy at Harvard College* 44, no. 1 (1912): 1–203.

———. "A Contribution to the Zoögeography of the West Indies, with Especial Reference to Amphibians and Reptiles." *Memoirs of the Museum of Comparative Zoölogy at Harvard College* 44, no. 2 (1914): 209–359.

———. "For Zoogeographers Only." In *Naturalist at Large*, 299–310. Boston, MA: Little, Brown, 1943.

———. "Marine Biological Laboratories." *Science* 98, no. 2537 (1943): 141–43.

———. "More about Harvard Biology." *Harvard Alumni Bulletin* 27, no. 8 (1924): 211–13.

———. *Naturalist at Large*. Boston, MA: Little, Brown, 1943.

———. *A Naturalist in Cuba*. Boston, MA: Little, Brown, 1945.

———. "Notes on Bermudan Fishes." *Bulletin of the Museum of Comparative Zoology* 46 (1905): 109–34.

———. "Notes on the Herpetology of Jamaica." *Bulletin of the Museum of Comparative Zoology at Harvard College* 52 (1910): 271–301.

———. "Some Remarks upon Matthew's 'Climate and Evolution' with Supplementary Note by W. D. Matthew." *Annals of the New York Academy of Sciences* 27 (1916): 1–15.

Barbour, Thomas, and Rosamond Barbour. *Letters Written While on a Collecting Trip in the East Indies*. Paterson, NJ: Chas. A. Shriner, 1913.

Barbour, Thomas, and Gladwyn Kingsley Noble. "A Revision of the Lizards of the Genus Ameiva." *Bulletin of the Museum of Comparative Zoology at Harvard College* 59, no. 6 (1915): 417–79.

Barbour, Thomas, and C. T. Ramsden. "The Herpetology of Cuba." *Memoirs of the Museum of Comparative Zoölogy at Harvard College* 47, no. 2 (1919): 69–213.

Barrow, Mark V. *Nature's Ghosts: Confronting Extinction from the Age of Jefferson to the Age of Ecology*. Chicago: University of Chicago Press, 2009.

———. *A Passion for Birds: American Ornithology after Audubon*. Princeton, NJ: Princeton University Press, 1998.

"Bartica Company Annual Report, January, 1912." In *Materials of Corporation Finance*, edited by Charles William Gerstenberg, 387–97. New York: Prentice-Hall, 1922.

"Bartica Rubber Company." *United States Investor* 26, no. 5 (1915): 33.
Bates, Marston. "The Advantage of the Tropical Environment for Studies on the Species Problem." In *Plants and Plant Science in Latin America*, edited by Frans Verdoorn, 235–36. Waltham, MA: Chronica Botanica, 1945.
———. *The Forest and the Sea: A Look at the Economy of Nature and the Ecology of Man*. New York: Random House, 1960.
Beard, J. S. "Some Ecological Work in the Caribbean." *Empire Forestry Journal* 24, no. 1 (1945): 40–46.
Beebe, William. *Edge of the Jungle*. New York: H. Holt, 1921.
———. "Fauna of Four Square Feet of Forest Debris." *Zoologica* 2 (1916): 107–19.
———. *Geographic Variation in Birds with Special Reference to the Effects of Humidity*. New York: New York Zoological Society, 1907.
———. "The Higher Vertebrates of British Guiana." *Zoologica* 2, nos. 7, 8, 9 (1919): 205–38.
———. *High Jungle*. New York: Duell, Sloan, and Pearce, 1949.
———. *Jungle Peace*. New York: H. Holt, 1918.
———. "Lizards of the Genus Ameiva in Bartica District: Notes on Their Color and Pattern Variation." *Zoologica* 2, no. 9 (1919): 235–38.
———. *Notes on the Birds of Pará, Brazil*. New York: New York Zoological Society, 1916.
———. *Studies of a Tropical Jungle: One Quarter of a Square Mile of Jungle at Kartabo, British Guiana*. New York: New York Zoological Society, 1925.
———. "A Yard of Jungle." *Atlantic Monthly* 117 (1916): 40–47.
Beebe, William, George Inness Hartley, and Paul Griswold Howes. *Tropical Wild Life in British Guiana: Zoological Contributions from the Tropical Research Station of the New York Zoological Society*. New York: New York Zoological Society, 1917.
Bemis, Samuel Flagg. " 'America' and Americans." *Yale Review* 57, no. 3 (1968): 321–36.
Benson, Etienne. "From Wild Lives to Wildlife and Back." *Environmental History* 16, no. 3 (2011): 418–22.
———. *Wired Wilderness: Technologies of Tracking and the Making of Modern Wildlife*. Baltimore, MD: Johns Hopkins University Press, 2010.
Benson, Keith R. "From Museum Research to Laboratory Research: The Transformation of Natural History into Academic Biology." In *The American Development of Biology*, edited by Ronald Rainger, Keith R. Benson, and Jane Maienschein, 49–83. Philadelphia: University of Pennsylvania Press, 1988.
———. "The Naples Stazione Zoologica and Its Impact on the Emergence of American Marine Biology." *Journal of the History of Biology* 21, no. 2 (1988): 331–41.
———. "Why American Marine Stations? The Teaching Argument." *American Zoologist* 28, no. 1 (1988): 7–14.
Bermuda Biological Station for Research. *The First Century: Celebrating 100 Years of Marine Science*. St. George's, Bermuda: Bermuda Biological Station for Research, 2003.
Bigelow, Henry B. "Thomas Barbour, 1884–1946." *National Academy of Sciences Biographical Memoirs* 27 (1952): 11–45.

Billick, Ian, and Mary V. Price. "The Ecology of Place." In *The Ecology of Place: Contributions of Place-Based Research to Ecological Understanding*, edited by Ian Billick and Mary V. Price, 1–10. Chicago: University of Chicago Press, 2010.

Bisby, G. R. "The Barro Colorado Island Laboratory." *Science* 62, no. 1596 (1925): 111.

Black, G. A., Theodosius Dobzhansky, and C. Pavan. "Some Attempts to Estimate Species Diversity and Population Density of Trees in Amazonian Forests." *Botanical Gazette* 111, no. 4 (1950): 413–25.

Blakeslee, Albert F. "A Paradise for Plant Lovers: The Biological Station at Alto Da Serra, Brazil." *Scientific Monthly* 25, no. 1 (1927): 5–18.

Board of Agriculture and Department of Public Gardens and Plantations. *Annual Report*. Kingston, Jamaica: Government Printing Office, 1901–5.

Board of Regents of the Smithsonian Institution. *Annual Report, 1921*. Washington, DC: U.S. Government Printing Office, 1922.

———. *Annual Report, 1947*. Washington, DC: U.S. Government Printing Office, 1948.

Bocking, Stephen. *Ecologists and Environmental Politics: A History of Contemporary Ecology*. New Haven, CT: Yale University Press, 1997.

Bond, Mary Wickham. *How 007 Got His Name*. London: Collins, 1966.

Boom, Brian M. "Botanical Expeditions of the New York Botanical Garden." *Brittonia* 48, no. 3 (1996): 297–307.

Booth, John A. *Costa Rica: Quest for Democracy*. Boulder, CO: Westview Press, 1999.

"Botanical Society of America." *Botanical Gazette* 24, no. 3 (1897): 179–86.

"A Botanist in Jamaica: A Young New Yorker's Travels in the Blue Mountains." *Daily Gleaner*, 28 August 1894, 7.

Bowers, Janice E. "Plant World and Its Metamorphosis from a Popular Journal into Ecology." *Bulletin of the Torrey Botanical Club* 119, no. 3 (1992): 333–41.

———. *A Sense of Place: The Life and Work of Forrest Shreve*. Tucson: University of Arizona Press, 1988.

Bowler, Peter J. *The Earth Encompassed: A History of the Environmental Sciences*. New York: Norton, 2000.

Bowman, Kirk S. "New Scholarship on Costa Rican Exceptionalism." *Journal of Interamerican Studies and World Affairs* 41, no. 2 (1999): 123–30.

Brackett, Mary M. "A Summer in the Tropics." *Plant World* 8 (1905): 6–12, 29.

Braker, Elizabeth. "The Changing Face of Tropical Biology?" *Tropinet* 11, no. 1 (2000): 1–2.

"Brazil-U.S. Agreement Is Part of Greater Plan." *Science News-Letter* 35, no. 12 (1939): 182–83.

Bridges, William. *Gathering of Animals: An Unconventional History of the New York Zoological Society*. New York: Harper and Row, 1974.

Briggs, Laura. *Reproducing Empire: Race, Sex, Science, and U.S. Imperialism in Puerto Rico*. Berkeley: University of California Press, 2002.

Britton, N. L. "Report of the Secretary and Director-in-Chief for the Year 1907." *Bulletin of the New York Botanical Garden* 6, no. 19 (1908): 1–21.

———. "The Tropical Station at Cinchona, Jamaica." *Journal of the New York Botanical Garden* 5, no. 49 (1904): 1–7.

Brock, Darryl Erwin. "American Empire and the Scientific Survey of Puerto Rico." PhD diss., Fordham University, 2014.

Brockway, Lucile. *Science and Colonial Expansion: The Role of the British Royal Botanic Gardens*. New York: Academic Press, 1979.

Brokaw, Nicholas, Todd A. Crowl, Ariel E. Lugo, William H. McDowell, Frederick N. Scatena, Robert B. Waide, and Michael R. Willig, eds. *A Caribbean Forest Tapestry: The Multidimensional Nature of Disturbance and Response*. New York: Oxford University Press, 2012.

Brooks, W. K. "Johns Hopkins Marine Laboratory." *Science* 19, no. 465 (1892): 10–11.

———. "Notes from the Biological Laboratory." *Johns Hopkins University Circulars* 17, no. 132 (1897): 131–32.

Brown, James H. "The Legacy of Robert MacArthur: From Geographical Ecology to Macroecology." *Journal of Mammalogy* 80, no. 2 (1999): 333–34.

Bu, Liping. "Educational Exchange and Cultural Diplomacy in the Cold War." *Journal of American Studies* 33, no. 3 (1999): 393–415.

Bucheli, Marcelo. *Bananas and Business: The United Fruit Company in Colombia, 1899–2000*. New York: New York University Press, 2005.

Buechner, Helmut K., and F. Raymond Fosberg. "A Contribution toward a World Program in Tropical Biology." *BioScience* 17, no. 8 (1967): 532–38.

Bull, John, and Dean Amadon. "In Memoriam: Eugene Eisenmann." *Auk* 100, no. 1 (1983): 188–91.

Burkhardt, Frederick, and Sydney Smith, eds. *The Correspondence of Charles Darwin*. Vol. 1, *1821–1836*. Cambridge: Cambridge University Press, 1985.

Burlingame, L. J. "Evolution of the Organization for Tropical Studies." *Revista de Biologia Tropical* 50, no. 2 (2002): 439–72.

Burnett, D. Graham. *Masters of All They Surveyed: Exploration, Geography, and a British El Dorado*. Chicago: University of Chicago Press, 2000.

Cahan, Marion D. "The Harvard Garden in Cuba: A Brief History." *Arnoldia: The Magazine of the Arnold Arboretum* 51, no. 3 (1991): 22–32.

Calhoun, C. H. "Jungle Folk Sanctuary Is Provided in Panama." *New York Times*, 6 October 1929, 13.

———. "Panama's Island Ark." *New York Times*, 31 October 1948, X17.

Camerini, Jane R. "Evolution, Biogeography, and Maps: An Early History of Wallace's Line." *Isis: International Review Devoted to the History of Science and Its Cultural Influences* 84 (1993): 700–727.

———. "Wallace in the Field." *Osiris* 11 (1996): 44–65.

Campbell, Douglas Houghton. "Botanical Aspects of Jamaica." *American Naturalist* 32, no. 373 (1898): 34–42.

Campbell, Neil A., and Ariel Lugo. "A Conversation with Ariel Lugo." *American Biology Teacher* 55, no. 5 (1993): 292–96.

Canfield, Michael R. *Theodore Roosevelt in the Field*. Chicago: University of Chicago Press, 2015.

Carey, Mark. "Inventing Caribbean Climates: How Science, Medicine, and Tourism Changed Tropical Weather from Deadly to Healthy." *Osiris* 26, no. 1 (2011): 129–41.

Carnegie Institution of Washington. *Year Book No. 1: 1902*. Vol. 1. Washington, DC: Carnegie Institution of Washington, 1903.

Carpenter, Clarence Ray. "Behavior and Social Relations of Free-Ranging Primates." *Scientific Monthly* 48, no. 4 (1939): 319–25.

———. "Behavior of Red Spider Monkeys in Panama." *Journal of Mammalogy* 16, no. 3 (1935): 171–80.

———. "A Field Study of the Behavior and Social Relations of Howling Monkeys." *Comparitive Psychology Monographs* 10, no. 2 (1934): 1–168.

Carse, Ashley. *Beyond the Big Ditch: Politics, Ecology, and Infrastructure at the Panama Canal*. Cambridge, MA: MIT Press, 2014.

———. "'Like a Work of Nature': Revisiting the Panama Canal's Environmental History at Gatun Lake." *Environmental History* 21, no. 2 (2016): 231–39.

Carter, Edward Carlos, ed. *Surveying the Record: North American Scientific Exploration to 1930*. Philadelphia: American Philosophical Society, 1999.

Castree, Noel. "The Geopolitics of Nature." In *A Companion to Political Geography*, edited by John A. Agnew, Katharyne Mitchell, and Gerard Toal, 423–39. Malden, MA: Blackwell, 2003.

Castro Herrera, Guillermo. "Isthmus in the World: Elements for an Environmental History of Panama." *Global Environment* 1 (2008): 10–55.

"Census of Rare Animals in Captivity." *International Zoo Yearbook* 36, no. 1 (1998): 575.

Chai, Shauna-Lee, and Edmund Tanner. "One-Hundred-Fifty-Year Legacy of Land Use on Tree Species Composition in Old-Secondary Forests of Jamaica." *Journal of Ecology* 99, no. 1 (2011): 113–21.

Champion, H. G. "The Tropical Rain Forest, Richards, P. W." *Empire Forestry Review* 33, no. 2 (1954): 134–35.

Chapman, Frank. "In an Eden Where Man Befriends Beast." *New York Times*, 4 December 1932, SM6, 16.

———. "An Island Ark: An Unusual Wildlife Refuge Near Panama." *World's Work* 53 (1926): 61–74.

———. *Life in an Air Castle: Nature Studies in the Tropics*. New York: D. Appleton-Century, 1938.

———. *My Tropical Air Castle: Nature Studies in Panama*. New York: Appleton, 1929.

———. "The Nesting Habits of Wagler's Oropendola (*Zarhynchus wagleri*) on Barro Colorado Island." *Bulletin of the American Museum of Natural History* 58, no. 3 (1928): 123–66.

———. "Who Treads Our Trails?" *National Geographic Magazine* 52, no. 3 (1927): 341–45.

Chazdon, Robin Lee, and Julie S. Denslow. "Floristic Composition and Species Richness." In *Foundations of Tropical Forest Biology*, edited by Robin Lee Chazdon and T. C. Whitmore, 513–22. Chicago: University of Chicago Press, 2002.

Chazdon, Robin Lee, and T. C. Whitmore, eds. *Foundations of Tropical Forest Biology: Classic Papers with Commentaries*. Chicago: University of Chicago Press, 2002.

Christen, Catherine. "At Home in the Field: Smithsonian Tropical Science Field Stations in the U.S. Panama Canal Zone and the Republic of Panama." *Americas* 58, no. 4 (2002): 537–75.

———. "Development and Conservation on Costa Rica's Osa Peninsula, 1937–1977: A Regional Case Study of Historical Land Use Policy and Practice in a Small Neotropical Country." PhD diss., Johns Hopkins University, 1994.

———. "Tropical Field Ecology and Conservation Initiatives on the Osa Peninsula, Costa Rica, 1962–1973." In *Les Sciences Hors d'Occident au XX Siècle*, edited by Yvon Chatelin and Christophe Bonneuil, 223–42. Paris: IRD, 1995.

Christen, Catherine, Selene Herculano, Kathryn Hochstetler, Renae Prell, Marie Price, and J. Timmons Roberts. "Latin American Environmentalism: Comparative Views." *Studies in Comparative International Development* 33, no. 2 (1998): 58.

Christian, Michelle. "'Latin America without the Downside': Racial Exceptionalism and Global Tourism in Costa Rica." *Ethnic and Racial Studies* 36, no. 10 (2013): 1599–1618.

Cittadino, Eugene. "Ecology and the Professionalization of Botany in America, 1890–1905." *Studies in the History of Biology* 4 (1980): 171–98.

———. *Nature as the Laboratory: Darwinian Plant Ecology in the German Empire, 1880–1900*. New York: Cambridge University Press, 1990.

Clark, David B. "Ecological Field Studies in the Tropics: Geographical Origin of Reports." *Bulletin of the Ecological Society of America* 66, no. 1 (1985): 6–9.

Clark, H. C. "The United Fruit Company's Cooperative Action in the Development of Antivenins." *Annual Report: United Fruit Company, Medical Department* (1927): 279–91.

Clement, Ian D., Vivian W. Clement, Frank G. Walsingham, John W. Weeks, and Katherine C. Weeks. *Guide to the Atkins Garden*. Central Soledad, Cienfuegos, Cuba: Harvard University Atkins Garden & Research Laboratory, 1954.

Cody, Martin L. "A General Theory of Clutch Size." *Evolution* (1966): 174–84.

Coker, W. C. "Professor Duncan Starr Johnson." *Science* 86, no. 2240 (1937): 510–12.

Colby, Jason M. *The Business of Empire: United Fruit, Race, and U.S. Expansion in Central America*. Ithaca, NY: Cornell University Press, 2011.

Coleman, David C. *Big Ecology: The Emergence of Ecosystem Science*. Berkeley: University of California Press, 2010.

Coleman, William. "Evolution into Ecology? The Strategy of Warming's Ecological Plant Geography." *Journal of the History of Biology* 19, no. 2 (1986): 181–96.

Collen, Ben, Mala Ram, Tara Zamin, and Louise McRae. "The Tropical Biodiversity Data Gap: Addressing Disparity in Global Monitoring." *Tropical Conservation Science* 1, no. 2 (2008): 75–88.

Collias, Nicholas E. "The Role of American Zoologists and Behavioural Ecologists in the Development of Animal Sociology, 1934–1964." *Animal Behaviour* 41, no. 4 (1991): 613–31.

Conference on Tropical Botany, Fairchild Tropical Garden, May 5–7, 1960. Washington, DC: Division of Biology and Agriculture, National Academy of Sciences and National Research Council, 1960.

Conniff, Michael L. *Black Labor on a White Canal: Panama, 1904–1981*. Pittsburgh: University of Pittsburgh Press, 1985.

———. *Panama and the United States: The End of the Alliance*. Athens, GA: University of Georgia Press, 2012.

Conte, Christopher A. "Imperial Science, Tropical Ecology, and Indigenous History: Tropical Research Stations in Northeastern German East Africa, 1896 to the Present." In *Colonialism and the Modern World: Selected Studies*, edited by Gregory Blue, Martin P. Bunton, and Ralph C. Croizier, 246–61. Armonk, NY: Sharpe, 2002.

Contreras, Fernando C. Agüero, and Cristóbal Ríos Albuerne. "El Jardín Botánico de Cienfuegos: Sintesis Histórica." *Islas* 111 (1995): 146–53.

Cooke, Mel. "Great House, Not So Great: Dilapidated Buildings Mar Gardens." *Gleaner*, 18 September 2011.

Coursen, Blair. "Jungle Laboratory: A Visit to Barro Colorado Island." *Turtox News* 34, no. 8 (1956): 138–46.

Cowles, Henry C. "Review: A New Treatise on Ecology." *Botanical Gazette* 27, no. 3 (1899): 214–16.

Craib, Raymond B., and D. Graham Burnett. "Insular Visions: Cartographic Imagery and the Spanish-American War." *Historian* 61 (1998): 100–118.

Craig, Patricia. *Centennial History of the Carnegie Institution of Washington*. Vol. 4. Cambridge: Cambridge University Press, 2004.

Croat, Thomas B. *Flora of Barro Colorado Island*. Stanford, CA: Stanford University Press, 1978.

Cronon, William. "The Trouble with Wilderness; or, Getting Back to the Wrong Nature." *Environmental History* 1, no. 1 (1996): 7–28.

Cruz, Felipe Fernandes. "Flight of the Steel Toucans: Aeronautics and Nation-Building in Brazil's Frontiers." PhD diss., University of Texas at Austin, 2016.

Cushman, Gregory T. *Guano and the Opening of the Pacific World: A Global Ecological History*. Cambridge: Cambridge University Press, 2013.

Da Cunha, A. Brito. "On Dobzhansky and His Evolution." *Biology and Philosophy* 13, no. 2 (1998): 289–300.

Da Cunha, A. Brito, Theodosius Dobzhansky, and A. Sokoloff. "On Food Preferences of Sympatric Species of Drosophila." *Evolution* (1951): 97–101.

Dammerman, K. W. "A History of the Visitors' Laboratory (Treub Laboratorium) of the Botanic Gardens, Buitenzorg, 1884–1934." In *Science and Scientists in the Netherlands Indies*, edited by Pieter Honig and Frans Verdoorn, 59–75. New York: Board for the Netherlands Indies, 1945.

Darlington, P. J. "Experiments on Mimicry in Cuba, with Suggestions for Future Study." *Transactions of the Royal Entomological Society of London* 87, no. 23 (1938): 681–95.

———. "The Origin of the Fauna of the Greater Antilles, with Discussion of Dispersal of Animals over Water and through the Air." *Quarterly Review of Biology* 13, no. 3 (1938): 274–300.

Darwin, Charles, and Franci Darwin. *The Power of Movement in Plants*. London: John Murray, 1880.

Dasmann, Raymond. *Called by the Wild: The Autobiography of a Conservationist*. Berkeley: University of California Press, 2002.

———. *A Different Kind of Country*. New York: Macmillan, 1968.

Davis, Frederick Rowe. *The Man Who Saved Sea Turtles: Archie Carr and the Origins of Conservation Biology*. Oxford: Oxford University Press, 2007.

Davis, T. A. W., and Paul W. Richards. "The Vegetation of Moraballi Creek, British Guiana: An Ecological Study of a Limited Area of Tropical Rain Forest. Part II." *Journal of Ecology* (1934): 106–55.

Davis, Wade. *One River: Explorations and Discoveries in the Amazon Rain Forest*. New York: Simon and Schuster, 1996.

Dean, Warren. *Brazil and the Struggle for Rubber: A Study in Environmental History*. Cambridge: Cambridge University Press, 1987.

———. *With Broadax and Firebrand: The Destruction of the Brazilian Atlantic Forest*. Berkeley: University of California Press, 1995.

De Bont, Raf. *Stations in the Field: A History of Place-Based Animal Research, 1870–1930*. Chicago: University of Chicago Press, 2015.

De Chadarevian, Soraya. "Laboratory Science versus Country-House Experiments: The Controversy between Julius Sachs and Charles Darwin." *British Journal for the History of Science* 29, no. 1 (1996): 17–41.

De Girolami, Ettore. "Reseña Histórica de la Fundación de la Revista de Biología Tropical." *Revista de Biología Tropical/International Journal of Tropical Biology and Conservation* 36, no. 2B (1988): 341–46.

Deikumah, Justus P., Clive A. McAlpine, and Martine Maron. "Biogeographical and Taxonomic Biases in Tropical Forest Fragmentation Research." *Conservation Biology* 28, no. 6 (2014): 1522–31.

De Souza, Vanderlei Sebastião, and Ricardo Ventura Santos. "The Emergence of Human Population Genetics and Narratives about the Formation of the Brazilian Nation (1950–1960)." *Studies in History and Philosophy of Science Part C: Studies in History and Philosophy of Biological and Biomedical Sciences* 47, Part A (2014): 97–107.

Diamond, Jared M. "The Island Dilemma: Lessons of Modern Biogeographic Studies for the Design of Natural Reserves." *Biological Conservation* 7, no. 2 (1975): 129–46.

Ditmars, Raymond L. "Antivenin Institute of America." *New York Zoological Society Bulletin* 29, no. 4 (1926): 134–35.

Dobzhansky, Theodosius. "Evolution in the Tropics." *American Scientist* 38 (1950): 209–21.

———. "Genetics of Natural Populations XXVI. Chromosomal Variability in Island and Continental Populations of *Drosophila willistoni* from Central America and the West Indies." *Evolution* (1957): 280–93.

———. *The Roving Naturalist: Travel Letters of Theodosius Dobzhansky*. Philadelphia, PA: American Philosophical Society, 1980.

Dobzhansky, Theodosius, and C. Pavan. "Local and Seasonal Variations in Relative Frequencies of Species of Drosophila in Brazil." *Journal of Animal Ecology* 19, no. 1 (1950): 1–14.

Doel, Ronald E., and Kristine C. Harper. "Prometheus Unleashed: Science as a Diplomatic Weapon in the Lyndon B. Johnson Administration." *Osiris* 21, no. 1 (2006): 66–85.

Domínguez Soto, Tania, Lázaro Jesús Ojeda Quintana, and Xiomara Moreno. "El Jardín Botánico de Cienfuegos y el Sistema Agrario Municipal, un Enfoque Territorial de Extensionismo para la Gestión Local." *Centro Agrícola* 43, no. 4 (2016): 88–92.

Donnelly, Maureen A., Megan H. Chen, and Graham G. Watkins. "Sampling Amphibians and Reptiles in the Iwokrama Forest Ecosystem." *Proceedings of the Academy of Natural Sciences of Philadelphia* 154 (2005): 55–69.

Donoghue, Michael E. *Borderland on the Isthmus: Race, Culture, and the Struggle for the Canal Zone.* Durham, NC: Duke University Press, 2014.

Douglas, Marjory Stoneman. *Adventures in a Green World: The Story of David Fairchild and Barbour Lathrop.* Coconut Grove, FL: Field Research Projects, 1973.

Dowler, C. C., and F. H. Tschirley. "Effect of Herbicides on a Puerto Rican Rain Forest." In *A Tropical Rain Forest: A Study of Irradiation and Ecology at El Verde, Puerto Rico*, edited by Howard T. Odum and Robert F. Pigeon, B315–24. Oak Ridge, TN: Division of Technical Information, U.S. Atomic Energy Commission, 1970.

Drayton, Richard Harry. *Nature's Government: Science, Imperial Britain, and the "Improvement" of the World.* New Haven, CT: Yale University Press, 2000.

Dritschilo, William. "A Mathematician Who Knew His Warblers." In *Earth Days: Ecology Comes of Age as a Science*, 126–41. New York: iUniverse, 2004.

Driver, Felix. "Imagining the Tropics: Views and Visions of the Tropical World." *Singapore Journal of Tropical Geography* 25, no. 1 (2004): 1–17.

Driver, Felix, and Luciana Martins, eds. *Tropical Visions in an Age of Empire.* Chicago: University of Chicago Press, 2005.

Dropkin, Victor H. "Host Specificity Relation of Termite Protozoa." *Ecology* 22, no. 2 (1941): 200–202.

———. "The Use of Mixed Colonies of Termites in the Study of Host-Symbiont Relations." *Journal of Parasitology* 32, no. 3 (1946): 247–51.

Drummond, Jose Augusto, Jose Luiz de Andrade Franco, and Alessandro Bortoni Ninis. "Brazilian Federal Conservation Units: A Historical Overview of Their Creation and of Their Current Status." *Environment and History* 15, no. 4 (2009): 463–91.

Dunlap, Thomas R. *In the Field, among the Feathered: A History of Birders and Their Guides.* New York: Oxford University Press, 2011.

Dupree, A. Hunter. *Science in the Federal Government: A History of Policies and Activities to 1940.* Cambridge, MA: Belknap Press, 1957.

East, Edward M., and William Henry Weston. *A Report on the Sugar Cane Mosaic Situation in February, 1924, at Soledad, Cuba.* Vol. 1, *Contributions from the Harvard Institute for Tropical Biology and Medicine.* Cambridge, MA: Harvard University Press, 1925.

Ebert, James D. "Carnegie Institution of Washington and Marine Biology: Naples, Woods Hole, and Tortugas." *Biological Bulletin* 168 (1985): 172–82.

"Editorial: Loss of Cuban Botanical Garden Throws a Bright Spotlight on FTG." *Fairchild Tropical Garden Bulletin* (1962): 49–50.

"Editorials: An American Tropical Laboratory." *Botanical Gazette* 22, no. 5 (1896): 415–16.

"Editorials: The Tropical Laboratory." *Botanical Gazette* 23, no. 3 (1897): 200–203.

"Editorials: Tropical Laboratory Commission." *Botanical Gazette* 23, no. 2 (1897): 126–27.

Edwards, Paul N. *The Closed World: Computers and the Politics of Discourse in Cold War America*. Cambridge, MA: MIT Press, 1996.

Edwards, Thera. "Towards an Historiography of the Hill Gardens at Cinchona, Jamaica." *Caribbean Geography* 19 (2014): 69–88.

Eggeling, W. J. "Review: The Tropical Rain Forest." *Empire Forestry Review* 31, no. 4 (1952): 329–31.

Ehrlich, Paul. *The Population Bomb*. New York: Ballantine, 1968.

Ehrlich, Paul R., and Peter H. Raven. "Butterflies and Plants: A Study in Coevolution." *Evolution* (1964): 586–608.

Eisner, Thomas, Hans Eisner, Jerrold Meinwald, Carl Sagan, Charles Walcott, Ernst Mayr, Edward O. Wilson, Peter H. Raven, Anne Ehrlich, Paul R. Ehrlich, Archie Carr, Eugene P. Odum, and Carl Gans. "Conservation of Tropical Forests." *Science New Series* 213, no. 4514 (1981): 1314.

Elliot, Nils Lindahl. "A Memory of Nature: Ecotourism on Panama's Barro Colorado Island." *Journal of Latin American Cultural Studies* 19, no. 3 (2010): 237–59.

Elton, Charles S. *Animal Ecology*. London: Sidgwick and Jackson, 1927.

Emerson, Alfred E. "The Jungle Laboratory of Tropical Biology Conducted by the University of Pittsburgh." *Science* 61, no. 1576 (1925): 281.

———. "Social Coordination and the Superorganism." *American Midland Naturalist* 21, no. 1 (1939): 182–209.

Enders, Robert K. "Changes Observed in the Mammal Fauna of Barro Colorado, 1929–1937." *Ecology* 20, no. 1 (1939): 104–6

———. "Mammalian Life Histories of Barro Colorado Island, Panama." *Bulletin of the Museum of Comparative Zoology Harvard* 78, no. 4 (1935): 385–502.

———. "Notes on Some Mammals from Barro Colorado Island, Canal Zone." *Journal of Mammology* 11, no. 3 (1930): 280–92.

Enright, Kelly. *The Maximum of Wilderness: The Jungle in the American Imagination*. Charlottesville: University of Virginia Press, 2012.

Erickson, Paul. *The World the Game Theorists Made*. Chicago: University of Chicago Press, 2015.

Espinosa, Mariola. *Epidemic Invasions: Yellow Fever and the Limits of Cuban Independence, 1878–1930*. Chicago: University of Chicago Press, 2009.

Estrada, Alejandro, R. Coates-Estrada, and M. Martínez-Ramos. "La Estación Biológica Tropical Los Tuxtlas: Un Recurso Para el Estudio y Conservación de las

Selvas del Trópico Húmedo." In *Investigaciones Sobre la Regeneración de Selvas Altas en Veracruz, México*, Vol. 2, 395–400. Alhambra, México: Instituto Nacional de Investigaciones sobre Recursos Bióticos, 1985.

Evans, Sterling. *The Green Republic: A Conservation History of Costa Rica*. Austin: University of Texas Press, 1999.

"Expedition of the Department of Tropical Research of the New York Zoological Society." *Science* 96, no. 2497 (1942): 422.

Fairchild, David. "The Jungles of Panama." *National Geographic Magazine* 41 (1922): 131–45.

———. "The Sensitive Plant as a Weed in the Tropics." *Botanical Gazette* 34, no. 3 (1902): 228–30.

———. *The World Was My Garden*. New York: Charles Scribner's Sons, 1938.

Fan, Fa-ti. *British Naturalists in Qing China: Science, Empire, and Cultural Encounter*. Cambridge, MA: Harvard University Press, 2004.

Farnham, Timothy J. *Saving Nature's Legacy: Origins of the Idea of Biological Diversity*. New Haven, CT: Yale University Press, 2007.

Fawcett, William. "The Public Gardens and Plantations of Jamaica." *Botanical Gazette* 24, no. 5 (1897): 345–69.

Fernández Prieto, Leida. "Azúcar y Ciencia en Cuba: 1878–1898." *Tzintzun* 31 (2000): 29–54.

———. *Espacio de Poder: Ciencia y Agricultura en Cuba: El Círculo de Hacendados, 1878–1917*. Madrid: Consejo Superior de Investigaciones Científicas; Universidad de Sevilla: Diputación de Sevilla, 2008.

———. "Islands of Knowledge: Science and Agriculture in the History of Latin America and the Caribbean." *Isis* 104, no. 4 (2013): 788–97.

———. "Saberes Híbridos: Las Sugar Companys y la Moderna Plantación Azucarera en Cuba." *Asclepio* 67, no. 1 (2015): 1–15.

Fernós, Rodrigo. *Amistad y Progreso: Los Congresos Cientificos Pan Americanos, 1898–1916*. Chula Vista, CA: Aventine Press, 2003.

Finlay, Mark R. *Growing American Rubber: Strategic Plants and the Politics of National Security*. New Brunswick, NJ: Rutgers University Press, 2009.

Fischer, Alfred G. "Latitudinal Variations in Organic Diversity." *Evolution* 14, no. 1 (1960): 64–81.

Fisher, Ronald A., A. Steven Corbet, and Carrington B. Williams. "The Relation between the Number of Species and the Number of Individuals in a Random Sample of an Animal Population." *Journal of Animal Ecology* (1943): 42–58.

Fitzgerald, Deborah. "Exporting American Agriculture: The Rockefeller Foundation in Mexico, 1943–53." *Social Studies of Science* 16, no. 3 (1986): 457–83.

Foster, Robin B., and Nicholas V. L. Brokaw. "Structure and History of the Vegetation of Barro Colorado Island." In *The Ecology of a Tropical Forest Seasonal Rhythms and Long-Term Changes*, edited by A. Stanley Rand, Donald M. Windsor, and Egbert Giles Leigh Jr., 67–82. Washington, DC: Smithsonian Institution Press, 1982.

Fraser, Cary. *Ambivalent Anti-Colonialism: The United States and the Genesis of West Indian Independence, 1940–1964*. Santa Barbara, CA: Praeger, 1994.

Fretwell, Stephen D. "The Impact of Robert MacArthur on Ecology." *Annual Review of Ecology and Systematics* (1975): 1–13.

Frodin, D. G. *Guide to Standard Floras of the World: An Annotated, Geographically Arranged Systematic Bibliography of the Principal Floras, Enumerations, Checklists, and Chorological Atlases of Different Areas*. 2nd ed. Cambridge: Cambridge University Press, 2001.

Frost, Stuart Ward. "Animal Life on Barro Colorado Island: A Lecture." *Journal of the Washington Academy of Science* 21, no. 8 (1931): 173–75.

———. "Collecting Leaf-Miners on Barro Colorado Island, Panama." *Scientific Monthly* 30, no. 5 (1930): 443–49.

Fuertes, Louis Agassiz. "Impressions of the Voices of Tropical Birds." *Bird-Lore* 16, nos. 1–6 (1914): 1–4, 96.

Fuller, Geo. D. "A Montane Rain-Forest." *Botanical Gazette* 60, no. 3 (1915): 237–40.

Funes Monzote, Reinaldo. *From Rainforest to Cane Field in Cuba: An Environmental History since 1492*. Chapel Hill: University of North Carolina Press, 2008.

Galloway, Beverly Thomas. "The Buitenzorg Gardens." *Botanical Gazette* 22, no. 6 (1896): 496–97.

Gálvez, Antonio, Mercedes Maqueda, Manuel Martínez-Bueno, and Eva Valdivia. "Scientific Publication Trends and the Developing World: What Can the Volume and Authorship of Scientific Articles Tell Us about Scientific Progress in Various Regions?" *American Scientist* 88, no. 6 (2000): 526–33.

Garfield, Seth. *In Search of the Amazon: Brazil, the United States, and the Nature of a Region*. Durham, NC: Duke University Press, 2013.

Garrison Wilkes, H. "Efraim Hernández Xolocotzi-Guzman 1913–1991." *Economic Botany* 45, no. 2 (1991): 301–2.

Gaspar, David Barry, and David Patrick Geggus. *A Turbulent Time: The French Revolution and the Greater Caribbean*. Bloomington: Indiana University Press, 1997.

Geissler, P. W., and A. H. Kelly. "Field Station as Stage: Re-enacting Scientific Work and Life in Amani, Tanzania." *Social Studies of Science* 46, no. 6 (2016): 912–37.

Gentry, Alwyn H. *Four Neotropical Rainforests*. New Haven, CT: Yale University Press, 1990.

"The George Mercer Award for 1959." *Bulletin of the Ecological Society of America* 40, no. 4 (1959): 112–11.

Gieryn, Thomas F. "City as Truth-Spot: Laboratories and Field-Sites in Urban Studies." *Social Studies of Science* 36, no. 1 (2006): 5–38.

Glick, Thomas F. "The Rockefeller Foundation and the Emergence of Genetics in Brazil, 1943–1960." In *Missionaries of Science: The Rockefeller Foundation and Latin America*, edited by Marcos Cueto, 149–64. Bloomington: Indiana University Press, 1994.

———. "Science and Society in Twentieth-Century Latin America." In *The Cambridge History of Latin America*. Vol. 6, *1930 to the Present*, edited by Leslie Bethell, 461–536. Cambridge: Cambridge University Press, 1995.

Goetzmann, William H. *Exploration and Empire: The Explorer and the Scientist in the Winning of the American West*. New York: Knopf, 1966.

Goldman, Edward Alphonso, and James Zetek. "Panama." In *Naturalist's Guide to the Americas*, edited by Victor E. Shelford, 612–22. Baltimore, MD: Williams and Wilkins, 1926.

Golley, Frank, Howard T. Odum, and Ronald F. Wilson. "The Structure and Metabolism of a Puerto Rican Red Mangrove Forest in May." *Ecology* (1962): 9–19.

Gómez, Luis Diego, Jay Savage, and Daniel Janzen. "Searchers on That Rich Coast: Costa Rican Field Biology, 1400–1980." In *Costa Rican Natural History*, edited by Daniel Janze, 1–22. Chicago: University of Chicago Press, 1983.

Gómez-Pompa, Arturo. *Mi Vida en las Selvas Tropicales: Memorias de un Botánico*. Ciudad de México: Arturo Gómez-Pompa, 2016.

Gómez-Pompa, Arturo, and Armando Butanda C. *Indice de Proyectos en Desarrollo en Ecologia Tropical: Index of Current Tropical Ecology Research*. Xalapa: Instituto de Investigaciones sobre Recursos Bioticos, 1977.

Gómez-Pompa, Arturo, C. Vazquez-Yanes, and S. Guevara. "The Tropical Rain Forest: A Nonrenewable Resource." *Science New Series* 177, no. 4051 (1972): 762–65.

Gooday, Graeme. "Placing or Replacing the Laboratory in the History of Science?" *Isis* 99, no. 4 (2008): 783–95.

Goodland, R. J. "The Tropical Origin of Ecology: Eugen Warming's Jubilee." *Oikos* 26, no. 2 (1975): 240–45.

Goss, Andrew. *The Floracrats: State-Sponsored Science and the Failure of the Enlightenment in Indonesia*. Madison: University of Wisconsin Press, 2011.

Gould, Carol Grant. *The Remarkable Life of William Beebe: Explorer and Naturalist*. Washington, DC: Island Press/Shearwater Books, 2004.

Gowricharn, Ruben S. *Caribbean Transnationalism: Migration, Pluralization, and Social Cohesion*. Lanham, MD: Lexington Books, 2006.

Grandin, Greg. *Fordlandia: The Rise and Fall of Henry Ford's Forgotten Jungle City*. New York: Metropolitan Books, 2009.

Graves, Gary R. "Historical Decline and Probable Extinction of the Jamaican Golden Swallow *Tachycineta euchrysea euchrysea*." *Bird Conservation International* 24, no. 2 (2014): 239–51.

Greene, Julie. *The Canal Builders: Making America's Empire at the Panama Canal*. New York: Penguin Press, 2009.

Gross, Alfred O. "A Jungle Laboratory: Companions of the Wild at Barro Colorado Island." *Nature Magazine* 15, no. 1 (1930): 11–15.

Grubb, Peter J., and Edmund Tanner. "The Montane Forests and Soils of Jamaica: A Reassessment." *Journal of the Arnold Arboretum* (1976): 313–68.

Guha, Ramachandra. "The Authoritarian Biologist and the Arrogance of Anti-humanism." *Ecologist* 27, no. 1 (1997): 14–20.

Haberlandt, Gottlieb Freidrich Johann. *Eine Botanische Tropenreise*. Leipzig: W. Engelmann, 1893.

Hagen, Joel B. *An Entangled Bank: The Origins of Ecosystem Ecology*. New Brunswick, NJ: Rutgers University Press, 1992.

———. "Problems in the Institutionalization of Tropical Biology: The Case of the Barro Colorado Island Biological Laboratory." *History and Philosophy of the Life Sciences* 12 (1990): 225–47.

"Handbook of the Ecological Society of America." *Bulletin of the Ecological Society of America* 1, no. 3 (1917): 9–56.

Haraway, Donna Jeanne. *Primate Visions: Gender, Race, and Nature in the World of Modern Science*. New York: Routledge, 1989.

Harmon, R. S. *The Río Chagres, Panama: A Multidisciplinary Profile of a Tropical Watershed*. Dordrecht: Springer, 2005.

Harris, Christopher. "Edwin F. Atkins and the Evolution of United States Cuba Policy, 1894–1902." *New England Quarterly* 78, no. 2 (2005): 202–31.

Harrison, Michelle. *King Sugar: Jamaica, the Caribbean, and the World Sugar Industry*. New York: New York University Press, 2001.

"Harvard, out of Cuban Garden Joins FTG in Plant Research." *Fairchild Tropical Garden Bulletin* (1962): 48–49.

Harvard University. *Reports of the President and the Treasurer of Harvard College, 1898–1899*. Cambridge, MA: Harvard University, 1900.

———. *Reports of the President and the Treasurer of Harvard College, 1919–1920*. Cambridge, MA: Harvard University, 1921.

———. *Reports of the President and the Treasurer of Harvard College, 1923–1924*. Cambridge, MA: Harvard University, 1925.

———. *Reports of the President and the Treasurer of Harvard College, 1924–1925*. Cambridge, MA: Harvard University, 1926.

———. *Reports of the President and the Treasurer of Harvard College, 1925–1926*. Cambridge, MA: Harvard University, 1927.

———. *Reports of the President and the Treasurer of Harvard College, 1926–1927*. Cambridge, MA: Harvard University, 1928.

———. *Reports of the President and the Treasurer of Harvard College, 1927–1928*. Cambridge, MA: Harvard University, 1929.

———. *Report of the President of Harvard College and the Reports of Departments for 1937–38: Official Register of Harvard University*. Cambridge, MA: Harvard University, 1939.

———. *Report of the President of Harvard College and the Reports of Departments for 1941–42: Official Register of Harvard University*. Cambridge, MA: Harvard University, 1943.

———. *Report of the President of Harvard College and the Reports of Departments for 1942–43: Official Register of Harvard University*. Cambridge, MA: Harvard University, 1944.

———. *Report of the President of Harvard College and Reports of Departments, 1947–1948*. Cambridge, MA: Harvard University, 1950.

———. *Report of the President of Harvard College and Reports of Departments, 1948–1949*. Cambridge, MA: Harvard University, 1952.

———. *Report of the President of Harvard College and Reports of Departments, 1949–1950.* Cambridge, MA: Harvard University, 1954.
———. *Report of the President of Harvard College and Reports of Departments, 1954–1955.* Cambridge, MA: Harvard University, 1956.
———. *Report of the President of Harvard College and Reports of Departments, 1955–1956.* Cambridge, MA: Harvard University, 1957.
———. *Report of the President of Harvard College and Reports of Departments, 1957–1958.* Cambridge, MA: Harvard University, 1959.
———. *Report of the President of Harvard College and Reports of Departments, 1958–1959.* Cambridge, MA: Harvard University, 1960.
———. *Report of the President of Harvard College and Reports of Departments, 1959–1960.* Cambridge, MA: Harvard University, 1961.
———. *Report of the President of Harvard College and Reports of Departments, 1960–1961.* Cambridge, MA: Harvard University, 1962.
———. *Report of the President of Harvard College and Reports of Departments, 1961–1962.* Cambridge, MA: Harvard University, 1963.
———. *Report of the President of Harvard College and Reports of Departments, 1970–1971.* Cambridge, MA: Harvard University, 1972.
Hazen, Dan. "Cienfuegos Botanical Garden: Harvard's Legacy, Cuba's Challenge." *DRCLAS News* Fall (1998): 6–8.
Heckadon-Moreno, Stanley. *Naturalistas de Istmo de Panamá: Un Siglo de Historia Natural sobre el Puente Biológico de Las Américas.* Balboa, Panamá: Instituto Smithsonian de Investigaciones Tropicales y Fundación Santillana para Iberoamérica, 1998.
Heggie, Vanessa. "Why Isn't Exploration a Science?" *Isis* 105, no. 2 (2014): 318–34.
Henderson, Robert W., and Robert Powell. "Thomas Barbour and the Utuwana Voyages (1929–1934) in the West Indies." *Bonner Zoologische Beiträge* 52, no. 3 (2004): 297–309.
Henson, Pamela M. "A Baseline Environmental Survey: The 1910–1912 Smithsonian Biological Survey of the Panama Canal Zone." *Environmental History* 21, no. 2 (2016): 222–30.
———. "Invading Arcadia: Women Scientists in the Field in Latin America, 1900–1950." *Americas* 58 (2002): 577–600.
———. "'What Holds the Earth Together': Agnes Chase and American Agrostology." *Journal of the History of Biology* 36, no. 3 (2003): 437–60.
Hillman, Richard S., and Thomas J. D'Agostino. *Understanding the Contemporary Caribbean.* Kingston, Jamaica: L. Rienner, 2003.
History of the Office of the Coordinator of Inter-American Affairs. Washington: U.S. Government Printing Office, 1947.
Hitchcock, A. S. "Biological Station of the Institution for Research in Tropical America." *Science Supplement* (29 February 1924): x.
Hodge, W. H. "Paul Hamilton Allen, 1911–1963." *Taxon* 13, no. 3 (1964): 73–77.
———. "Review: The Tropical Rain Forest." *Geographical Review* 1 (1954): 160.
Hodge, Walter H., and David D. Keck. "Biological Research Centres in Tropical America." *Bulletin of the Association for Tropical Biology* 1 (1962): 107–20.

Hogue, John S. "Cruise Ship Diplomacy: Making U.S. Leisure and Power in the Anglophone Caribbean, 1900–1973." PhD diss., University of Wisconsin-Madison, 2013.

Holdridge, Leslie R. "Determination of World Plant Formations from Simple Climatic Data." *Science* 105 (1947): 267–68.

———. "Forestry in Puerto Rico." *Caribbean Forester* 1, no. 1 (1940): 7–11.

———. *Life Zone Ecology*. Provisional ed. San José, Costa Rica: Tropical Science Center, 1964.

Hopkins, Jack W. *Policymaking for Conservation in Latin America: National Parks, Reserves, and the Environment*. Westport, CT: Praeger, 1995.

Howard, Richard A. "The Role of Botanists during World War II in the Pacific Theatre." *Botanical Review* 60, no. 2 (1994): 197–257.

———. "The Role of Botanists during World War II in the Pacific Theatre." In *Science and the Pacific War: Science and Survival in the Pacific, 1939–1945*, edited by Roy M. MacLeod, 83–118. Boston: Kluwer, 2000.

Hubbell, Stephen P. "To Know a Tropical Forest: What Mechanisms Maintain High Tree Diversity on Barro Colorado Island, Panama?" In *The Ecology of Place: Contributions of Place-Based Research to Ecological Understanding*, edited by Ian Billick and Mary V. Price, 327–51. Chicago: University of Chicago Press, 2010.

———. *The Unified Neutral Theory of Biodiversity and Biogeography*. Princeton, NJ: Princeton University Press, 2001.

Hubbell, Theodore H. "The Organization for Tropical Studies." *BioScience* 17, no. 4 (1967): 236–40.

Huettmann, Falk. "Field Schools and Research Stations in a Global Context: La Suerte (Costa Rica) and Ometepe (Nicaragua) in a Wider Perspective." In *Central American Biodiversity*, 175–98. New York: Springer, 2015.

Humphrey, James Ellis. "Botany in Jamaica." *Science* 22, no. 550 (1893): 85.

———. "The Tropical Laboratory." *Botanical Gazette* 23, no. 1 (1897): 50–51.

Hunter, George W. "Tropical Wild Life in British Guiana." *Zoological Society Bulletin* 22, no. 1 (1919): 21–26.

Hutchinson, G. Evelyn. "Homage to Santa Rosalia; or, Why Are There So Many Kinds of Animals?" *American Naturalist* 93, no. 870 (1959): 145–59.

Hyles, Joshua R. *Guiana and the Shadows of Empire: Colonial and Cultural Negotiations at the Edge of the World*. Lanham, MD: Lexington Books, 2014.

"Inter-American Institute of Agricultural Sciences." *Science* 97, no. 2521 (1943): 371.

Ishida, Yoichi. "Patterns, Models, and Predictions: Robert MacArthur's Approach to Ecology." *Philosophy of Science* 74, no. 5 (2007): 642–53.

Ishmael, Odeen. *The Guyana Story: From Earliest Times to Independence*. Bloomington, IN: Xlibris, 2013.

Ives, Anthony R. "Community Diversity and Stability: Changing Perspectives and Changing Definitions." In *Ecological Paradigms Lost: Routes of Theory Change*, edited by Beatrix Beisner, 159–82. Amsterdam: Elsevier, 2005.

Jack, Homer A. "Biological Field Stations of the World." *Chronica Botanica* 9, no. 1 (1945): 1–73.

Jacobs, Nancy J. "The Intimate Politics of Ornithology in Colonial Africa." *Comparative Studies in Society and History* 48, no. 3 (2006): 564–603.

Jacoby, Karl. *Crimes against Nature: Squatters, Poachers, Thieves, and the Hidden History of American Conservation*. Berkeley: University of California Press, 2001.

Janzen, Daniel. "The Future of Tropical Ecology." *Annual Review of Ecology and Systematics* 17 (1986): 305–24.

———. "Herbivores and the Number of Tree Species in Tropical Forests." *American Naturalist* (1970): 501–28.

———. "The Impact of Tropical Studies on Ecology." In *Changing Scenes in Natural Sciences, 1776–1976*, edited by Clyde E. Goulden, 159–87. Philadelphia: Academy of Natural Sciences, 1977.

———. "Uncertain Future of Tropics." *Natural History* 81, no. 9 (1972): 80.

———. "Why Mountain Passes Are Higher in the Tropics." *American Naturalist* 101, no. 919 (1967): 233–49.

Jimenez Saa, Humberto. "Dr. Leslie R. Holdridge: La Capacidad de Crear a Partir de Lo Cotidiano." *Manejo Integrado de Plagas y Agroecología (Costa Rica)* 75 (2005): 1–6.

Johnson, Duncan S. "The Cinchona Botanical Station." *Popular Science Monthly* 85, no. 36 (1914): 521–30.

———. "Invasion of Virgin Soil in the Tropics." *Botanical Gazette* 72, no. 5 (1921): 305–12.

———. "An Opportunity to Study the Origin and Development of a Tropical Forest." *Journal of the New York Botanical Garden* 11, no. 132 (1910): 271–72.

———. "Revegetation of a Denuded Tropical Valley." *Botanical Gazette* 84, no. 3 (1927): 294–306.

Johnson, Duncan S., Douglas Houghton Campbell, A. W. Evans, C. H. Farr, and Forrest Shreve. "Cinchona as a Tropical Station for American Botanists." *Science* 43, no. 1122 (1916): 917–19.

Johnson, Kristin. *Ordering Life: Karl Jordan and the Naturalist Tradition*. Baltimore, MD: Johns Hopkins University Press, 2012.

Jones, Jeffrey R. "Colonization in Central America." In *Agricultural Expansion and Pioneer Settlements in the Humid Tropics*, edited by William B. Morgan Walther Manshard, 241–65. Hong Kong: United Nations University Press, 1988.

Jordan, John M. "A Small World of Little Americans: The $1 Diplomacy of Wendell Willkie's One World." *Indiana Magazine of History* 88, no. 3 (1992): 173–204.

Josiah, Barbara P. *Migration, Mining, and the African Diaspora: Guyana in the Nineteenth and Twentieth Centuries*. New York: Palgrave Macmillan, 2011.

Joyce, Christopher. "Species Are the Spice of Life." *New Scientist* 112, no. 1529 (1986): 20–21.

Justus, James Robert. "The Stability-Diversity-Complexity Debate of Theoretical Community Ecology: A Philosophical Analysis." PhD diss., University of Texas at Austin, 2007.

Kalamandeen, Michelle. "Indigenous Rights, Conservation, and Climate Change Strategies in Guyana." In *Conservation Biology: Voices from the Tropics*, edited by N. S. Sodhi, L. G. Gibson, and Peter H. Raven. Hoboken, NJ: Wiley Blackwell, 2013.

Kalko, Elisabeth K. V., and Charles O. Handley Jr. "Neotropical Bats in the Canopy: Diversity, Community Structure, and Implications for Conservation." *Plant Ecology* 153, nos. 1–2 (2001): 319–33.

Kannowski, Paul B. "Tropical Radioecology." *Ecology* 52, no. 6 (1971): 1131–32.

Karr, James R. "Avian Extinction on Barro Colorado Island, Panama: A Reassessment." *American Naturalist* 119, no. 2 (1982): 220–39.

Kaspari, Michael. "Knowing Your Warblers: Thoughts on the 50th Anniversary of MacArthur (1958)." *Bulletin of the Ecological Society of America* 89, no. 4 (2008): 448–58.

Kay, Lily E. *The Molecular Vision of Life: Caltech, the Rockefeller Foundation, and the Rise of the New Biology*. New York: Oxford University Press, 1993.

Keiner, Christine. "The Panatomic Canal and the U.S. Environmental Management State, 1964–1978." *Environmental History* 21, no. 2 (2016): 278–87.

———. "A Two-Ocean Bouillabaisse: Science, Politics, and the Central American Sea-Level Canal Controversy." *Journal of the History of Biology* (2017): 1–53.

Kennedy, Dane Keith. *The Magic Mountains: Hill Stations and the British Raj*. Berkeley: University of California Press, 1996.

Kenny, Judith T. "Claiming the High Ground: Theories of Imperial Authority and the British Hill Stations in India." *Political Geography* 16, no. 8 (1997): 655–73.

———. "Climate, Race, and Imperial Authority: The Symbolic Landscape of the British Hill Station in India." *Annals of the Association of American Geographers* 85, no. 4 (1995): 694–714.

Kenoyer, Leslie A. "Botanizing on Barro Colorado Island, Panama." *Scientific Monthly* 27 (1928): 322–36.

———. "Fern Ecology of Barro Colorado Island, Panama Canal Zone." *American Fern Journal* 18, no. 1 (1928): 6–13.

———. "General and Successional Ecology of the Lower Tropical Rain-Forest at Barro Colorado Island, Panama." *Ecology* 10, no. 2 (1929): 201–22.

———. "A Study of Raunkaier's Law of Frequence." *Ecology* 8, no. 3 (1927): 341–49.

Kier, Gerold, Jens Mutke, Eric Dinerstein, Taylor H. Ricketts, Wolfgang Küper, Holger Kreft, and Wilhelm Barthlott. "Global Patterns of Plant Diversity and Floristic Knowledge." *Journal of Biogeography* 32, no. 7 (2005): 1107–16.

Kinchy, Abby J. "On the Borders of Post-war Ecology: Struggles over the Ecological Society of America's Preservation Committee, 1917–1946." *Science as Culture* 15, no. 1 (2006): 23–44.

Kingsland, Sharon. "The Battling Botanist: Daniel Trembly MacDougal, Mutation Theory, and the Rise of Experimental Evolutionary Biology in America, 1900–1912." *Isis* 82, no. 3 (1991): 479–509.

———. "Creating a Science of Nature Reserve Design: Perspectives from History." *Environmental Modeling and Assessment* 7, no. 2 (2002): 61–69.

———. *The Evolution of American Ecology, 1890–2000*. Baltimore, MD: Johns Hopkins University Press, 2005.

———. "Frits Went's Atomic Age Greenhouse: The Changing Labscape on the Lab-Field Border." *Journal of the History of Biology* 42, no. 2 (2009): 289–324.

———. *Modeling Nature: Episodes in the History of Population Ecology*. Chicago: University of Chicago Press, 1985.

———. "The Role of Place in the History of Ecology." In *The Ecology of Place: Contributions of Place-Based Research to Ecological Understanding*, edited by Ian Billick and Mary V. Price, 15–39. Chicago: University of Chicago Press, 2010.

Klein, Kerwin Lee. "Reclaiming the 'F' Word, or Being and Becoming Postwestern." *Pacific Historical Review* 65, no. 2 (1996): 179–215.

Klopfer, Peter H. *Politics and People in Ethology: Personal Reflections on the Study of Animal Behavior*. Lewisburg, PA: Bucknell University Press, 1999.

Klopfer, Peter H., and R. H. MacArthur. "Niche Size and Faunal Diversity." *American Naturalist* 94, no. 877 (1960): 293–300.

———. "On the Causes of Tropical Species Diversity: Niche Overlap." *American Naturalist* 95, no. 883 (1961): 223–26.

Knight, Franklin W., and Colin A. Palmer. *The Modern Caribbean*. Chapel Hill: University of North Carolina Press, 1989.

Kohler, Robert E. *All Creatures: Naturalists, Collectors, and Biodiversity, 1850–1950*. Princeton, NJ: Princeton University Press, 2006.

———. *Landscapes and Labscapes: Exploring the Lab-Field Border in Biology*. Chicago: University of Chicago Press, 2002.

———. *Lords of the Fly: Drosophila Genetics and the Experimental Life*. Chicago: University of Chicago Press, 1994.

———. "Paul Errington, Aldo Leopold, and Wildlife Ecology: Residential Science." *Historical Studies in the Natural Sciences* 41, no. 2 (2011): 216–54.

Kramer, Paul A. *The Blood of Government: Race, Empire, the United States, and the Philippines*. Chapel Hill: University of North Carolina Press, 2006.

———. "Empires, Exceptions, and Anglo-Saxons: Race and Rule between the British and United States Empires, 1880–1910." *Journal of American History* 88, no. 4 (2002): 1315–53.

———. "Power and Connection: Imperial Histories of the United States in the World." *American Historical Review* 116, no. 5 (2011): 1348–91.

Kroll, Gary. *America's Ocean Wilderness: A Cultural History of Twentieth-Century Exploration*. Lawrence: University Press of Kansas, 2008.

Kucera, Thomas E., and Reginald H. Barrett. "A History of Camera Trapping." In *Camera Traps in Animal Ecology*, edited by Allan F. O'Connell, James D. Nichols, and Ullas K. Karanth, 9–26. New York: Springer, 2010.

Kwa, Chung Lin. *Mimicking Nature: The Development of Systems Ecology in the United States, 1950–1975*. PhD diss., University of Amsterdam, 1989.

"Laboratories." *Bulletin of the New York Botanical Garden* 7, no. 25 (1911): 284.

"Laboratories for Botanical Research." *Nature* 69, no. 1797 (1904): 538–39.

Lachenal, Guillaume. "At Home in the Postcolony: Ecology, Empire and Domesticity at the Lamto Field Station, Ivory Coast." *Social Studies of Science* 46, no. 6 (2016): 877–93.

LaFeber, Walter. *The Panama Canal: The Crisis in Historical Perspective*. Updated ed. New York: Oxford University Press, 1989.

Langley, Lester D. *The Banana Wars: United States Intervention in the Caribbean, 1898–1934*. Wilmington, DE: SR Books, 2002.

Lapp, Michael. "The Rise and Fall of Puerto Rico as a Social Laboratory, 1945–1965." *Social Science History* 19, no. 2 (1995): 169–99.

Lasso, Marixa. "From Citizens to 'Natives': Tropical Politics of Depopulation at the Panama Canal Zone." *Environmental History* 21, no. 2 (2016): 240–49.

"The Latin American Committee on National Parks: One Year Old." *Bulletin (Association for Tropical Biology)* 4 (1965): 20–18.

Lawrence, Mark Atwood. "Exception to the Rule? The Johnson Administration and the Panama Canal." In *Looking Back at LBJ: White House Politics in a New Light*, 20–47. Lawrence: University Press of Kansas, 2005.

Ledesma-Mateos, Ismael, and Ana Barahona. "The Institutionalization of Biology in Mexico in the Early 20th Century: The Conflict between Alfonso Luis Herrera (1868–1942) and Isaac Ochoterena (1885–1950)." *Journal of the History of Biology* 36, no. 2 (2003): 285–307.

LeGrand, Catherine. "Living in Macondo: Economy and Culture in a United Fruit Company Banana Enclave in Colombia." In *Close Encounters of Empire: Writing the Cultural History of U.S.–Latin American Relations*, edited by G. M. Joseph, Catherine LeGrand, and Ricardo Donato Salvatore, 333–68. Durham, NC: Duke University Press, 1998.

Leigh, Egbert Giles. *Tropical Forest Ecology: A View from Barro Colorado Island*. New York: Oxford University Press, 1999.

Leigh, Egbert Giles, A. Stanley Rand, and Donald M. Windsor. *The Ecology of a Tropical Forest: Seasonal Rhythms and Long-Term Changes*. Washington, DC: Smithsonian Institution Press, 1982.

Leigh, Egbert Giles, Jr. "Barro Colorado." In *Encyclopedia of Islands*, edited by Rosemary G. Gillespie and D. A. Clague, 88–91. Berkeley: University of California Press, 2009.

Lévêque, Christian, and Jean-Claude Mounolou. *Biodiversity*. Chichester: Wiley, 2003.

Levins, Richard, and Robert MacArthur. "The Maintenance of Genetic Polymorphism in a Spatially Heterogeneous Environment: Variations on a Theme by Howard Levene." *American Naturalist* 100, no. 916 (1966): 585–89.

Levy, Daniel C. *To Export Progress: The Golden Age of University Assistance in the Americas*. Bloomington: Indiana University Press, 2005.

Lewis, Michael. *Inventing Global Ecology: Tracking the Biodiversity Ideal in India, 1947–1997*. Athens: Ohio University Press, 2004.

———. "Scientists or Spies? Ecology in a Climate of Cold War Suspicion." *Economic and Political Weekly* (2002): 2323–32.

Limerick, Patricia Nelson. *The Legacy of Conquest: The Unbroken Past of the American West*. New York: Norton, 1987.

Lindsay-Poland, John. *Emperors in the Jungle: The Hidden History of the U.S. in Panama*. Durham, NC: Duke University Press, 2003.

Lindsay-Poland, John, and Fellowship of Reconciliation (U.S.) Chemical Weapons Working Group. *Test Tube Republic: Chemical Weapons Tests in Panama and U.S.*

Responsibility: A Report. San Francisco, CA: Fellowship of Reconciliation Panama Campaign, 1998.
Lisboa, Pedro L. B., and Samuel Soares de Almeida. "Vida e Obra de João Murça Pires (1917–1994)." *Acta Botanica Brasilica* 9 (1995): 303–14.
Livingston, George, Bonnie Waring, Luis F. Pacheco, Damayanti Buchori, Yuexin Jiang, Lawrence Gilbert, and Shalene Jha. "Perspectives on the Global Disparity in Ecological Science." *BioScience* (2016): 147–55.
Livingstone, David N. *Putting Science in Its Place: Geographies of Scientific Knowledge*. Chicago: University of Chicago Press, 2003.
———. "Tropical Climate and Moral Hygiene: The Anatomy of a Victorian Debate." *British Journal for the History of Science* 32, no. 1 (1999): 93–110.
Locher, Uli. "Migration and the Costa Rican Environment since 1900." In *The Costa Rica Reader: History, Culture, Politics*, edited by Steven Paul Palmer and Iván Molina Jiménez, 284–92. Durham, NC: Duke University Press, 2004.
Lomolino, Mark V., Dov F. Sax, and James H. Brown, eds. *Foundations of Biogeography: Classic Papers with Commentaries*. Monograph ed. Chicago: University of Chicago Press, 2004.
Longley, Kyle. *Sparrow and the Hawk: Costa Rica and the United States during the Rise of Jose Figueres*. Tuscaloosa: University of Alabama Press, 1997.
Lot-Helgueras, Antonio. "La Estación Biológica Tropical Los Tuxtlas: Pasado, Presente y Futuro." In *Investigaciones Sobre la Regeneracion de Selvas Altas en Veracruz, Mexico*. Vol. 1. Mexico City: Compania Editorial Continental, 1976.
Loveridge, Arthur. "Thomas Barbour, Herpetologist: 1884–1946." *Herpetologica* 3, no. 2 (1946): 33–39.
Lucier, Paul. "The Origins of Pure and Applied Science in Gilded Age America." *Isis* 103, no. 3 (2012): 527–36.
Lugo, Ariel E. "Contributions of H. T. Odum to Tropical Ecology." In *Maximum Power: The Ideas and Applications of H. T. Odum*, edited by Howard Thomas Odum and Charles A. S. Hall, 23–24. Niwot: University Press of Colorado, 1995.
Maasen, Sabine, Everett Mendelsohn, and Peter Weingart. *Biology as Society, Society as Biology: Metaphors*. Dordrecht: Kluwer, 1995.
MacArthur, Robert. "Fluctuations of Animal Populations and a Measure of Community Stability." *Ecology* 36, no. 3 (1955): 533–36.
———. *Geographical Ecology: Patterns in the Distribution of Species*. New York: Harper and Row, 1972.
———. "On the Relative Abundance of Bird Species." *Proceedings of the National Academy of Sciences of the United States of America* 43, no. 3 (1957): 293–95.
———. "On the Relative Abundance of Species." *American Naturalist* 94, no. 874 (1960): 25–36.
———. "Population Ecology of Some Warblers of Northeastern Coniferous Forests." *Ecology* 39, no. 4 (1958): 599–619.
MacArthur, Robert H., and Joseph H. Connell. *The Biology of Populations*. New York: Wiley, 1966.

MacArthur, Robert, and Richard Levins. "Competition, Habitat Selection, and Character Displacement in a Patchy Environment." *Proceedings of the National Academy of Sciences of the United States of America* 51, no. 6 (1964): 1210–17.

———. "The Limiting Similarity, Convergence, and Divergence of Coexisting Species." *American Naturalist* 101, no. 921 (1967): 377–85.

MacArthur, Robert H., and John W. MacArthur. "On Bird Species Diversity." *Ecology* 42, no. 3 (1961): 594–98.

MacArthur, Robert H., John W. MacArthur, and James Preer. "On Bird Species Diversity II. Prediction of Bird Census from Habitat Measurements." *American Naturalist* 96, no. 888 (1962): 167–74.

MacArthur, Robert H., Harry Recher, and Martin Cody. "On the Relation between Habitat Selection and Species Diversity." *American Naturalist* 100, no. 913 (1966): 319–32.

MacArthur, Robert H., and Edward O. Wilson. "An Equilibrium Theory of Insular Zoogeography." *Evolution* 17, no. 4 (1963): 373–87.

———. *The Theory of Island Biogeography*. Princeton, NJ: Princeton University Press, 1967.

MacDougal, D. T. "Botanic Gardens I." *Popular Science Monthly* 50 (December 1896): 172–86.

———. "The Mechanism of Movement and Transmission of Impulses in Mimosa and Other 'Sensitive' Plants: A Review with Some Additional Experiments." *Botanical Gazette* 22, no. 4 (1896): 293–300.

———. "Mimosa: A Typical Sensitive Plant." In *Living Plants and Their Properties: A Collection of Essays*, edited by Joseph Charles Arthur and D. T. MacDougal. New York: Baker and Tayler, 1898.

———. "The Movements of Plants." *Bulletin of the Botanical Department of Jamaica* 4 (1897): 217–27.

———. "A New Botanical Laboratory in the American Tropics." *Science New Series* 5, no. 114 (1897): 395–96.

———. "Plant Zones of Arizona." *Proceedings of the Indiana Academy of Science* (1891): 97.

———. "Recent Botanical Explorations in Idaho." *Science* 20, no. 513 (1892): 311–13.

———. "The Tendrils of Passiflora Caerulea." *Botanical Gazette* 18, no. 4 (1893): 123–30.

———. "A Tropical Laboratory." *Botanical Gazette* 22, no. 6 (1896): 496.

———. "The Tropical Laboratory Commission." *Botanical Gazette* 23, no. 2 (1897): 129.

———. "The Tropical Laboratory Commission." *Botanical Gazette* 23, no. 3 (1897): 207–8.

Magnus, David. "Down the Primrose Path: Competing Epistemologies in Early Twentieth-Century Biology." In *Biology and Epistemology*, edited by Richard Creath and Jane Maienschein, 91–121. Cambridge: Cambridge University Press, 1999.

Magurran, Anne E. *Measuring Biological Diversity*. Malden, MA: Blackwell, 2004.

Maienschein, Jane. *One Hundred Years Exploring Life, 1888–1988: The Marine Biological Laboratory at Woods Hole*. Boston: Jones and Bartlett, 1989.

———. *Transforming Traditions in American Biology, 1880–1915*. Baltimore, MD: Johns Hopkins University Press, 1991.

Malhado, Ana C. M. "Amazon Science Needs Brazilian Leadership." *Science*, n.s., 331, no. 6019 (2011): 857.

Malhado, Ana C. M., Rafael S. D. de Azevedo, Peter A. Todd, Ana M. C. Santos, Nídia N. Fabré, Vandick S. Batista, Leonardo J. G. Aguiar, and Richard J. Ladle. "Geographic and Temporal Trends in Amazonian Knowledge Production." *Biotropica* 46, no. 1 (2014): 6–13.

Mancilla, Francisco Javier Dosil. *Faustino Miranda, una Vida Dedicada a la Botánica*. Madrid: Editorial CSIC, 2007.

Manganaro, Christine Leah. "Assimilating Hawaiʻi: Racial Research in a Colonial 'Laboratory,' 1919–1939." PhD diss., University of Minnesota, 2012.

Manning, Kenneth R. *Black Apollo of Science: The Life of Ernest Everett Just*. New York: Oxford University Press, 1983.

Martin, Laura, Bernd Blossey, and Erle Ellis. "Mapping Where Ecologists Work: Biases in the Global Distribution of Terrestrial Ecological Observations." *Frontiers in Ecology and the Environment* 10 (2012): 195–201.

Matthew, William Diller. *Climate and Evolution*. New York: New York Academy of Sciences, 1915.

Matthews, Robert W., and Janice R. Matthews. "Adaptive Aspects of Insular Evolution: A Symposium." *Biotropica* 2, no. 1 (1970): 1–2.

Maurer, Noel, and Carlos Yu. *The Big Ditch: How America Took, Built, Ran, and Ultimately Gave Away the Panama Canal*. Princeton, NJ: Princeton University Press, 2011.

Mayfield, Harold. "In Memoriam: Josselyn Van Tyne." *Auk* 74 (1957): 322–32.

Mayr, Ernst, and William B. Provine, eds. *The Evolutionary Synthesis: Perspectives on the Unification of Biology*. Cambridge, MA: Harvard University Press, 1980.

Mazur, Allan, and Jinling Lee. "Sounding the Global Alarm: Environmental Issues in the U.S. National News." *Social Studies of Science* 23, no. 4 (1993): 681–720.

McConnell, Ro. *Land of Waters: Explorations in the Natural History of Guyana, South America*. Sussex, UK: Book Guild, 2000.

McCook, Stuart. "Global Currents in National Histories of Science: The 'Global Turn' and the History of Science in Latin America." *Isis* 104, no. 4 (2013): 773–76.

———. "The Neo-Columbian Exchange: The Second Conquest of the Greater Caribbean, 1720–1930." *Latin American Research Review* 46 (2011): 11–31.

———. *States of Nature: Science, Agriculture, and Environment in the Spanish Caribbean, 1760–1940*. Austin: University of Texas Press, 2002.

———. " 'The World Was My Garden': Tropical Botany and Cosmopolitanism in American Science, 1898–1935." In *Colonial Crucible: Empire in the Making of the Modern American State*, edited by Alfred W. McCoy and Francisco A. Scarano, 499–507. Madison: University of Wisconsin Press, 2009.

McCoy, Alfred W., and Francisco A. Scarano, eds. *Colonial Crucible: Empire in the Making of the Modern American State*. Madison: University of Wisconsin Press, 2009.

McIntosh, Robert P. *The Background of Ecology: Concept and Theory*. Cambridge: Cambridge University Press, 1985.
———. "An Index of Diversity and the Relation of Certain Concepts to Diversity." *Ecology* 48, no. 3 (1967): 392–404.
———. "Pioneer Support for Ecology." *BioScience* 33, no. 2 (1983): 107–12.
McNeill, John Robert. *Mosquito Empires: Ecology and War in the Greater Caribbean, 1620–1914*. New York: Cambridge University Press, 2010.
Medina, Eden, Ivan da Costa Marques, Christina Holmes, and Marcos Cueto. *Beyond Imported Magic: Essays on Science, Technology, and Society in Latin America*. Cambridge, MA: MIT Press, 2014.
Meine, Curt. "Conservation Biology: Past and Present." In *Conservation Biology for All*, edited by Navjot S. Sodhi and Paul R. Ehrlich, 7–26. Oxford: Oxford University Press, 2010.
Meine, Curt, Michael Soulé, and Reed F. Noss. " 'A Mission-Driven Discipline': The Growth of Conservation Biology." *Conservation Biology* 20, no. 3 (2006): 631–51.
Mertz, David B. "Thomas Park, 1908–1992." *Bulletin of the Ecological Society of America* 73, no. 4 (1992): 248–51.
Mickulas, Peter Philip. *Britton's Botanical Empire: The New York Botanical Garden and American Botany, 1888–1929*. Bronx: New York Botanical Garden, 2007.
Millard, Candice. *River of Doubt: Theodore Roosevelt's Darkest Journey*. New York: Doubleday, 2005.
Miller, Clark A. " 'An Effective Instrument of Peace': Scientific Cooperation as an Instrument of U.S. Foreign Policy, 1938–1950." *Osiris* 21, no. 1 (2006): 133–60.
Mintz, Sidney Wilfred, and Sally Price, eds. *Caribbean Contours*. Baltimore, MD: Johns Hopkins University Press, 1985.
Mitman, Gregg. *The State of Nature: Ecology, Community, and American Social Thought, 1900–1950*. Chicago: University of Chicago Press, 1992.
———. "When Nature *Is* the Zoo: Vision and Power in the Art and Science of Natural History." *Osiris* 11 (1996): 117–43.
Mitman, Gregg, and Richard W. Burkhardt Jr. "Struggling for Identity: The Study of Animal Behavior in America, 1930–1945." In *The Expansion of American Biology*, edited by Keith Benson, Jane Maienschein, and Ronald Rainger, 164–94. New Brunswick, NJ: Rutgers University Press, 1991.
Monge-Nájera, Julian, and Lizeth Diaz. "Thirty-Five Years of Tropical Biology: A Quantitative History." *Revista Biología Tropical* 36, no. 2B (1988): 347–59.
Montgomery, Georgina M. "Place, Practice and Primatology: Clarence Ray Carpenter, Primate Communication and the Development of Field Methodology, 1931–1945." *Journal of the History of Biology* 38, no. 3 (2005): 495–533.
———. *Primates in the Real World: Escaping Primate Folklore and Creating Primate Science*. Charlottesville: University of Virginia Press, 2015.
———. "Primates in the Real World: Place, Practice and the History of Primate Field Studies, 1924–1970." PhD diss., University of Minnesota, 2005.

Moon, Suzanne. *Technology and Ethical Idealism: A History of Development in the Netherlands East Indies*. Leiden: CNWS Publications, 2007.
Mordecai, Martin. "Cinchona: 'The Nearest Place to Heaven.'" *Jamaica Journal* 17, no. 2 (1984): 2–9.
Moynihan, Martin. "Report on the Canal Zone Biological Area." *Annual Report of the Smithsonian Institution* (1958): 180–87.
———. "Report on the Canal Zone Biological Area." *Annual Report of the Smithsonian Institution* (1960): 172–76.
———. "Report on the Canal Zone Biological Area." *Annual Report of the Smithsonian Institution* (1964): 231–35.
———. "Report of the Canal Zone Biological Area." *Annual Report of the Smithsonian Institution* (1965): 307–13.
———. "Smithsonian Tropical Research Institute." *Smithsonian Year* (1967): 171–82.
"Mr. James Zetek." In *El "Libro Azul" de Panamá: Relato e Historia Sobre la Vida de Las Personas Más Prominentes*, edited by William T. Scoullar, 173. Panama City, Panama: Bureau de Publicidad de la América Latina, 1917.
Muka, Samantha K. "The Right Tool and the Right Place for the Job: The Importance of the Field in Experimental Neurophysiology, 1880–1945." *History and Philosophy of the Life Sciences* 38, no. 3 (2016): 7.
———. "Working at Water's Edge: Life Sciences at American Marine Stations, 1880–1930." PhD diss., University of Pennsylvania, 2014.
Mukerji, Chandra. *A Fragile Power: Scientists and the State*. Princeton, NJ: Princeton University Press, 1989.
Murça Pires, J., Theodosius Dobzhansky, and G. A. Black. "An Estimate of the Number of Species of Trees in an Amazonian Forest Community." *Botanical Gazette* 114, no. 4 (1953): 467–77.
Myers, Norman. *Conversion of Tropical Moist Forests: A Report Prepared by Norman Myers for the Committee on Research Priorities in Tropical Biology of the National Research Council*. Washington, DC: National Academy of Sciences, 1980.
———. *The Sinking Ark: A New Look at the Problem of Disappearing Species*. Oxford: Pergamon Press, 1979.
National Science Foundation. *Tenth Annual Report, 1960*. Washington, DC: U.S. Government Printing Office, 1961.
———. *Eleventh Annual Report, 1961*. Washington, DC: U.S. Government Printing Office, 1962.
A Naturalist in the Rainforest: A Portrait of Alexander Skutch. Directed by Paul Feyling. Oley, PA: Bullfrog Films, Paul Feyling Productions, 1995. VHS, 54 min.
Navia, Juan M., Hady López, Margarita Cimadevilla, Edelmira Fernández, Angel Valiente, I. D. Clement, and Robert S. Harris. "Nutrient Composition of Cuban Foods." *Journal of Food Science* 20, no. 2 (1955): 97–113.
New York Zoological Society. *Annual Report*. Vol. 20. New York: New York Zoological Society, 1915.
———. *Annual Report*. Vol. 21. New York: New York Zoological Society, 1916.

———. *Annual Report*. Vol. 24. New York: New York Zoological Society, 1919.
———. *Annual Report*. Vol. 25. New York: New York Zoological Society, 1920.
———. *Annual Report*. Vol. 26. New York: New York Zoological Society, 1921.
Nicolson, Malcolm. "Humboldtian Plant Geography after Humboldt: The Link to Ecology." *British Journal for the History of Science* 29, no. 3 (1996): 289–310.
Nikisianis, Nikos, and George P. Stamou. "Quantifying Nature: Ideological Representations in the Concept of Diversity." *History and Philosophy of the Life Sciences* 33, no. 3 (2011): 365–88.
Niles, Blair, and William Beebe. *Our Search for a Wilderness: An Account of Two Ornithological Expeditions to Venezuela and to British Guiana*. New York: Holt, 1910.
"Noah's Ark in Jungle Isle." *Los Angeles Times*, 15 August 1924, 7.
"Notes, News and Comment." *Journal of the New York Botanical Garden* 4 (1903): 138–40.
"Notes, News and Comment." *Bulletin of the New York Botanical Garden* 5 (1904): 151–52.
Odenbaugh, Jay. "Searching for Patterns, Hunting for Causes: Robert MacArthur, the Mathematical Naturalist." In *Outsider Scientists: Routes to Innovation in Biology*, edited by Oren Solomon Harman and Michael R. Dietrich, 181–200. Chicago: University of Chicago Press, 2013.
Odum, Eugene P. *Fundamentals of Ecology*. Philadelphia: Saunders, 1953.
Odum, Eugene P., and Howard T. Odum. *Fundamentals of Ecology*. 2nd ed. Philadelphia: Saunders, 1959.
Odum, Howard T. "Ecological Tools and Their Use: Man and the Ecosystem." In *Proceedings of the Lockwood Conference on the Suburban Forest and Ecology*, edited by P. E. Waggoner and J. O. Ovington, 57–75. New Haven: Connecticut Agricultural Experiment Station, 1962.
———. "The Possible Effect of Cloud Cover on Bird Migration in Central America." *Auk* (1947): 316–17.
———. "The Rain Forest and Man: An Introduction." In *A Tropical Rain Forest: A Study of Irradiation and Ecology at El Verde, Puerto Rico*, edited by Howard T. Odum and Robert F. Pigeon, A5–11. Oak Ridge, TN: Division of Technical Information, U.S. Atomic Energy Commission, 1970.
———. "Summary: An Emerging View of the Ecological System at El Verde." In *A Tropical Rain Forest: A Study of Irradiation and Ecology at El Verde, Puerto Rico*, edited by Howard T. Odum and Robert F. Pigeon, I191–281. Oak Ridge, TN: Division of Technical Information, U.S. Atomic Energy Commission, 1970.
Odum, Howard T., W. Abbott, and R. Selander. "Studies on the Productivity of the Lower Montane Rain Forest of Puerto Rico." *Bulletin of the Ecological Society of America* 39 (1958): 85.
Odum, Howard T., Paul R. Burkholder, and Juan Rivero. "Measurements of Productivity of Turtle Grass Flats, Reefs, and the Bahia Fosforescente of Southern Puerto Rico." *Publications of the Institute of Marine Science of the University of Texas* 6 (1959): 159–70.
Odum, Howard T., B. J. Copeland, and Robert Z. Brown. "Direct and Optical Assay of Leaf Mass of the Lower Montane Rain Forest of Puerto Rico." *Proceedings of the National Academy of Sciences of the United States of America* 49, no. 4 (1963): 429.

Odum, Howard T., William McConnell, and Walter Abbott. "The Chlorophyll 'A' of Communities." *Publications of the Institute of Marine Science of the University of Texas* 5 (1958): 65–96.
Odum, Howard T., and Eugene P. Odum. "Trophic Structure and Productivity of a Windward Coral Reef Community on Eniwetok Atoll." *Ecological Monographs* 25, no. 3 (1955): 291–320.
Odum, Howard T., and Robert F. Pigeon, eds. *A Tropical Rain Forest: A Study of Irradiation and Ecology at El Verde, Puerto Rico*. Oak Ridge, TN: Division of Technical Information, U.S. Atomic Energy Commission, 1970.
Odum, Howard T., and Richard C. Pinkerton. "Time's Speed Regulator: The Optimum Efficiency for Maximum Power Output in Physical and Biological Systems." *American Scientist* (1955): 331–43.
Ojeda Quintana, Lázaro J., Cristóbal Ríos Albuerne, Ileana Fernández Santana, and Félix Pazos Sánchez. "El Jardín Botánico de Cienfuegos: Ciento Cinco Años en la Conservación de la Diversidad Biológica Vegetal." *Centro Agrícola* 34, no. 1 (2007): 51–55.
Okihiro, Gary Y. *Pineapple Culture: A History of the Tropical and Temperate Zones*. Berkeley: University of California Press, 2009.
Oksanen, Markku, and Juhani Pietarinen. *Philosophy and Biodiversity*. Cambridge: Cambridge University Press, 2004.
Orians, Gordon H., and E. W. Pfeiffer. "Ecological Effects of the War in Vietnam." *Science* 168, no. 3931 (1970): 544–54.
Osborne, Michael A. "Acclimatizing the World: A History of the Paradigmatic Colonial Science." *Osiris* 15 (2000): 135–51.
Overfield, Richard A. "The Agricultural Experiment Station and Americanization: The Hawaiian Experience, 1900–1910." *Agricultural History* 60, no. 2 (1986): 256–66.
———. "Science Follows the Flag: The Office of Experiment Stations and American Expansion." *Agricultural History* 64, no. 2 (1990): 31–40.
Palmer, Colin A. *Cheddi Jagan and the Politics of Power: British Guiana's Struggle for Independence*. Chapel Hill: University of North Carolina Press, 2010.
Palmer, Steven Paul. *Launching Global Health: The Caribbean Odyssey of the Rockefeller Foundation*. Ann Arbor: University of Michigan Press, 2010.
Palmer, Steven Paul, and Iván Molina Jiménez, eds. *The Costa Rica Reader: History, Culture, Politics*. Durham, NC: Duke University Press, 2004.
Park, Orlando. "Observations Concerning the Future of Ecology." *Ecology* 26, no. 1 (1945): 1–9.
———. "Studies in Nocturnal Ecology, VII. Preliminary Observations on Panama Rain Forest Animals." *Ecology* 19, no. 2 (1938): 208–23.
Park, Thomas. "Analytical Population Studies in Relation to General Ecology." *American Midland Naturalist* 21, no. 1 (1939): 235–55.
———. "Some Observations on the History and Scope of Population Ecology." *Ecological Monographs* 16, no. 4 (1946): 313–20.
Parker, Jason. *Brother's Keeper: The United States, Race, and Empire in the British Caribbean, 1937–1962*. New York: Oxford University Press, 2008.

———. "Remapping the Cold War in the Tropics: Race, Communism, and National Security in the West Indies." *International History Review* 24, no. 2 (2002): 318–47.

Parkes, Kenneth C. "In Memoriam: James Bond." *Auk* 106, no. 4 (1989): 718–20.

Paton, R. F. "The Illinois State Academy of Science." *Science New Series* 93, no. 2426 (1941): 621–22.

Pauly, Philip J. *Biologists and the Promise of American Life: From Meriwether Lewis to Alfred Kinsey*. Princeton, NJ: Princeton University Press, 2000.

———. *Fruits and Plains: The Horticultural Transformation of America*. Cambridge, MA: Harvard University Press, 2007.

———. "Summer Resort and Scientific Discipline: Woods Hole and the Structure of American Biology, 1882–1925." In *The American Development of Biology*, edited by Ronald Rainger, Keith R. Benson, and Jane Maienschein, 121–50. Philadelphia: University of Pennsylvania Press, 1988.

Pavan, Crodowaldo, and Antonio Brito da Cunha. "Theodosius Dobzhansky and the Development of Genetics in Brazil." *Genetics and Molecular Biology* 26 (2003): 387–95.

Peard, Julyan G. *Race, Place, and Medicine: The Idea of the Tropics in Nineteenth-Century Brazilian Medicine*. Durham, NC: Duke University Press, 1999.

Peleggi, Maurizio. "The Social and Material Life of Colonial Hotels: Comfort Zones as Contact Zones in British Colombo and Singapore, ca. 1870–1930." *Journal of Social History* 46, no. 1 (2012): 124–53.

Penard, Thomas E. "Remarks on Beebe's 'Tropical Wild Life.'" *Auk* 36, no. 2 (1919): 217–25.

Pennisi, Elizabeth. "Panama Lab Overcomes Political Turmoil." *Scientist*, 13 November 1989.

Pérez, Louis A. *Cuba and the United States: Ties of Singular Intimacy*. Athens: University of Georgia Press, 2003.

———. *Cuba: Between Reform and Revolution*. Oxford: Oxford University Press, 2014.

———. *Cuba in the American Imagination: Metaphor and the Imperial Ethos*. Chapel Hill: University of North Carolina Press, 2008.

———. *Intervention, Revolution, and Politics in Cuba, 1913–1921*. Pittsburgh: University of Pittsburgh Press, 1978.

———. *The War of 1898: The United States and Cuba in History and Historiography*. Chapel Hill: University of North Carolina Press, 1998.

Pianka, Eric R. "Latitudinal Gradients in Species Diversity: A Review of Concepts." *American Naturalist* 100, no. 910 (1966): 33–46.

Pianka, Eric R., and Henry S. Horn. "Ecology's Legacy from Robert MacArthur." In *Ecological Paradigms Lost: Routes of Theory Change*, 213–34. Burlington, MA: Elsevier, 2005.

Pielou, E. C. *Ecological Diversity*. New York: Wiley, 1975.

Pierce, S. "Environmental History of La Selva Biological Station: How Colonization and Deforestation of Sarapiquí Canton, Costa Rica, Have Altered the Ecological Context of the Station." In *Changing Tropical Forests: Historical Perspectives on*

Today's Challenges in Central and South America, edited by Harold K. Steen and Richard P. Tucker, 40–57. Durham, NC: Forest History Society, 1992.

Pite, Rebekah E. "The Force of Food: Life on the Atkins Family Sugar Plantation in Cienfuegos, Cuba, 1884–1900." *Massachusetts Historical Review* 5 (2003): 59–94.

Pitman, Nigel. "Research in Biodiversity Hotspots Should Be Free." *Trends in Ecology and Evolution* 25, no. 7 (2010): 381.

Popenoe, John. "News of the Garden." *Fairchild Tropical Garden Bulletin* (1980): 4.

Prance, Ghillean T. *Biological Diversification in the Tropics: Proceedings of the Fifth International Symposium of the Association for Tropical Biology, Macuto Beach, Caracas, Venezuela, February 8–13, 1979*. New York: Columbia University Press, 1982.

———. "João Murça Pires (1917–1994)." *Taxon* 44, no. 4 (1995): 653–55.

Pratt, Mary Louise. *Imperial Eyes: Travel Writing and Transculturation*. London: Routledge, 1992.

Preston, F. W. "The Commonness, and Rarity, of Species." *Ecology* 29, no. 3 (1948): 254–83.

———. "Time and Space and the Variation of Species." *Ecology* 41, no. 4 (1960): 611–27.

Provine, William B. *Sewall Wright and Evolutionary Biology*. Chicago: University of Chicago Press, 1989.

Pruna, Pedro. *Ciencia y Científicos en Cuba Colonial: La Real Academia de Ciencias de la Habana, 1861–1898*. La Habana: Editorial Academia, 2001.

———. *Historia de la Ciencia y la Tecnología en Cuba*. La Habana: Editorial Científico Técnica, 2006.

———. "National Science in a Colonial Context: The Royal Academy of Sciences of Havana, 1861–1898." *Isis* 85, no. 3 (1994): 412–26.

Pugliese, David. "Panama: Bombs on the Beach." *Bulletin of the Atomic Scientists* 58, no. 4 (2002): 55–60.

Pyenson, Lewis. *Empire of Reason: Exact Sciences in Indonesia, 1840–1940*. Leiden: Brill, 1989.

Quammen, David. *The Song of the Dodo*. New York: Scribner, 1996.

Quintero Toro, Camilo. *Birds of Empire, Birds of Nation: A History of Science, Economy, and Conservation in United States–Colombia Relations*. Bogotá: Universidad de los Andes, 2012.

———. "¿En Qué Anda la Historia de la Ciencia y el Imperialismo? Saberes Locales, Dinámicas Coloniales y el Papel de Los Estados Unidos en la Ciencia en el Siglo XX." *Historia Crìtica* (2006): 151–72.

———. "Trading in Birds: A History of Science, Economy, and Conservation in United States–Colombia Relations." PhD diss., University of Wisconsin-Madison, 2007.

———. "Trading in Birds: Imperial Power, National Pride, and the Place of Nature in U.S.–Colombia Relations." *Isis* 102, no. 3 (2011): 421–45.

Rabe, Stephen G. *U.S. Intervention in British Guiana: A Cold War Story*. Chapel Hill: University of North Carolina Press, 2005.

Raby, Megan. "Ark and Archive: Making a Place for Long-Term Research on Barro Colorado Island, Panama." *Isis* 106, no. 4 (2015): 798–824.

———. "'The Jungle at Our Door': Panama and American Ecological Imagination in the 20th Century." *Environmental History* 21, no. 2 (2016): 260–69.

———. "A Laboratory for Tropical Ecology: Colonial Models and American Science at Cinchona, Jamaica." In *Spatializing the History of Ecology: Sites, Journeys, Mappings*, edited by Raf de Bont and Jens Lachmund, 56–78. New York: Routledge, 2017.

———. "Making Biology Tropical: American Science in the Caribbean, 1898–1963." PhD diss., University of Wisconsin-Madison, 2013.

Raffles, Hugh. "Intimate Knowledge." *International Social Science Journal* 54, no. 3 (2003): 325–35.

"Raid Men Who Sold $10,000,000 in Stock." *New York Times*, 21 December 1912, 1.

Rainger, Ronald. *An Agenda for Antiquity: Henry Fairfield Osborn and Vertebrate Paleontology at the American Museum of Natural History, 1890–1935*. Tuscaloosa: University of Alabama Press, 1991.

Raj, Kapil. *Relocating Modern Science: Circulation and the Construction of Knowledge in South Asia and Europe, 1650–1900*. Basingstoke: Palgrave Macmillan, 2007.

Randall, Stephen, and Graeme Mount. *The Caribbean Basin: An International History*. New York: Routledge, 2013.

Rankin, Monica A. *The History of Costa Rica*. Santa Barbara, CA: ABC-CLIO, 2012.

Rau, Phil. *Jungle Bees and Wasps of Barro Colorado Island (Panama)*. Kirkwood, MO: P. Rau, 1933.

Raunkiær, Christen. "Formationsundersøgelse Og Formationsstatistik." *Botanisk Tidsskrift* 30 (1909): 20–132.

Raven, Peter H., Peter S. Ashton, Gerardo Budowski, Arturo Gómez-Pompa, Daniel H. Janzen, William M. Lewis Jr., Harold Mooney, Paulo Nogueira-Neto, Gordon H. Orians, Harald Sioli, Hilgard O'Reilly Sternberg, John J. Terborgh, Frank H. Wadsworth, and Paul J. Zinke. *Research Priorities in Tropical Biology*. Washington, DC: National Academy of Sciences, 1980.

Recher, Harry. "Population Density and Seasonal Changes of the Avifauna in a Tropical Forest before and after Gamma Radiation." In *A Tropical Rain Forest: A Study of Irradiation and Ecology at El Verde, Puerto Rico*, edited by Howard T. Odum and Robert F. Pigeon, E69–93. Oak Ridge, TN: Division of Technical Information, U.S. Atomic Energy Commission, 1970.

Recher, Harry, and Judy Recher. "A Contribution to the Knowledge of the Avifauna of the Sierra de Luquillo, Puerto Rico." *Caribbean Journal of Science* 6 (1966): 151–61.

Rees, Amanda. *The Infanticide Controversy: Primatology and the Art of Field Science*. Chicago: University of Chicago Press, 2009.

———. "A Place That Answers Questions: Primatological Field Sites and the Making of Authentic Observations." *Studies in History and Philosophy of Science Part C: Studies in History and Philosophy of Biological and Biomedical Sciences* 37, no. 2 (2006): 311–33.

Renda, Mary A. *Taking Haiti: Military Occupation and the Culture of U.S. Imperialism, 1915–1940*. Chapel Hill: University of North Carolina Press, 2001.

"Review: Montane Rain-Forest of Jamaica." *Journal of Ecology* 3, no. 4 (1915): 244–46.

Richards, P. W. *The Tropical Rain Forest: An Ecological Study*. Cambridge: Cambridge University Press, 1952.

Riehl, Herbert. *Tropical Meteorology*. New York: McGraw-Hill, 1954.

Ripley, S. Dillon. "Perspectives in Tropical Biology." *BioScience* 17, no. 8 (1967): 538–40.

Rivas, Darlene. *Missionary Capitalist: Nelson Rockefeller in Venezuela*. Chapel Hill: University of North Carolina Press, 2002.

"Robert H. MacArthur, 1930–1972." *Ecology* 53, no. 6 (1972): 995–96.

Roberts, Lissa. "Situating Science in Global History: Local Exchanges and Networks of Circulation." *Itinerario* 33, no. 01 (2009): 9–30.

Robin, Libby. "Biological Diversity as a Political Force in Australia." In *A History of Environmentalism: Local Struggles, Global Histories*, edited by Marco Armiero and Lise Sedrez, 39–56. New York: Bloomsbury, 2014.

———. "The Rise of the Idea of Biodiversity: Crises, Responses and Expertise." *Quaderni* 76 (2011): 25–37.

Robinson, Michael H. "Is Tropical Biology Real?" *Tropical Ecology* 19, no. 1 (1978): 30–50.

Robinson, Winifred J. "A Visit to the Botanical Laboratory at Cinchona, Jamaica." *Journal of the New York Botanical Garden* 5, no. 58 (1904): 187–94.

Rodgers, Andrew Denny. *John Merle Coulter: Missionary in Science*. Princeton, NJ: Princeton University Press, 1944.

Rodriguez, Julia. "Beyond Prejudice and Pride: The Human Sciences in Nineteenth- and Twentieth-Century Latin America." *Isis* 104, no. 4 (2013): 807–17.

Roosevelt, Theodore. "A Naturalists' Tropical Laboratory." *Scribner's Magazine* 61, no. 1 (1917): 46–64.

———. "Where the Steady Trade-Winds Blow." *Scribner's Magazine* 61, no. 2 (1917): 169–88.

Ropp, Steve C. "Beyond United States Hegemony: Colombia's Persistent Role in the Shaping and Reshaping of Panama." *Journal of Caribbean History* 39, no. 2 (2005): 140–72.

Rosen, Walter G. "What's in a Name?" *BioScience* 47, no. 10 (1997): 708–12.

Rosenberg, Emily S. "Transnational Currents in a Shrinking World." In *A World Connecting, 1870–1945*, edited by Emily S. Rosenberg, 815–998. Cambridge, MA: Belknap Press, 2012.

Rowland, H. A. "A Plea for Pure Science." *Science* 2, no. 29 (1883): 242–50.

Royte, Elizabeth. *The Tapir's Morning Bath: Mysteries of the Tropical Rain Forest and the Scientists Who Are Trying to Solve Them*. Boston: Houghton Mifflin, 2001.

Rubinoff, Ira. "Mixing Oceans and Species." *Natural History* 74, no. 7 (1965): 69–72.

———. "The Sea-Level Canal Controversy." *Biological Conservation* 3, no. 1 (1970): 33–36.

———. "Tax Rich Nations, Save the Jungle." *International Wildlife* 18, no. 3 (1988): 24.

Rubinoff, Ira, and Egbert Giles Leigh Jr. "Dealing with Diversity: The Smithsonian Tropical Research Institute and Tropical Biology." *Trends in Ecology and Evolution* 5, no. 4 (1990): 115–18.

Ruiz Cedillo, Julieta, and Leticia Durand. "La Estación Biológica Tropical 'Los Tuxtlas' (Veracruz: México): Conviene una Actitud Preservacionista." Paper presented at Commons in an Age of Global Transition: Challenges, Risks and Opportunities, the Tenth Biennial Conference of the International Association for the Study of Common Property, Oaxaca, México, 9–13 August 2004.

Rumore, Gina. "Preservation for Science: The Ecological Society of America and the Campaign for Glacier Bay National Monument." *Journal of the History of Biology* 45, no. 4 (2012): 1–38.

Safier, Neil. "Global Knowledge on the Move: Itineraries, Amerindian Narratives, and Deep Histories of Science." *Isis* 101, no. 1 (2010): 133–45.

Salt, George W., and John Golding Myers. *Report on Sugar-Cane Borers at Soledad, Cuba*. Vol. 3, *Harvard Institute for Tropical Biology and Medicine*. Cambridge, MA: Harvard University Press, 1926.

Sandoval-García, Carlos. *Shattering Myths on Immigration and Emigration in Costa Rica*. Lanham, MD: Lexington Books, 2011.

Sapp, Jan. *Coexistence: The Ecology and Evolution of Tropical Biodiversity*. New York: Oxford University Press, 2016.

Sarkar, Sahotra. *Biodiversity and Environmental Philosophy: An Introduction*. Cambridge: Cambridge University Press, 2005.

———. "Complementarity and the Selection of Nature Reserves: Algorithms and the Origins of Conservation Planning, 1980–1995." *Archive for History of Exact Sciences* 66, no. 4 (2012): 397–426.

———. "From Ecological Diversity to Biodiversity." In *The Cambridge Companion to the Philosophy of Biology*, edited by David L. Hull and Michael Ruse, 388–409. Cambridge: Cambridge University Press, 2007.

Sastre-D. J., Inés, and Eugenio Santiago-Valentín. "Botanical Explorations of Puerto Rico by N. L. Britton and E. G. Britton: Their Significance in Plant Conservation, Horticulture, and Education." *Brittonia* 48, no. 3 (1996): 322–36.

Schiebinger, Londa. *Plants and Empire: Colonial Bioprospecting in the Atlantic World*. Cambridge, MA: Harvard University Press, 2004.

Schiebinger, Londa, and Claudia Swan. *Colonial Botany: Science, Commerce, and Politics in the Early Modern World*. Philadelphia: University of Pennsylvania Press, 2005.

Schimper, A. F. W. *Pflanzen-Geographie Auf Physiologischer Grundlage*. Jena: G. Fischer, 1898.

———. *Plant-Geography upon a Physiological Basis*. Translated by William Rogers Fisher. Oxford: Clarendon Press, 1903.

Schmidt, Karl P. "Warder Clyde Allee, 1885–1955." *Biographical Memoirs of the National Academy of Sciences* 30 (1957): 3–40.

Schneider, Daniel W. "Local Knowledge, Environmental Politics, and the Founding of Ecology in the United States: Stephen Forbes and "the Lake as a Microcosm" (1887)." *Isis* 91, no. 4 (2000): 681–705.

Schwartzman, Simon. *Space for Science: The Development of the Scientific Community in Brazil*. University Park: Penn State Press, 1991.

"Science News." *Science* 89, no. 2308 (1939): 6–8.

Scott, James C. *Seeing Like a State: How Certain Schemes to Improve the Human Condition Have Failed*. New Haven, CT: Yale University Press, 1998.

Seifriz, William. "The Length of the Life Cycle of a Climbing Bamboo: A Striking Case of Sexual Periodicity in *Chusquea abietifolia* Griseb." *American Journal of Botany* 7, no. 3 (1920): 83–94.

"Seventeen Years of Camera-Trapping." *STRI News*, 26 April 2010.

Shelford, Victor E. *Naturalist's Guide to the Americas*. Baltimore, MD: Williams and Wilkins, 1926.

Sheller, Mimi. *Consuming the Caribbean: From Arawaks to Zombies*. New York: Routledge, 2003.

Shreve, Forrest. "A Collecting Trip at Cinchona." *Torreya* 6, no. 5 (1906): 81–84.

———. "The Direct Effects of Rainfall on Hygrophilous Vegetation." *Journal of Ecology* 2, no. 2 (1914): 82–98.

———. *A Montane Rain-Forest: A Contribution to the Physiological Plant Geography of Jamaica*. Washington, DC: Carnegie Institution of Washington, 1914.

———. "A Winter at the Tropical Station of the Garden." *Journal of the New York Botanical Garden* 7, no. 80 (1906): 193–96.

Simberloff, Daniel S., and Edward O. Wilson. "Experimental Zoogeography of Islands: The Colonization of Empty Islands." *Ecology* 50, no. 2 (1969): 278–96.

Simonian, Lane. *Defending the Land of the Jaguar: A History of Conservation in Mexico*. Austin: University of Texas Press, 1995.

Singerman, David Roth. "Inventing Purity in the Atlantic Sugar World, 1860–1930." PhD diss., Massachusetts Institute of Technology, 2014.

"Sixty Birds in New Group Exhibited at Museum." *New York Times*, 9 December 1927, 22.

Skutch, Alexander F. *A Bird Watcher's Adventures in Tropical America*. Austin: University of Texas Press, 1977.

———. "Helpers at the Nest." *Auk* 52, no. 3 (1935): 257–73.

———. "Life Histories of Central American Birds." *Pacific Coast Avifauna* 31 (1954): 1–448.

Slack, Nancy G. *G. Evelyn Hutchinson and the Invention of Modern Ecology*. New Haven, CT: Yale University Press, 2010.

Slater, Candace. "Amazonia as Edenic Narrative." In *Uncommon Ground: Rethinking the Human Place in Nature*, edited by William Cronon, 114–31. New York: Norton, 1996.

———, ed. *In Search of the Rain Forest*. Durham, NC: Duke University Press, 2003.

Sluga, Glenda. "UNESCO and the (One) World of Julian Huxley." *Journal of World History* 21, no. 3 (2010): 393–418.

Smith, Neil. *American Empire: Roosevelt's Geographer and the Prelude to Globalization*. *California Studies in Critical Human Geography*. Berkeley: University of California Press, 2003.

Smith, Robert E. "Bundy Denies Reports of Cuban Interference in Botanical Gardens." *Harvard Crimson*, 1 December 1960.

———. "Clement Tells of Cuban Research." *Harvard Crimson*, 12 January 1962.
———. "Corporation Vote Ends Cuban Research Project." *Harvard Crimson*, 2 December 1961.
Smithsonian Institution. "Smithsonian Biological Survey of the Panama Canal Zone." *Smithsonian Miscellaneous Collections* 59, no. 11 (1912): 15–27.
———. *Smithsonian Year, 1965*. Washington, DC: U.S. Government Printing Office, 1965.
———. *Smithsonian Year, 1966*. Washington, DC: U.S. Government Printing Office, 1966.
———. *Smithsonian Year, 1969*. Washington, DC: U.S. Government Printing Office, 1969.
———. *Smithsonian Year, 1970*. Washington, DC: U.S. Government Printing Office, 1970.
Snyder, Thomas E., Alexander Wetmore, and Bennet A. Porter. "James Zetek, 1886–1959." *Journal of Economic Entomology* 52, no. 6 (1959): 1230–32.
Snyder, Thomas E., and James B. Zetek. "Damage by Termites in the Canal Zone and Panama and How to Prevent It." *U.S. Department of Agricultural Bulletin* 1232 (1924): 1–26.
———. "Inspection of Test Building Treated for Termites on Barro Colorado Island, Canal Zone, Panama." *Pests and Their Control* 16, no. 3 (1948): 8–14.
———. "Test House on Barro Colorado Island Resists Termites and Decay." *Wood Presreving News* 19, no. 7 (1941): 4–5, 13.
———. "Test of Wood Preservatives to Prevent Termite Attack, Conducted by the Bureau of Entomology, U.S. Department of Agriculture." In *Termites and Termite Control*, edited by Charles A. Kofoid, 404–17. Berkeley: University of California Press, 1934.
Soluri, John. *Banana Cultures: Agriculture, Consumption, and Environmental Change in Honduras and the United States*. Austin: University of Texas Press, 2005.
Soto Laveaga, Gabriela. *Jungle Laboratories: Mexican Peasants, National Projects, and the Making of the Pill*. Durham, NC: Duke University Press, 2009.
Sponsel, Alistair. "Coral Reefs and the Concept of Fragility in the Age of Ecology." Paper presented at the Ecological Society of America Annual Meeting, Baltimore, Maryland, 10 August 2015.
Standley, Paul C. "The Flora of Barro Colorado Island, Panama." *Smithsonian Miscellaneous Collection* 78, no. 8 (1927): 1–32.
———. "Flora of the Panama Canal Zone." *Contributions from the U.S. National Herbarium* 27 (1928): 1–416.
Stanley, Philip E., George C. G. Argent, and Harold L. K. Whitehouse. "A Botanical Biography of Professor Paul Richards C. B. E." *Journal of Bryology* 20 (1998): 323–70.
Stark, James H. *Stark's Illustrated Bermuda Guide*. Boston: James H. Stark, 1897.
———. *Stark's Jamaica Guide*. Boston: James H. Stark, 1898.
Starry, David Edward. "A Jungle Island in Panama: The Strange Pageant of Primeval Life." *Travel* 63, no. 2 (1934): 15–19, 58.
Steen, Harold K. *An Interview with Frank H. Wadsworth*. Durham, NC: Forest History Society, 1993.

Stepan, Nancy. *Picturing Tropical Nature*. Ithaca, NY: Cornell University Press, 2001.
"Sterling Debenture Arrests." *Printer's Ink* 81, no. 13 (1912): 52–53.
Stiles, F. Gary. "In Memoriam: Alexander F. Skutch, 1904–2004." *Auk* 122, no. 2 (2005): 710–18.
Stocks, Gabriela, Lisa Seales, Franklin Paniagua, Erin Maehr, and Emilio M. Bruna. "The Geographical and Institutional Distribution of Ecological Research in the Tropics." *Biotropica* 40, no. 4 (2008): 397–404.
Stone, Donald. "The Organization for Tropical Studies (OTS): A Success Story in Graduate Training and Research." In *Tropical Rainforests: Diversity and Conservation*, edited by Frank Almeda and Catherine M. Pringle, 143–87. San Francisco: California Academy of Sciences, 1988.
Striffler, Steve. *In the Shadows of State and Capital: The United Fruit Company, Popular Struggle, and Agrarian Restructuring in Ecuador, 1900–1995*. Durham, NC: Duke University Press, 2002.
Striffler, Steve, and Mark Moberg. *Banana Wars: Power, Production, and History in the Americas*. Durham, NC: Duke University Press, 2003.
Sutter, Paul. " 'The First Mountain to Be Removed': Yellow Fever Control and the Construction of the Panama Canal." *Environmental History* 21, no. 2 (2016): 250–59.
———. "Nature's Agents or Agents of Empire? Entomological Workers and Environmental Change during the Construction of the Panama Canal." *Isis* 98, no. 4 (2007): 724–54.
———. "The Tropics: A Brief History of an Environmental Imaginary." In *Oxford Handbook of Environmental History*, edited by Andrew C. Isenberg, 178–204. Oxford: Oxford University Press, 2014.
Swanson, Rebecca Cowles. "Bridging the Gap between the Naturalist Tradition and Ecological Science: Robert H. MacArthur, Edward O. Wilson and the Theory of Island Biogeography." PhD diss., North Carolina State University, 2000.
Swindle, William S. "Notes from a Marine Biological Laboratory." *Popular Science Monthly*, February (1894): 449–58.
Takacs, David. *The Idea of Biodiversity: Philosophies of Paradise*. Baltimore, MD: Johns Hopkins University Press, 1996.
Tangley, Laura. "Biological Diversity Goes Public." *BioScience* 36, no. 11 (1986): 708–15.
———. "Studying (and Saving) the Tropics: At Age 25, the Organization for Tropical Studies Is Adding More Conservation Activities to Its Successful Teaching and Research Agenda." *BioScience* 38, no. 6 (1988): 375–85.
Tanner, E. V. J. "Four Montane Rain Forests of Jamaica: A Quantitative Characterization of the Floristics, the Soils and the Foliar Mineral Levels, and a Discussion of the Interrelations." *Journal of Ecology* 65, no. 3 (1977): 883–918.
Tansley, A. G. "Review: The Tropical Rain Forest." *Journal of Ecology* 41, no. 2 (1953): 393–98.
Taylor, Anna Heyward, Edmund R. Taylor, and Alexander Moore. *Selected Letters of Anna Heyward Taylor: South Carolina Artist and World Traveler*. Columbia: University of South Carolina Press, 2010.

Taylor, Peter J. "How Do We Know We Have Global Environmental Problems? Undifferentiated Science-Politics and Its Potential Reconstruction." In *Changing Life: Genomes, Ecologies, Bodies, Commodities*, edited by Peter J. Taylor, Saul E. Halfon, and Paul N. Edwards, 149–74. Minneapolis: University of Minnesota Press, 1997.

———. "Technocratic Optimism, H. T. Odum, and the Partial Transformation of Ecological Metaphor after World War II." *Journal of the History of Biology* 21, no. 2 (1988): 213–44.

———. *Unruly Complexity: Ecology, Interpretation, Engagement*. Chicago: University of Chicago Press, 2010.

Terborgh, John. "Preservation of Natural Diversity: The Problem of Extinction Prone Species." *BioScience* 24, no. 12 (1974): 715–22.

Theobald, E. J., A. K. Ettinger, H. K. Burgess, L. B. DeBey, N. R. Schmidt, H. E. Froehlich, C. Wagner, J. Hille Ris Lambers, J. Tewksbury, M. A. Harsch, and J. K. Parrish. "Global Change and Local Solutions: Tapping the Unrealized Potential of Citizen Science for Biodiversity Research." *Biological Conservation* 181 (2015): 236–44.

Thompson, Alvin O. *Flight to Freedom: African Runaways and Maroons in the Americas*. Kingston, Jamaica: University of the West Indies Press, 2006.

Thompson, Krista A. *An Eye for the Tropics: Tourism, Photography, and Framing the Caribbean Picturesque*. Durham, NC: Duke University Press, 2006.

Thoreau, Henry David. *Walden; or, Life in the Woods*. Boston, MA: Ticknor and Fields, 1854.

Tilley, Helen. *Africa as a Living Laboratory: Empire, Development, and the Problem of Scientific Knowledge, 1870–1950*. Chicago: University of Chicago Press, 2011.

———. "Global Histories, Vernacular Science, and African Genealogies; or, Is the History of Science Ready for the World?" *Isis* 101, no. 1 (2010): 110–19.

Tjossem, Sara Fairbank. "Preservation of Nature and Academic Respectability: Tensions in the Ecological Society of America, 1915–1979." PhD diss., Cornell University, 1994.

Tobey, Ronald C. *Saving the Prairies: The Life Cycle of the Founding School of American Plant Ecology, 1895–1955*. Berkeley: University of California Press, 1981.

Tobin, Beth Fowkes. *Colonizing Nature: The Tropics in British Arts and Letters, 1760–1820*. Philadelphia: University of Pennsylvania Press, 2005.

Townsend, Charles W. "An Island in the Tropics." *Harvard Graduates Magazine* 34, no. 133 (1924): 36–45.

"Training Reactor in Puerto Rico." *Science* 128, no. 3316 (1958): 133.

Tucker, Richard P. *Insatiable Appetite: The United States and the Ecological Degradation of the Tropical World*. Berkeley: University of California Press, 2000.

Tydecks, Laura, Vanessa Bremerich, Ilona Jentschke, Gene E. Likens, and Klement Tockner. "Biological Field Stations: A Global Infrastructure for Research, Education, and Public Engagement." *BioScience* 66, no. 2 (2016): 164–71.

Underwood, Lucien Marcus. "Account by Professor Underwood of Explorations in Jamaica." *Journal of the New York Botanical Garden* 4, no. 43 (1903): 109–19.

Vadrot, Alice B. M. *The Politics of Knowledge and Global Biodiversity*. New York: Routledge, 2014.

Van Dyke, John C. *In the West Indies: Sketches and Studies in Tropic Seas and Islands*. New York: Charles Scribner's Sons, 1932.

Van Tyne, Josselyn. "The Barro Colorado Laboratory as a Station for Ornithological Research." *Wilson Bulletin* 42 (1930): 225–32.

Vernberg, F. John. "Field Stations of the United States." *American Zoologist* 3, no. 3 (1963): 245–386.

Vessuri, Hebe M. C. "Academic Science in Twentieth-Century Latin America." In *Science in Latin America: A History*, edited by Bernabé Madrigal and Juan José Saldaña, 197–230. Austin: University of Texas Press, 2009.

Vetter, Jeremy. *Field Life: Science in the American West during the Railroad Era*. Pittsburgh: University of Pittsburgh Press, 2016.

———. "Labs in the Field? Rocky Mountain Biological Stations in the Early Twentieth Century." *Journal of the History of Biology* 45, no. 4 (2012): 587–611.

———. "Rocky Mountain High Science." In *Knowing Global Environments: New Historical Perspectives on the Field Sciences*, edited by Jeremy Vetter, 108–34. New Brunswick, NJ: Rutgers University Press, 2011.

———. "Science Along the Railroad: Expanding Field Work in the U.S. Central West." *Annals of Science* 61 (2004): 187–211.

———. "Wallace's Other Line: Human Biogeography and Field Practice in the Eastern Colonial Tropics." *Journal of the History of Biology* 39 (2006): 89–123.

von Ihering, H. "The History of the Neotropical Region." *Science* (1900): 857–64.

Vuilleumier, François. "Dean of American Ornithologists: The Multiple Legacies of Frank M. Chapman of the American Museum of Natural History." *Auk* 122, no. 2 (2005): 389–402.

Vuilleumier, François, and Allison V. Andors. "Origins and Development of North American Avian Biogeography." In *Contributions to the History of North American Ornithology*, edited by William E. Davis Jr. and Jerome A. Jackson, 387–428. Cambridge, MA: Nuttall Ornithological Club, 1995.

Wadsworth, Frank H. "A Forest Research Institution in the West Indies: The First 50 Years." In *Tropical Forests: Management and Ecology*, edited by Ariel E. Lugo and Carol Lowe, 33–56. New York: Springer, 1995.

———. "Review of Past Research in the Luquillo Mountains of Puerto Rico." In *A Tropical Rain Forest: A Study of Irradiation and Ecology at El Verde, Puerto Rico*, edited by Howard T. Odum and Robert F. Pigeon, B33–46. Oak Ridge, TN: Division of Technical Information, U.S. Atomic Energy Commission, 1970.

Waide, Robert B., and Ariel E. Lugo. *Long-Term Ecological Research on the Luquillo Experimental Forest: A Proposal to the National Science Foundation*. 1988. http://luq.lternet.edu/sites/default/files/proposals/1988LUQLTER1Proposal.pdf (accessed 15 April, 2017).

Wakild, Emily. *Revolutionary Parks: Conservation, Social Justice, and Mexico's National Parks, 1910–1940*. Tucson: University of Arizona Press, 2011.

Wallace, Alfred Russel. *The Geographical Distribution of Animals with a Study of the Relations of Living and Extinct Faunas as Elucidating the Past Changes of the Earth's Surface*. New York: Harper and Brothers, 1876.

———. *Tropical Nature and Other Essays*. London: Macmillan, 1878.

Warming, Eugenius. *Lagoa Santa: Et Bidrag Til Den Biologiske Plantegeografi*. Kjøbenhavn: B. Luno, 1892.

———. "On the Vegetation of Tropical America." *Botanical Gazette* 27, no. 1 (1899): 1–18.

———. *Plantesamfund: Grundtræk Af Den Økologiske Plantegeografi*. Kjøbenhavn: P. G. Philipsen, 1895.

Warming, Eugenius, and E. F. Knoblauch. *Lehrbuch der Ökologischen Pflanzengeographie: Eine Einführung in die Kenntnis der Pflanzenvereine*. Berlin: Gebrüder Borntraeger, 1896.

Warming, Eugenius, and Martin Vahl. *Oecology of Plants*. Oxford: Oxford University Press, 1909.

Warren, Louis S. *The Hunter's Game: Poachers and Conservationists in Twentieth-Century America*. New Haven, CT: Yale University Press, 1997.

Warren, Wini. *Black Women Scientists in the United States*. Bloomington: Indiana University Press, 1999.

Watkins, Graham G., and Maureen A. Donnelly. "Biodiversity Research in the Neotropics: From Conflict to Collaboration." *Proceedings of the Academy of Natural Sciences of Philadelphia* 154 (2005): 127–36.

Weed, Walter Harvey. *The Mines Handbook*. International ed. Vol. 15, Tuckahoe, NY: Mines Handbook Company, 1922.

West, Terry. "USDA Forest Service Involvement in Post World War II International Forestry." In *Changing Tropical Forests: Historical Perspectives on Today's Challenges in Central and South America*, edited by Harold K. Steen and Richard P. Tucker, 277–91. Durham, NC: Forest History Society, 1992.

Weston, William H., Jr. "The Biological Station on Barro Colorado Island." *Harvard Alumni Bulletin* (1929): 376–83.

Wheeler, William Morton. "The Organization of Research." *Science* 53, no. 1360 (1921): 53–67.

Whippo, Craig W., and Roger P. Hangarter. "The 'Sensational' Power of Movement in Plants: A Darwinian System for Studying the Evolution of Behavior." *American Journal of Botany* 96, no. 12 (2009): 2115–27.

Whitesell, Stephen, Robert J. Lilieholm, and Terry L. Sharik. "A Global Survey of Tropical Biological Field Stations." *BioScience* 52, no. 1 (2002): 55–64.

Whittaker, R. H. "Vegetation of the Siskiyou Mountains, Oregon and California." *Ecological Monographs* 30, no. 3 (1960): 279–338.

Wilcox, Uthai Vincent. "A Paradise for Zoologists." *New York Times*, 11 November 1928, 98.

Wilkinson, Xenia V. "Tapping the Amazon for Victory: Brazil's 'Battle for Rubber' of World War II." PhD diss., Georgetown University, 2009.

Wille, Robert-Jan. "The Coproduction of Station Morphology and Agricultural Management in the Tropics: Transformations in Botany at the Botanical Garden at Buitenzorg, Java 1880–1904." In *New Perspectives on the History of Life Sciences and Agriculture*, edited by Denise Phillips and Sharon Kingsland, 253–75. New York: Springer, 2015.

———. "De Stationisten: Laboratoriumbiologie, Imperialisme en de Lobby Voor Nationale Wetenschapspolitiek, 1871–1909." PhD diss., Radboud University, 2015.

Willis, Arthur J. "Obituary: Paul Westamacott Richards, CBE (1908–95)." *Journal of Ecology* 84, no. 5 (1996): 795–98.

Willis, Edwin O. "Populations and Local Extinctions of Birds on Barro Colorado Island, Panama." *Ecological Monographs* 44, no. 2 (1974): 153–69.

Wilson, Edward O., ed. *BioDiversity*. Washington, DC: National Academy Press, 1988.

———. *The Diversity of Life*. Cambridge, MA: Belknap Press, 1992.

———. "Eminent Ecologist, 1973, Robert Helmer MacArthur." *Bulletin of the Ecological Society of America* 54, no. 3 (1973): 11–12.

———. *Naturalist*. Washington, DC: Island Press, 1994.

———. "Philip Jackson Darlington, Jr." *Biographical Memoirs of the National Academy of Sciences* (1991): 33–44.

Wilson, Edward O., and Edwin O. Willis. "Applied Biogeography." In *Ecology and Evolution of Communities*, edited by Martin L. Cody and Jared M. Diamond, 522–53. Cambridge, MA: Belknap Press, 1975.

Wilson, Kerrie A., Nancy A. Auerbach, Katerina Sam, Ariana G. Magini, Alexander St. L. Moss, Simone D. Langhans, Sugeng Budiharta, Dilva Terzano, and Erik Meijaard. "Conservation Research Is Not Happening Where It Is Most Needed." *PLoS Biology* 14, no. 3 (2016): e1002413.

Winsor, Mary P. *Reading the Shape of Nature: Comparative Zoology at the Agassiz Museum*. Chicago: University of Chicago Press, 1991.

Woodwell, George M., and H. H. Smith, eds. *Diversity and Stability in Ecological Systems: Report of a Symposium Held May 26–28, 1969*. Upton, NY: Brookhaven National Laboratory, 1969.

World Heritage Committee. 2015 *Decisions Adopted by the World Heritage Committee at Its 39th Session (Bonn, 2015)*. United Nations Educational, Scientific and Cultural Organization. http://whc.unesco.org/document/137710 (accessed 18 February, 2016).

Worster, Donald. *Nature's Economy*. San Francisco: Sierra Club Books, 1977.

———. *Nature's Economy: A History of Ecological Ideas*. 2nd ed. Cambridge: Cambridge University Press, 1994.

Wright, Jonathan Jeffrey, and Diarmid Finnegan. *Spaces of Global Knowledge: Exhibition, Encounter and Exchange in an Age of Empire*. Burlington, VT: Ashgate, 2015.

Wyman, Richard L., Eugene Wallensky, and Mark Baine. "The Activities and Importance of International Field Stations." *BioScience* 59, no. 7 (2009): 584–92.

Yantko, Joan A., and Frank B. Golley. "A Census of Tropical Ecologists." *BioScience* 28, no. 4 (1978): 260–64.

Young, Alvin L. *The History of the U.S. Department of Defense Programs for the Testing, Evaluation, and Storage of Tactical Herbicides*. Contract DAAD19-02-D-0001, U.S. Army Research Office, P.O. Box 12211, Research Triangle Park NC 27709, U.S.A. 2006. http://www.dtic.mil/docs/citations/ADA534602 (accessed 15 April, 2017).

"Zetek: Canal Zone Biologist." *Illinois Alumni News* April (1948): 10–11.

Zimmerer, Carl S. "Biodiversity." In *A Companion to Environmental Geography*, edited by Noel Castree, 50–65. Chichester, UK: Wiley-Blackwell, 2009.

Zuk, Marlene. "Temperate Assumptions: How Where We Work Influences How We Think." *American Naturalist* 188, no. S1 (2016): S1–S7.

Index

Note: Illustrations and photographs are indicated by page numbers in italics.